Statistical Thermodynamics

STUDIES IN CHEMICAL PHYSICS

General Editor

A.D. Buckingham, Professor of Chemistry,
University of Cambridge

Series Foreword

The field of science known as 'Chemical Physics' has greatly expanded in recent years. It is an essential part of both physics and chemistry and now impinges on biology, crystallography, the science of materials and even on astronomy. The aim of this series is to present short, authoritative and readable books on different topics in chemical physics at a level that is appreciated by the non-specialist and yet is of prime interest to the expert–in fact, the type of book that we all welcome and enjoy.

I was grateful to be given the opportunity to help plan this series, and warmly thank the authors and publishers whose efforts have brought it into being.

A.D. Buckingham,
University Chemical Laboratory,
Cambridge, U.K.

Other titles in the series

Advanced Molecular Quantum Mechanics, R.E. Moss
Chemical Applications of Molecular Beam Scattering, K.P. Lawley and M.A.D. Fluendy
Electronic Transitions and the High Pressure Chemistry and Physics of Solids, H.G. Drickamer and C.W. Frank
Aqueous Dielectrics, J.B. Hasted

Statistical Thermodynamics

B.J. McCLELLAND

Dept. of Chemistry and Applied Chemistry,
University of Salford

CHAPMAN AND HALL & SCIENCE PAPERBACKS
LONDON

First published 1973
by Chapman and Hall Ltd.
11 New Fetter Lane, London EC4P 4EE

Set by E.W.C. Wilkins Ltd, London N12 0EH
and printed in Great Britain by
Fletcher & Son Ltd,
Norwich

SBN *412 10350 8 (cased edition)*
SBN *412 20780 X (paperback edition)*

Distributed in the U.S.A.
by Halsted Press, a Division
of John Wiley & Sons, Inc., New York

Library of Congress Catalog Card Number 73-13384

Contents

Preface

This book is addressed to students of physical chemistry; I hope that it will also be found of use to those studying physics and related subjects. It is based upon some thirty courses on statistical mechanics which I have given during the past twelve years, and is intended, as were most of my lectures, for those with little or no prior knowledge of the subject; the only assumption made is that the reader is conversant with the basic principles of thermodynamics and the simple concepts of molecular energy levels.

I have attempted, within the limitations of a very elementary treatment, to introduce a modest degree of rigour, because I believe that oversimplification ultimately causes more difficulties than it avoids. It is hoped, therefore, that the book will provide, firstly, a working knowledge of statistical thermodynamics, applicable to straightforward practical problems of physical chemistry, and secondly, a direct route to advanced work.

Choice of subject matter has, of necessity, been somewhat arbitrary. We are here more concerned with principles and methods of calculation than with a detailed comparison with experiment: the book would need to be considerably lengthened to accommodate the latter if it were not to degenerate into a 'mere gabble of facts'. I hope, however, that sufficient ground is covered to enable the reader who requires further information to make an intelligent search for it in the advanced works. On one or two occasions, the title of this work has been interpreted rather loosely: it was considered advisable to include certain items which are, perhaps, more a part of statistical mechanics than statistical thermodynamics.

The first part of the book (Chapters One to Ten) is devoted to what might be called the naïve statistical theory: the classical Maxwell–Boltzmann method expressed in terms of quantized molecular energies. An alternative and, in many ways, a commendable mode of approach would have been to discard the Maxwell–Boltzmann method outright, in favour of the ensemble theories of Gibbs. Nevertheless, experience has shown that beginners are more able to appreciate the significance of Gibbsian statistics *after* exploring the applications and limitations of the naïve theory; moreover, the Maxwell–Boltzmann theory is adequate for much of the routine work of physical chemistry. Refinement of the treatment and discussion of the canonical ensemble is, accordingly, left until the later part of the book.

A few comments must be made concerning numerical work.

In the first place, the successful calculation, from spectroscopic data, of thermodynamic functions of gaseous substances remains one of the major achievements of statistical mechanics, and it is important that the physical chemist be able to carry through such computations efficiently. Hence, although this has necessitated a little repetition, in Chapter Ten important results are collected together in the form of tables; these make the calculations for quite complicated molecules a rather simple matter.

Secondly, it is unlikely that there will be, for some time, a universally agreed system of physical units. Consequently, wherever convenient, dimensionless expressions have been used; for example, entropies and heat capacities are often expressed in terms of their natural unit R, the molar gas constant. Otherwise, the SI system is mostly used in the first nine chapters and calculations involving common non-SI units are discussed in Chapter 10.

In teaching statistical mechanics, I have inevitably been influenced by many existing texts; a regrettably very short and incomplete bibliography is given below. Although this book contains a few details of treatment which I think are new, it is undoubtedly true that the greater number of the ideas introduced have their origins, directly or otherwise, in the important works of Fowler and Guggenheim, and Mayer and Mayer. It is with pleasure that I recall, too, the extent to which my initial approach to teaching the subject was determined in the mid-nineteen-fifties by the lectures of the then Professor F.S. Dainton, F.R.S.

Lastly, it is my pleasant duty to record my thanks to my wife for her continued help during the preparation of the manuscript.

B.J. McClelland

Salford
July, 1972

Bibliography

The books listed below are standard and are referred to in the text by the authors' names only. The list is not in any way intended to represent a survey of the existing literature.

(i) Statistical Mechanics and Thermodynamics

J.G. Aston, Chapter 4 in *Treatise on Physical Chemistry*, edited by H.S. Taylor and S. Glasstone (van Nostrand, 1942).

R.H. Fowler, *Statistical Mechanics* (Cambridge, 1966).

R.H. Fowler and E.A. Guggenheim, *Statistical Thermodynamics* (Cambridge, 1952).

T.L. Hill, *Statistical Thermodynamics* (Addison-Wesley, 1962).

G.N. Lewis and M. Randall (revised by K.S. Pitzer and L. Brewer), *Thermodynamics* (McGraw-Hill, 1961).

J.E. Mayer and M.G. Mayer, *Statistical Mechanics* (Wiley, 1940).

G.S. Rushbrooke, *Statistical Mechanics* (Oxford, 1949).

E. Schrödinger, *Statistical Thermodynamics* (Cambridge, 1967).

R.C. Tolman, *The Principles of Statistical Mechanics* (Oxford, 1938).

(ii) Other Sources

G. Herzberg, I *The Spectra of Diatomic Molecules* (van Nostrand, 1950).

II *Infra-Red and Raman Spectra* (van Nostrand, 1945).

H. Jeffreys and B.S. Jeffreys, *Methods of Mathematical Physics* (Cambridge, 1962).

H. Margenau and G.M. Murphy, *The Mathematics of Physics and Chemistry* (van Nostrand, 1956).

L. Pauling and E.B. Wilson, *Introduction to Quantum Mechanics* (McGraw-Hill, 1935).

Units, Notation and Physical Constants

Units

The basic SI units of interest in the present work are: the metre (m), kilogramme (kg), second (s), ampere (A), and temperature degree Kelvin (K).
Derived units are:

Physical Quantity	*Name and Symbol*	*Definition of Unit*
energy	joule (J)	$kg\,m^2\,s^{-2}$
force	newton (N)	$kg\,m\,s^{-2} = J\,m^{-1}$
electric charge	coulomb (C)	$A\,s$
electric potential	volt (V)	$kg\,m^2\,s^{-3}\,A^{-1} = J\,C^{-1}$

The mole (mol) of X is defined as the amount of X which contains the same number of molecules of X as there are atoms in 0·012 kg of the nuclide ^{12}C.
The unit of volume (V) used in this book is the litre (l) = $10^{-3}\,m^3$.
The SI unit of pressure (P) is Nm^{-2}, but we shall sometimes refer to pressures in atmospheres (1 atmosphere = $101{\cdot}325 \times 10^3\,Nm^{-2}$).

Notation

Notation for the quantities of thermodynamics is:
E Internal energy
H Enthalpy, or heat content

S Entropy
A Work function, or Helmholtz free energy
G Gibbs free energy
C_P, C_V Heat capacities at constant pressure and constant volume.

Properties *per mole* are indicated by placing a tilde over the appropriate symbol, thus: $\tilde{E}$, $\tilde{H}$, $\tilde{S}$, etc. (an exception is made for the gas constant R, which is traditionally the symbol for the molar quantity); correspondingly, the Avogadro number is denoted by $\tilde{N}$. The mass of one mole is denoted by $\tilde{M}$, and the molecular weight (a dimensionless number) by M. In c.g.s. units $\tilde{M} = M$ g, but in SI units $\tilde{M} = 10^{-3} M$ kg.

Physical Constants

Quantity	*Symbol*	*Value*
Avogadro number	$\tilde{N}$	$6{\cdot}023 \times 10^{23}$ mol^{-1}
Gas constant	R	$8{\cdot}314$ J K^{-1} mol^{-1}
Boltzmann constant	$\boldsymbol{k}$	$1{\cdot}381 \times 10^{-23}$ J K^{-1}
Thermochemical calorie	cal	$4{\cdot}184$ J
Planck's constant	h	$6{\cdot}626 \times 10^{-34}$ J s
Mass of the hydrogen atom	m_{H}	$1{\cdot}673 \times 10^{-27}$ kg
Mass of atom having unit atomic weight (atomic mass unit)	u	$1{\cdot}661 \times 10^{-27}$ kg
Speed of light	c	$2{\cdot}998 \times 10^{8}$ m s^{-1}
Elementary charge	e	$1{\cdot}602 \times 10^{-19}$ C

CHAPTER ONE

Introduction

The laws of thermodynamics are broad, empirical generalizations which allow correlations to be made among the various properties of any *macroscopic* (literally, 'large enough to be seen by the unaided eye') portion of matter in equilibrium. The number of atoms in such an amount of matter is of the order of 10^{20} or more, and is so large that the behaviour of the individual atoms or molecules is not explicitly considered; indeed, from the point of view of thermodynamics, the atomic constitution of matter is of little interest.

In contrast, the aim of *statistical mechanics* is to deduce the behaviour of matter in bulk, from a knowledge of the quantal laws which govern the individual molecules. In consequence of this more fundamental approach, statistical mechanics leads to a deeper insight into the theoretical significance of the laws of thermodynamics, and permits the calculation, in terms of atomic and molecular parameters, of quantities, such as heat capacity and entropy, which, in thermodynamics, are left to be determined experimentally. This particular aspect of statistical mechanics is called *statistical thermodynamics*.

The following is a summary of our initial approach to the subject. Many different energies are possible for the individual molecules of a substance. At any instant in time, it may happen that a certain number n_1 of molecules have the same energy ϵ_1, n_2 have the energy ϵ_2, n_3 the energy ϵ_3, and so on. We use the elementary theory of permutations and combinations to work out the most probable, or average, values of $n_1, n_2, n_3, \ldots$, and find that they are given by a simple equation, called the Maxwell–Boltzmann distribution law. In Chapter 3, we show how the values so obtained are related to the energy E and entropy S of the substance; it is then found that E and S, and hence all the other thermodynamic properties, can be expressed in terms of a function f, known as the partition function. Finally, the calculation of f in terms of atomic and molecular parameters is discussed.

We begin by introducing, in this chapter, a few basic notions about molecular energy levels.

1.1 Quantum states and complexions

For an isolated molecule in a confined space, the Schrödinger equation:

$$H\psi = \epsilon\psi \tag{1.1}$$

is satisfied for certain functions $\psi_1, \psi_2, \ldots, \psi_j, \ldots$ (eigenfunctions) and for certain values of ϵ, say $\epsilon_1, \epsilon_2, \ldots, \epsilon_i, \ldots$ (eigenvalues). Thus, such a molecule can exist in certain *quantum states*, each of which is characterized by a wave function ψ_j. The discrete values ϵ_i are referred to as *energy levels*, and we often speak of 'a molecule in the level ϵ_i', meaning that the molecule is in a quantum state for which the energy is ϵ_i. It may happen that a number, say g_i with $g_i > 1$, of states (i.e., wave functions ψ) have the same energy ϵ_i, and we then say that this energy level is g_i-*fold-degenerate*.

The Schrödinger equation can also be applied to macroscopic systems* of particles. This is probably a somewhat novel idea to the physical chemist, who is accustomed to working in terms of the quantum states of single atoms or molecules; there is, however, no restriction upon the size of the systems to which Equation 1.1 might be applied. It is convenient to use the symbols E and Ψ for the energy and wave function of a macroscopic system. For example, the quantum states Ψ_j and energy levels E_i of one mole of a substance, containing about 10^{25} particles (nuclei and electrons), would, in principle, be found by solving the equation:

* Throughout this book, the term *system* or *macroscopic system* is used to mean a macroscopic assembly of atoms, molecules, or ions; or, less specifically, a macroscopic portion of matter (cf. Tolman, p.43; Lewis and Randall, p.6). Fowler (p.8) and others use *system* in a different sense to mean one of the atoms etc. (or statistical elements) of which the assembly is composed.

$$H\Psi = E\Psi \tag{1.2}$$

a differential equation involving 3×10^{25} variables (e.g., the x, y, and z coordinates for each of the particles). It is quite unnecessary (even if it were possible), for the purposes of statistical mechanics, to seek a rigorous solution of this equation; the most important thing is the fact that such solutions do exist.

Not all the Ψ_j thus found may correspond to states which can actually be realized by the macroscopic system. For example, if we could solve Equation 1.2 for the particles which constitute a gaseous mixture of hydrogen and oxygen, we should find some quantum states which correspond to the presence of H_2O molecules. However, since hydrogen and oxygen, in the absence of a catalyst, do not react at lower temperatures, such states are *inaccessible* to the system at lower temperatures; they do not correspond to physical actuality. Another type of inaccessibility, resulting from symmetry restrictions on permitted wave functions, is discussed in Chapter 13. It is convenient to refer to any one of the accessible (physically possible) quantum states of a macroscopic system as a *complexion* of that system.

1.2 Independent molecules: distributions

The simplest systems for the application of statistical mechanics are those in which intermolecular forces can be neglected (e.g., perfect gas systems). The total energy of such a system is just the sum of the individual molecular energies; each of these will be one or other of the ϵ_i values derived from Equation 1.1.

As before, suppose that, for some particular quantum state of the whole system, there are n_1 molecules in the level ϵ_1, n_2 molecules in the level ϵ_2, and so on. The condition of the system which is characterized by the numbers $n_1, n_2, \ldots, n_i, \ldots$ is called a *distribution* (of molecules among the energy levels, or, equivalently, of energy among the molecules), and the numbers $n_1, n_2, \ldots, n_i, \ldots$ are themselves called *distribution numbers*.

It is intuitively evident (and will be proved in Section 13.6) that each complexion is *one physically distinct way of assigning the molecules to the energy levels*. The following are four complexions of a system of four distinguishable* molecules, a, b, c, d:

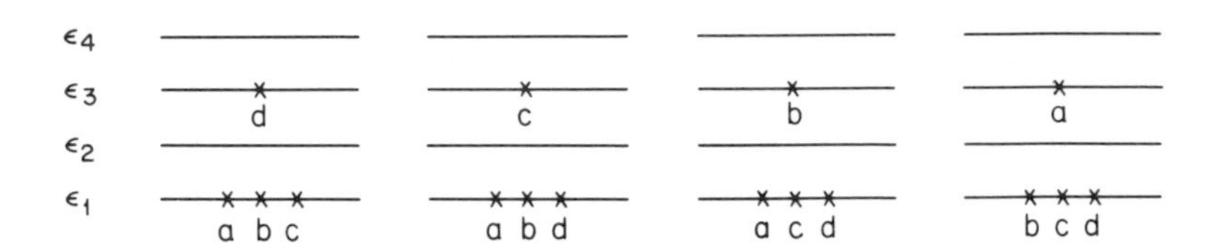

It will be seen that each of these complexions has the same values of the distribution numbers:

$$n_1 = 3; \quad n_3 = 1; \quad n_2, n_4, \ldots = 0\,,$$

and each has the same associated energy:

$$E = 3\epsilon_1 + \epsilon_3\;.$$

The number of complexions associated with a distribution will be denoted by Ω_D, and we shall find that the calculation of this number is an essential step in our statistical method. An equation for this purpose is obtained straightforwardly from an algebraic formula known as the *combinatory rule*; this is discussed in the next section.

1.3 Permutations and combinations: the combinatory rule

Consider the following problem. In how many different *orders* can three letters (a, b, c) be arranged?

The answer could be obtained by writing down all the different arrangements: (a, b, c), (a, c, b), (b, a, c), (b, c, a), (c, a, b), (c, b, a), and six different sequences are found. These different orders are called *permutations* of the group of letters. In the general case, the number of permutations of N things is easily calculated by the following procedure.

The first member of the sequence can be chosen in N ways. The second can be chosen in $(N - 1)$ ways because, after the first has been taken, there are only $(N - 1)$ things from which to choose. Similarly, the third can be chosen in $(N - 2)$ ways, and so on. The total number of ways of obtaining the sequence, i.e., the number

* The important question of whether it can be physically meaningful to label identical molecules in this way is discussed in Section 4.1.

of permutations, is therefore:

$$N \times (N-1) \times (N-2) \times \ldots \times 3 \times 2 \times 1 = N!$$

This product, which is denoted by $N!$, is called *factorial N*. For three letters, the number of permutations is $3! = 3 \times 2 \times 1 = 6$, which is the answer that we obtained in our earlier example.

A related problem is the following. In how many different ways can four molecules a, b, c, d be allocated to two energy levels ϵ_1 and ϵ_2, such that each level is occupied by two molecules? As before, the answer can be obtained by writing down all the different possibilities:

ϵ_1	ϵ_2
a b	c d
a c	b d
a d	c b
b c	a d
b d	a c
c d	a b

and there are six ways of allocating the molecules to the energy levels in this manner. Each of these ways is an example of what is known in mathematics as a *combination*; combinations are defined in general as partitions of a number of things into smaller groups. The calculation of the number of different combinations possible in a given case is accomplished by means of the *combinatory rule*. Before deriving this, let us illustrate the calculation by reference to the above example.

Suppose that the four letters are written down in some order, and the first two assigned to level ϵ_1, the second two to level ϵ_2. The $4! = 24$ ways in which this may be done are shown below:

ϵ_1	ϵ_2	ϵ_1	ϵ_2	ϵ_1	ϵ_2
a b	c d	a d	c b	b d	a c
b a	c d	d a	c b	d b	a c
a b	d c	a d	b c	b d	c a
b a	d c	d a	b c	d b	c a
a c	b d	b c	a d	c d	a b
c a	b d	c b	a d	d c	a b
a c	d b	b c	d a	c d	b a
c a	d b	c b	d a	d c	b a

The arrangements are grouped together in sets of four. Each such set corresponds to only one combination because, within the set, the same pairs of molecules occupy the same levels. There are thus $24/4 = 6$ combinations. The derivation of a general formula is now straightforward.

Suppose that N molecules are grouped into k levels, such that the first level contains n_1 molecules, the second n_2 molecules, and so on, as represented below:

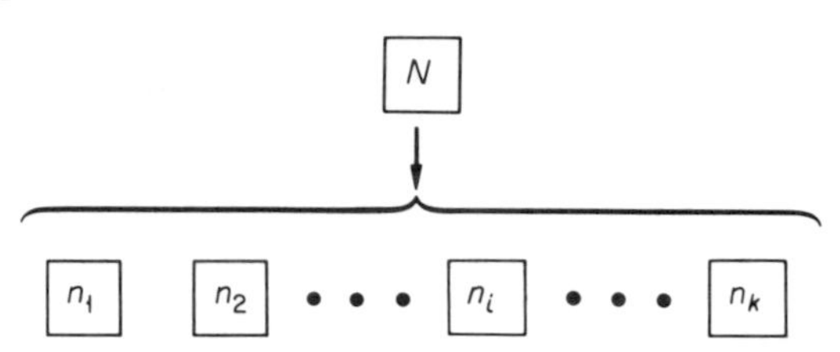

First let the N molecules (or their labels a, b, etc.) be placed in an ordered sequence, and from this sequence place the first n_1 molecules in level 1, the next n_2 in level 2, and so on. Since there are $N!$ ways of obtaining the ordered sequence (i.e., permutations), the above procedure can be carried out in $N!$ ways. Now, in determining the number of combinations, we are not concerned with the order *within* the various levels, and so we must divide $N!$ by a factor which is the number of equivalent ways in which each combination can be achieved. The number of ways in which a *given* n_i molecules can be arranged in sequence in the i-th level is $n_i!$, and therefore the number of ways (differing only in the ordering *within* the levels) of obtaining one combination is the product of such terms:

$$n_1!\, n_2! \ldots n_k! = \prod_{i=1}^{k} n_i!.$$

The symbol Π is an abbreviation meaning that the product of terms like $n_i!$ is to be taken; it may be compared with the abbreviation for a sum:

$$x_1 + x_2 + \ldots + x_k = \sum_{i=1}^{k} x_i.$$

The number of *non-equivalent* arrangements, i.e., the number C of combinations is thus:

$$C = \frac{N!}{\prod n_i!}. \tag{1.3}$$

Equation 1.3 is the combinatory rule.

1.4 Systems with definite total energy

Let us suppose that the total energy E of a system has some fixed and definite value. If intermolecular forces are neglected, the total energy is:

$$E = n_1 \epsilon_1 + n_2 \epsilon_2 + \ldots = \sum_i n_i \epsilon_i \tag{1.4}$$

where $n_1, n_2, \ldots$ are the distribution numbers defined in Section 1.2, and the summation is over all energy levels ϵ_i.

In real physical systems, there are always many distributions which have the *same* total energy E. Consider the following example.

> *Example 1.4*
>
> Suppose that, for a certain type of molecule, the allowed energy levels ϵ_i are $0, \omega, 2\omega, 3\omega, \ldots$, where ω is some unit of energy. Enumerate the complexions of a system of four such molecules having the total energy 2ω.

By inspection, it is immediately found that only two distributions (say distribution 1 and distribution 2) can have the total energy 2ω:

	Distribution 1	Distribution 2
$\epsilon_4 = 3\omega$		
$\epsilon_3 = 2\omega$	×	
$\epsilon_2 = \omega$		× ×
$\epsilon_1 = 0$	× × ×	× ×

Distribution 1: $n_1 = 3, n_2 = 0, n_3 = 1, n_4 \text{ etc.} = 0$

$$E = 3n_1 \epsilon_1 + n_3 \epsilon_3 = 2\omega$$

Distribution 2: $n_1 = 2, n_2 = 2, n_3 \text{ etc.} = 0$

$$E = 2n_1 \epsilon_1 + 2n_2 \epsilon_2 = 2\omega$$

According to the combinatory rule, the number of complexions associated with distribution 1 is:

$$\Omega_{D1} = \frac{4!}{3! \times 1!} = \frac{4 \times 3 \times 2}{3 \times 2} = 4 .$$

(note that $0! = 1$). The four complexions are those represented on p. 4. Similarly, the number of complexions associated with distribution 2 is:

$$\Omega_{D_2} = \frac{4!}{2! \times 2!} = 6 .$$

□ □ □

The total number of complexions (denoted by Ω), associated with a given total energy, is obtained by adding together the Ω_D for all the possible distributions:

$$\Omega = \Omega_{D_1} + \Omega_{D_2} + \Omega_{D_3} + \ldots . \tag{1.5}$$

Thus, in the above example, the value of Ω for $E = 2\omega$ is:

$$\Omega = 6 + 4 = 10 .$$

1.5 Degeneracy of energy levels

A slight complication arises when the energy levels are degenerate. This is demonstrated by the following example.

Example 1.5

Enumerate the complexions of the system in Example 1.4, when the levels $\epsilon_1 = 0$ and $\epsilon_2 = \omega$ are respectively 2- and 3-fold degenerate.

One of the complexions belonging to distribution 2 in Example 1.4 is that in which the molecules a and b are placed in the level ϵ_2, and the molecules c and d are placed in ϵ_1. However, if the levels are degenerate, this arrangement can be realized in more than one distinct way, as follows:

$\epsilon_2 = \omega$ (3 levels)	3 places for a 3 places for b	$3 \times 3 = 9$ ways of placing a and b in this level
$\epsilon_1 = 0$ (2 levels)	2 places for c 2 places for d	$2 \times 2 = 4$ ways of placing c and d in this level

Since a complexion of the assembly is obtained by taking one of the arrangements of a and b with one of the arrangements of c and d, there are 9 × 4 = 36 complexions with this allocation of a, b, c, and d to the two levels. It was shown in Example 1.4 that there are 6 different ways of allocating a pair of molecules to each of the two levels; hence, now:

$$\Omega_{D_2} = 36 \times 6 = 216 .$$

Similarly:

$$\Omega_{D_1} = 36 \times 4 = 144 .$$

□ □ □

The generalization of the procedure of the above example is rather obvious. If the level ϵ_i is g_i-fold degenerate, and is occupied by n_i molecules, then the g_i wave functions can be allocated to the n_i molecules in $(g_i)^{n_i}$ ways. Hence the occurrence of degeneracy increases Ω_D by the product of terms like $(g_i)^{n_i}$, one term for each level. Therefore, the number of ways of obtaining a distribution is:

$$\Omega_D = \frac{N! \prod_i (g_i)^{n_i}}{\prod n_i!} = N! \prod \frac{(g_i)^{n_i}}{n_i!} \qquad (1.6)$$

where the product is taken over all possible energy levels.

1.6 Probability and the most probable distribution

The word 'probability' is used familiarly and qualitatively to designate the extent to which some event is likely to happen. Mathematically, however, the term is defined in the following quantitative way. (For a fuller account see, e.g., Margenau and Murphy, p. 435).

Suppose that, in making an experimental observation, one might obtain any one of r possible results, and suppose that there is no known reason why some of these should be more likely than others. If, out of all the results, just c have in common a particular property X (and so r − c have not the property X), then the *probability of* X *is defined to be*:

$$P(X) = \frac{c}{r} . \qquad (1.7)$$

The meaning of this definition may be rendered clearer by consideration of a few trivial examples.

(i) Flip a coin to obtain heads or tails. Two results are possible, and one of these is heads, so:

$$P(\text{heads}) = \tfrac{1}{2} .$$

(ii) Throw a die with faces numbered from 1 to 6. The probability of getting the value 2 is:

$$P(2) = 1/6 .$$

The probability of getting a number which is even (i.e., 2, 4, or 6) is:

$$P(\text{even number}) = 3/6 = 1/2 .$$

because three of the six possible results have the desired property.

Now let us apply these ideas to systems of molecules, considering first a simple example.

Example 1.6

What is the probability of distribution 1 in Example 1.4. assuming that all complexions associated with a given total energy are equally likely?

If observations are made on the system having the known energy 2ω, any one out of the total of Ω complexions may be found. Of these complexions, Ω_{D_1} have the distribution numbers:

$$n_1 = 3; \; n_2 = 0; \; n_3 = 1; \; n_4 \text{ etc.} = 0 .$$

which characterize the distribution labelled 1. Hence, assuming that all the Ω complexions are equally likely, and writing Ω_{D_1} and Ω in place of c and r, respectively, we obtain from Equation 1.7:

$$P(\text{distribution 1}) = \frac{\Omega_{D_1}}{\Omega} = \frac{4}{10} = \frac{2}{5}$$

□ □ □

Generalizing this example a little, we can see that, if all the complexions associated with some total energy E are equally likely, then the probability of any distribution in a system having energy E is proportional to the corresponding value of Ω_D, and that the distribution with the *greatest* value of Ω_D is the *most probable*. In the next chapter, we will use this fact to determine the most probable values of the distribution numbers n_i, and will then show how these values determine the macroscopic properties of the system.

It may be noted in passing that probability, as defined by Equation 1.7, must be a proper fraction ($P \leqslant 1$, since $r \geqslant c$). Some authors refer to the value of Ω, or the ratio Ω/Ω_0 where Ω_0 is the lowest value of Ω that can be realized physically (e.g., by lowering the temperature towards 0K; see Section 8.3), as the *thermodynamic probability* of the system. However, these quantities are *not* probabilities in the usual mathematical sense.

1.7 Stirling's theorem

In applying equations such as 1.6 to assemblies of molecules, we shall often need to evaluate $N!$ or $n_i!$, where N and n_i are large numbers, perhaps of the order of 10^{24}. It is fortunate that a simple relationship, known as Stirling's approximation, exists (see Appendix 2). Thus:

$$\ln n! = n \ln n - n \,. \tag{1.8}$$

an approximation valid for large values of n.

By this means, Equation 1.6 can be put into a more convenient form. Taking logarithms:

$$\begin{aligned}\ln\Omega_D &= \ln N! + \ln[(g_1)^{n_1}(g_2)^{n_2}\ldots] - \ln[(n_1!)(n_2!)\ldots]\\ &= \ln N! + [n_1\ln g_1 + n_2\ln g_2 + \ldots] - [\ln(n_1!) + \ln(n_2!) + \ldots]\\ &= \ln N! + \sum_i n_i\ln g_i - \sum_i \ln(n_i!)\,.\end{aligned}$$

Applying Stirling's approximation:

$$\begin{aligned}\ln\Omega_D &= N\ln N - N + \sum n_i\ln g_i - \sum n_i\ln n_i + \sum n_i\\ &= N\ln N + \sum n_i\ln g_i - \sum n_i\ln n_i \qquad (1.9)\end{aligned}$$

since $N = \Sigma\, n_i$, the total number of molecules.

Problems

1.1 Suppose that, for a certain kind of molecule, the allowed energy levels are 0, ω, 2ω, 3ω, . . . , and are non-degenerate. Consider a system of six such molecules. What distributions are associated with a total energy 3ω of the system? Evaluate Ω_D for each of these distributions from Equation 1.6, and hence determine the probabilities of the distributions.

1.2 Repeat problem 1.1 for the case in which the energy levels 0 and ω are non-degenerate and levels 2ω and 3ω are respectively 6 – and 10-fold degenerate.

1.3 Consider a mixed crystal containing N_A molecules of A and N_B molecules of B arranged randomly in the crystal lattice sites. Use Equation 1.3 to show that the number of ways in which the molecules can occupy the sites is:

$$W = \frac{(N_A + N_B)!}{N_A!\,N_B!}$$

1.4 Use Stirling's theorem to show that if $N_A = N_B = N/2$ in problem 1.3, then:

$$W = 2^N.$$

1.5 If $N_A = N_B = 2$ in problem 1.3, show that $W = 6$. According to problem 1.4, the result would be $W = 2^4 = 16$. Why the discrepancy?

1.6 Show that if not more than one molecule could occupy any given quantum state, Equation 1.6 would be replaced by:

$$\Omega_D = N! \prod \frac{g_i!}{(g_i - n_i)!\,n_i!}$$

(This result is closely related to the basic equation of Fermi–Dirac statistics, discussed in Chapter 13).

CHAPTER TWO

Maxwell–Boltzmann Statistics

According to Section 1.6, the probability of a distribution is proportional to the corresponding value of Ω_D. Hence we shall obtain (the Maxwell–Boltzmann law for) the most probable distribution by evaluating those $n_1, n_2, \ldots, n_i, \ldots$ which lead to the greatest value of Ω_D for a system in a given state. Before carrying through this procedure, we summarize the assumptions which are implicit in our treatment; a more powerful method, in which most of the simplifying assumptions are avoided, is discussed in Chapter 14.

2.1 Nature of the Systems considered

Present consideration is restricted to systems which have the following characteristics.

(1) *The thermodynamic system is an assembly of N distinguishable and independent particles, in some state defined by the volume V and internal energy E.* By *independent*, it is meant that the forces of interaction between the particles are not explicitly considered when the total energy is calculated. Thus, the total energy is the sum of the energies of the individual molecules, as in Equation 1.4:

$$E = \sum n_i \epsilon_i \, .$$

The molecules of a perfect gas are independent in this sense; we shall, however, find it possible to treat solids and liquids as assemblies of approximately independent particles, even though the intermolecular forces are not negligible in these cases.

By *distinguishable*, it is meant that it is possible to identify individual molecules by labelling them (e.g., a, b, c, . . .), as we did in the previous chapter; in fact, this point requires careful discussion, and it is dealt with in Chapter 4.

(2) *The energies of the individual particles are eigenvalues of the Schrödinger equation* 1.1. Much of statistical mechanics (e.g., the old kinetic theory of gases) could be developed quite correctly by assuming that the molecular energies are those calculated by the methods of classical mechanics; but it is more precise, and often simpler, to use quantal energies throughout. Certain problems are, however, more easily solved by using classical dynamics as an approximation; some of these are considered in Chapter 11.

(3) *There is no restriction upon the ways in which energies ϵ_i can be allocated to the individual particles, other than that the sum of the energies of the particles is E.* In other words, any number of molecules can occupy any particular energy level ϵ_i, and all the different ways of allocating the molecules to energy levels, which result in the total energy E, are possible. This assumption is not always valid, and in Chapter 13 we shall study systems in which the occupation of a particular state is restricted by the Pauli principle.

(4) *All complexions associated with a given E and V are equally likely.* It was shown in Section 1.6 that a consequence of this assumption is that the probability of a distribution is proportional to the corresponding value of Ω_D. Although statistical thermodynamics can be developed without making the previous three assumptions, assumption (4) or some equivalent postulate is *always* necessary.

2.2 The most probable distribution: the Maxwell–Boltzmann distribution law

If all *complexions* associated with a given thermodynamic state (i.e., value of E and V) are equally probable, then the most probable *distribution* for that state is, as we have said, the one for which Ω_D has the largest value. Thus, it is required to find the values of $n_1, n_2, \ldots, n_i, \ldots$ which make Ω_D, calculated from Equation 1.6, a maximum.

It might be expected that these values would be the ones for which:

$$\frac{\partial \Omega_D}{\partial n_i} = 0 \qquad (2.1)$$

for all n_i. This would indeed be the case if all the n_i values could be varied independently, i.e., if there were *no restrictions upon the permitted values of* n_i. In the present case, however, a somewhat different maximization procudure is needed, because we require the largest value of Ω_D associated with a given thermodynamic state, in which the values of E, N, and V are already determined. Thus, the n_i-values finally obtained *must satisfy the following restrictive conditions*:

(i) $$n_1 + n_2 + \ldots \equiv \sum n_i = N, \text{ a constant ;} \qquad (2.2)$$

(ii) $$n_1\epsilon_1 + n_2\epsilon_2 + \ldots \equiv \sum n_i\epsilon_i = E, \text{ a constant .} \qquad (2.3)$$

Suppose that the values of $n_1, n_2, \ldots, n_i, \ldots$ are altered by small amounts* $\delta n_1, \delta n_2, \ldots, \delta n_i, \ldots$. Then the value of Ω_D is changed by the small amount $\delta\Omega_D$:

$$\delta\Omega_D = \left(\frac{\partial \Omega_D}{\partial n_1}\right)\delta n_1 + \left(\frac{\partial \Omega_D}{\partial n_2}\right)\delta n_2 + \ldots + \left(\frac{\partial \Omega_D}{\partial n_i}\right)\delta n_i + \ldots \qquad (2.4)$$

According to condition (1) above:

$$\delta n_1 + \delta n_2 + \ldots + \delta n_i + \ldots = \delta N = 0 \qquad (2.5)$$

and similarly, condition (2) yields:

$$\epsilon_1\delta n_1 + \epsilon_2\delta n_2 + \ldots + \epsilon_i\delta n_i + \ldots = \delta E = 0 \qquad (2.6)$$

For reasons which are shortly explained, we now introduce two constants, γ and β (the values of which, at this stage, are completely arbitrary). Multiplying Equation 2.5 by γ, and Equation 2.6 by β, and subtracting the results from Equation 2.4, one obtains:

* The n_i are regarded as continuous variables, for the purpose of maximizing the *function* Ω_D given by Equation 1.6. Consequently, non-integer values will usually be found for the n_i values at the maximum (these values can actually be regarded as *time averages* of distribution numbers – see Chapter 14). Since, for a macroscopic system, only very large n_i values have any physical importance, their fractional parts can be ignored.

$$\delta\Omega_D = \left(\frac{\partial\Omega_D}{\partial n_1} - \gamma - \beta\epsilon_1\right)\delta n_1 + \left(\frac{\partial\Omega_D}{\partial n_2} - \gamma - \beta\epsilon_2\right)\delta n_2 + \ldots + \left(\frac{\partial\Omega_D}{\partial n_i} - \gamma - \beta\epsilon_i\right)\delta n_i + \ldots \quad (2.7)$$

When Ω_D has a maximum value:

$$\delta\Omega_D = 0 .$$

The values of $n_1, n_2, \ldots, n_i, \ldots$ must therefore be such that the right-hand side of Equation 2.7 vanishes for any small values of $\delta n_1, \delta n_2, \ldots, \delta n_i, \ldots$ *which are consistent with Equations* 2.5 and 2.6. Suppose that we assign arbitrary values to $\delta n_3, \delta n_4, \ldots, \delta n_i, \ldots$. Then, δn_1 and δn_2 must be chosen such that Equations 2.5 and 2.6 hold. In other words, *all but two* of the δn values can be chosen independently.

Since Equation 2.7 holds for any γ and β, let us choose values of γ and β such that:

$$\frac{\partial\Omega_D}{\partial n_1} - \gamma - \beta\epsilon_1 = 0$$
$$\frac{\partial\Omega_D}{\partial n_2} - \gamma - \beta\epsilon_2 = 0 \quad (2.8)$$

Then: $$\delta\Omega_D = \left(\frac{\partial\Omega_D}{\partial n_3} - \gamma - \beta\epsilon_3\right)\delta n_3 + \left(\frac{\partial\Omega_D}{\partial n_4} - \gamma - \beta\epsilon_4\right)\delta n_4 + \ldots + \left(\frac{\partial\Omega_D}{\partial n_i} - \gamma - \beta\epsilon_i\right)\delta n_i + \ldots = 0 \quad (2.9)$$

where the values of $\delta n_3, \delta n_4, \ldots, \delta n_i, \ldots$ *can be chosen independently*.

The only way of ensuring that the right-hand side of Equation 2.9 is zero for any (small) values of $\delta n_3, \delta n_4, \ldots, \delta n_i, \ldots$ is to equate all the quantities in brackets to zero; i.e.:

$$\frac{\partial\Omega_D}{\partial n_3} - \gamma - \beta\epsilon_3 = 0$$

$$\frac{\partial\Omega_D}{\partial n_4} - \gamma - \beta\epsilon_4 = 0$$

$$\frac{\partial\Omega_D}{\partial n_i} - \gamma - \beta\epsilon_i = 0 \quad (2.10)$$

Combining Equations 2.8 and 2.10, we conclude that the values of all n_i which make Ω_D a maximum must satisfy the conditions:

$$\frac{\partial \Omega_D}{\partial n_i} - \gamma - \beta \epsilon_i = 0 \quad (i = 1, 2, 3, \ldots) \tag{2.11}$$

The above method of maximizing Ω_D subject to restrictive conditions is an example of *Lagrange's method of undetermined multipliers*. The parameters β and γ are the undetermined multipliers; however, their values are not indeterminate, because the n_i values must satisfy, in addition to Equation 2.11, the Equations 2.2 and 2.3. We shall later find that β is related in a simple way to the temperature of the system, and γ to a quantity known as the *partition function* (which is itself related to the free energy and temperature of the assembly).

The final part of the problem, namely, the determination of the n_i values which satisfy Equations 2.11, is solved by substituting for Ω_D in these equations. In practice, it is easier to find the maximum of the *logarithm* of Ω_D than of Ω_D itself; the values of n_i which maximize $\ln \Omega_D$ are, of course, just those which maximize Ω_D itself.

We have, therefore, instead of Equation 2.11, the following:

$$\frac{\partial \ln \Omega_D}{\partial n_i} - \gamma - \beta \epsilon_i = 0 \quad (i = 1, 2, 3, \ldots) \tag{2.12}$$

where *now* β and γ are chosen so as to satisfy the equations obtained by replacing Ω_D by $\ln \Omega_D$ in 2.8.

According to Equation 1.9:

$$\ln \Omega_D = N \ln N + \sum n_i \ln g_i - \sum n_i \ln n_i$$

Hence*:

$$\frac{\partial \ln \Omega_D}{\partial n_i} = \ln g_i - (\ln n_i + 1) \tag{2.13}$$

Substituting Equation 2.13 into 2.12:

$$\ln g_i - \ln n_i - 1 - \gamma - \beta \epsilon_i = 0$$

or

$$\ln n_i = \ln g_i - \alpha - \beta$$

where

$$\alpha = 1 + \gamma; \text{i.e.:}$$

$$n_i = g_i e^{-\alpha} e^{-\beta \epsilon_i} \tag{2.14}$$

* A further differentiation gives $\dfrac{\partial^2 \ln \Omega_D}{\partial n_i^2} = -\dfrac{1}{n_i}$ a negative number, and so equation (2.12) is indeed the condition for a maximum, not a minimum Ω_D.

Equation 2.14 is the *Maxwell–Boltzmann distribution law*. It gives the values of $n_1, n_2, \ldots, n_i, \ldots$ for the most probable distribution of molecules among the energy levels. We shall henceforth denote these *most probable distribution numbers* by $\overline{n_1}, \overline{n_2}, \ldots, \overline{n_i}, \ldots$.

2.3 Distribution among quantum states

The numbers $\overline{n_i}$ are the most probable numbers of molecules in the corresponding energy levels ϵ_i. For some purposes, it is useful to introduce numbers $\overline{m_1}, \overline{m_2}, \ldots, \overline{m_j}, \ldots$, which are the most probable numbers of molecules in the *quantum states* determined by the wave functions $\psi_1, \psi_2, \ldots, \psi_j, \ldots$. The distinction between the labelling of the energy levels and that of the quantum states can be illustrated by reference to the energy levels of Example 1.5:

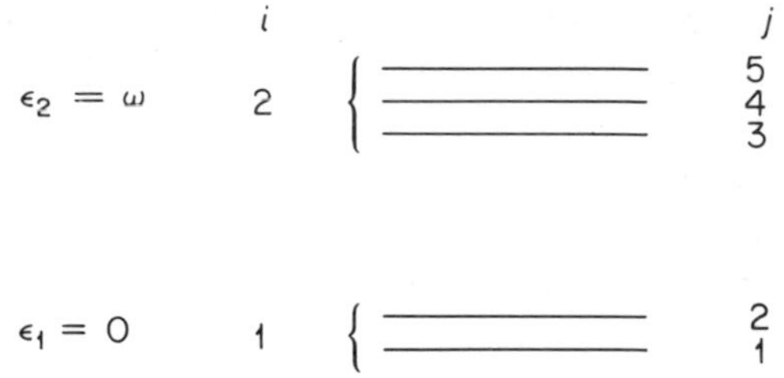

	i		j
$\epsilon_2 = \omega$	2	―――――― ―――――― ――――――	5 4 3
$\epsilon_1 = 0$	1	―――――― ――――――	2 1

Here, the quantum states 1 and 2 have energy ϵ_1, and the states 3, 4, and 5 have energy ϵ_2.

If ϵ_i is g_i-fold degenerate, then the $\overline{n_i}$ molecules are divided amongst g_i quantum states, so that the most probable average number of molecules in each quantum state associated with the level ϵ_i is:

$$\overline{m_j} = \frac{\overline{n_i}}{g_i} = e^{-\alpha} e^{-\beta\epsilon_i}$$

Thus:

$$\overline{m_j} = e^{-\alpha} e^{-\beta\epsilon_j} \tag{2.15}$$

where ϵ_j is the energy of the quantum state ψ_j.

2.4 The molecular partition function

Equations 2.14 and 2.15 contain the undetermined multipliers α and β, and these must be evaluated before the calculation of $\overline{n_i}$ or

$\overline{m_j}$ is possible. It was mentioned in Section 2.2 that the values of β and γ (= $\alpha + 1$) are not indeterminate because Equations 2.2 and 2.3 must hold for a system with a given number of particles N and total energy E. Thus:

$$\sum \overline{n_i} = N,$$

that is:

$$\begin{aligned} N &= g_1 e^{-\alpha} e^{-\beta\epsilon_1} + g_2 e^{-\alpha} e^{-\beta\epsilon_2} + \ldots + g_i e^{-\alpha} e^{-\beta\epsilon_i} + \ldots. \\ &= e^{-\alpha}(g_1 e^{-\beta\epsilon_1} + g_2 e^{-\beta\epsilon_2} + \ldots + g_i e^{-\beta\epsilon_i} + \ldots) \\ &= e^{-\alpha} \sum_i g_i e^{-\beta\epsilon_i} \end{aligned}$$

Hence:

$$e^{-\alpha} = \frac{N}{\sum_i g_i e^{-\beta\epsilon_i}} \tag{2.16}$$

The quantity appearing in the denominator on the right-hand side of Equation 2.16 is denoted by f:

$$f = \sum_i g_i e^{-\beta\epsilon_i} \tag{2.17}$$

and is called the *molecular partition function*, or *sum-over-states.** The adjective 'molecular' is used merely to distinguish f from another 'partition function' which is defined in Chapter 14.

Alternatively, one can sum over the $\overline{m_j}$ values given by Equation 2.15:

$$\begin{aligned} N &= \sum_j m_j \\ &= e^{-\alpha} \sum_j e^{-\beta\epsilon_j} \\ &= e^{-\alpha} f, \end{aligned}$$

where:

$$f = \sum_j e^{-\beta\epsilon_j}. \tag{2.18}$$

It is very important to observe that the definitions 2.17 and 2.18 are completely equivalent, because the summation has a different meaning in the two cases, In Equation 2.17, Σ means the sum over the different *energy levels* i; in Equation 2.18, Σ means the sum

* The expression *sum-over-states* is a translation by Tolman (p.532) of the term *Zustandsumme* introduced by Planck. Actually, Tolman's term has a more general significance than that indicated above, which we discuss in Chapter 14. Strictly speaking, the name *partition function* applies to a quantity introduced by Darwin and Fowler (see Section 2.5, and Fowler, p.38), but the latter is equivalent to f.

over different quantum states j. To the i-th term in Equation 2.17 there correspond g_i separate terms in Equation 2.18, and so:*

$$f = \sum_{\substack{\text{states} \\ j}} e^{-\beta\epsilon_j} = \sum_{\substack{\text{levels} \\ i}} g_i \, e^{-\beta\epsilon_i} \tag{2.19}$$

The origin of the name 'sum-over-states' is now obvious. The alternative, and less obvious, name 'partition function' is used for the following reason. The most probable numbers of molecules in the levels ϵ_k and ϵ_l are:

$$\overline{n_k} = g_k \, e^{-\alpha} \, e^{-\beta\epsilon_k}$$

$$\overline{n_l} = g_l \, e^{-\alpha} \, e^{-\beta\epsilon_l}$$

Hence:

$$\frac{\overline{n_k}}{\overline{n_l}} = \frac{g_k \, e^{-\beta\epsilon_k}}{g_l \, e^{-\beta\epsilon_l}} \tag{2.20}$$

In other words, for a given assembly of particles, the individual terms in the partition function are proportional to the corresponding (most probable) distribution numbers. Hence, the partition function shows the way in which the molecules are allocated to, or partitioned among, the energy levels (or, equivalently, how the total energy is partitioned among the molecules), in the most probable distribution.

The term $e^{-\alpha}$ can now be eliminated from Equations 2.14 and 2.15, giving the distribution numbers $\overline{n_i}$ and $\overline{m_j}$ in terms of f:

$$\begin{aligned} \overline{n_i} &= \frac{N}{f} g_i \, e^{-\beta\epsilon_i} \\ \overline{m_j} &= \frac{N}{f} e^{-\beta\epsilon_j} \end{aligned} \tag{2.21}$$

2.5 Average values of molecular properties

Consider any molecular property $\boldsymbol{p}$ (e.g., energy, dipole moment, etc.). In general, this property may have different values $\boldsymbol{p}_1, \boldsymbol{p}_2, \ldots, \boldsymbol{p}_j, \ldots$ respectively, for the various quantum states $1, 2, \ldots, j, \ldots$.

* The following analogy may also be helpful: $x_1 + x_1 + x_2 + x_2 + x_2$ has the same value as $2x_1 + 3x_2$.

The *average* value of $\boldsymbol{p}$, associated with any given way of allocating the molecules to quantum states, is defined to be:

$$\overline{\boldsymbol{p}} = \frac{\sum_j m_j \boldsymbol{p}_j}{\sum_j m_j} \tag{2.22}$$

The *most probable* value (denoted by $\overline{\overline{\boldsymbol{p}}}$) of this average is obtained by putting the most probable values of m_j in Equation 2.22. Thus:

$$\overline{\overline{\boldsymbol{p}}} = \frac{\sum_j \overline{m_j} \boldsymbol{p}_j}{\sum_j \overline{m_j}} \tag{2.23a}$$

$$= \frac{\sum_j \mathrm{e}^{-\beta \epsilon_j} \boldsymbol{p}_j}{f} \tag{2.23b}$$

This means that, if $\overline{\boldsymbol{p}}$ is determined experimentally for some system, the value which is obtained will most probably be the one given by Equation 2.23. In principle, other values of $\overline{\boldsymbol{p}}$, corresponding to other distributions, *might* be found, but, as is demonstrated in the following note, this possibility can be entirely neglected; i.e., it can be shown that, for macroscopic systems, *the experimental value of $\overline{\boldsymbol{p}}$ is identical with the value $\overline{\overline{\boldsymbol{p}}}$ calculated from Equation 2.23.*

There are two ways in which this can be done. The first is to make the reasonable assumption that the experimental value of $\overline{\boldsymbol{p}}$ should actually be calculated by averaging over *all* complexions (i.e., not just those belonging to the most probable distribution). The averaging can be carried out by a statistical method due to Darwin and Fowler, which uses the theory of the complex variable (Schrödinger, Ch. 6, especially p.37; Fowler, Ch.2), or, as we show in Section 14.7, by means of the canonical ensemble theory. This average value is still given by Equation 2.23a, provided that $\overline{m_j}$ *now* means the value of m_j *averaged over all complexions*. Remarkably, however, it is found that these average values are also given by Equation 2.21, and hence 2.23b follows. Instead of going into details here, we discuss the second, simpler demonstration.

Let D_0 be the most probable distribution and D_1 be any other distribution obtained from D_0 by rearranging the molecules among the energy levels in such a way that *the total energy is unchanged*. The probabilities of D_1 and D_0 are in the ratio:

$$\frac{\Omega_{D_1}}{\Omega_{D_0}}$$

and the physical importance of D_1 is determined by this ratio. The distribution numbers for D_0 are $\overline{n_1}, \overline{n_2}, ..., \overline{n_i}, ...$, and the corresponding values for D_1 can be written as:

$$(\overline{n_1} + \Delta n_1), (\overline{n_2} + \Delta n_2), ..., (\overline{n_i} + \Delta n_i), ...,$$

where $\Delta n_1, \Delta n_2, ..., \Delta n_i, ...$, are finite (positive or negative) numbers.

Now:

$$\ln \frac{\Omega_{D_1}}{\Omega_{D_0}} = \ln \Omega_{D_1} - \ln \Omega_{D_0} = \Delta \ln \Omega_D$$

Taylor series expansion of $\ln \Omega_D$ about $\ln \Omega_{D_0}$ yields:

$$\Delta \ln \Omega_D = \sum_i \left(\frac{\partial \ln \Omega_D}{\partial n_i} \right) \Delta n_i + \frac{1}{2} \sum_i \left(\frac{\partial^2 \ln \Omega_D}{\partial n_i^2} \right) (\Delta n_i)^2 + ... \qquad (2.24)$$

(Note that according to Equation 2.13 the mixed derivatives are zero)
From Equation 2.12:

$$\frac{\partial \ln \Omega_D}{\partial n_i} = \gamma + \beta \epsilon_i.$$

Hence:

$$\sum_i \left(\frac{\partial \ln \Omega_D}{\partial n_i} \right) \Delta n_i = \gamma \sum_i \Delta n_i + \beta \sum_i \epsilon_i \Delta n_i = 0,$$

the zero value arising because of the constancy of the total number of molecules ($\Sigma \Delta n_i = 0$) and total energy ($\Sigma \epsilon_i \Delta n_i = 0$). Differentiating Equation 2.13:

$$\frac{\partial^2 \ln \Omega_D}{\partial n_i^2} = -\frac{1}{n_i}$$

Therefore, Equation 2.24 becomes:

$$\Delta \ln \Omega_D = -\frac{1}{2} \sum_i \frac{(\Delta n_i)^2}{\overline{n_i}},$$

that is:

$$\frac{\Omega_{D_1}}{\Omega_{D_0}} = \exp \left(-\frac{1}{2} \sum_i \frac{(\Delta n_i)^2}{\overline{n_i}} \right).$$

For macroscopic systems, this is extremely small if D_1 differs at all significantly from D_0. To illustrate this, let us suppose that the distribution numbers of D_1 all differ from the corresponding numbers of D_0 by approximately 0·1% of the latter (we may assume that the signs and exact numerical values of Δn_i are chosen so that $\Sigma \Delta n_i$ and $\Sigma \epsilon_i \Delta n_i$ are both zero), i.e.:

$$\frac{|\Delta n_i|}{\overline{n_i}} \approx \frac{1}{10^3}$$

so that:

$$\frac{(\Delta n_i)^2}{n_i} = \frac{n_i}{10^6}$$

Then:

$$-\frac{1}{2}\sum\frac{(\Delta n_i)^2}{n_i} \approx -\frac{1}{2 \times 10^6}\sum n_i = -\frac{N}{2 \times 10^6}$$

For 1 mole, $N = 6{\cdot}023 \times 10^{23}$. Hence:

$$\frac{\Omega_{D_1}}{\Omega_{D_0}} = e^{-10^{17}},$$

which is an extremely small number. The same is true *a fortiori* for larger values of Δn_i. This justifies our neglect of distributions other than D_0.

A system is said to be in *equilibrium* when every one of its observable properties exhibits no change with time. At *any* instant, the observed value of any property $\bar{p}$ will almost certainly be that given by Equation 2.23, and, since this value does not change with time, it is a property of the system at equilibrium. The same is true of the distribution numbers $\bar{n_i}$. For these reasons, the most probable distribution may be referred to as the equilibrium distribution.

2.6 The parameter β

In Section 2.4, it was shown that, because the total number of particles N of the assembly is given, it is possible to eliminate the undetermined multiplier α from our equations. It remains to evaluate β, and unfortunately this is a less straightforward matter. However, it is possible to draw the following conclusions about the properties of this parameter.

(i) *β must be positive* because, with a negative or zero value of β, f, as given by Equation 2.17, would diverge. Thus, if β were zero, f would be the sum of an infinite number of terms g_i; if β were negative, the position would be even worse, for then f would be the sum of an infinite number of progressively increasing terms.

(ii) *β is related to the temperature of the system.* Consider two assemblies of, respectively, N' and N'' particles, having total energies E' and E'', the most probable distribution numbers being, correspondingly:

$$\begin{aligned} \overline{n'_i} &= g'_i\,\mathrm{e}^{-\alpha'}\,\mathrm{e}^{-\beta'\epsilon'_i} \\ \overline{n''_i} &= g''_i\,\mathrm{e}^{-\alpha''}\,\mathrm{e}^{-\beta''\epsilon''_i} \end{aligned} \tag{2.25}$$

Suppose that the assemblies are placed in thermal contact (but are otherwise isolated), *so that they can interchange energy* until equilibrium is reached. In the new equilibrium state, the energies of the assemblies need no longer be E' and E''; in fact, all that is known *a priori* is that the total energy of the *combined system* is:

$$E = E' + E''.$$

Equations 2.25 are obtained for the initial states of the separate systems *because* the following conditions hold (cf. Section 2.2):

$$\sum_i n'_i = N'; \quad \sum_i n''_i = N'' \tag{2.26}$$

$$\sum_i n'_i\epsilon'_i = E'; \quad \sum_i n''_i\epsilon''_i = E'' \tag{2.27}$$

Now, in the final state of equilibrium:

$$\sum_i n'_i = N'; \quad \sum_i n''_i = N'' \tag{2.26'}$$

but, instead of the *two* equations 2.27, we have the *one* condition:

$$\sum_i n'_i\epsilon'_i + \sum_i n''_i\epsilon''_i = E,$$

and it is easily shown (problem 2.5) that the method of undetermined multipliers *now* gives:

$$\begin{aligned} \overline{n'_i} &= g'_i\,\mathrm{e}^{-\alpha'}\,\mathrm{e}^{-\beta\epsilon'_i} \\ \overline{n''_i} &= g''_i\,\mathrm{e}^{-\alpha''}\,\mathrm{e}^{-\beta\epsilon''_i} \end{aligned} \tag{2.28}$$

where β must have the *same* value for both systems. Hence, if *any* two assemblies for which the Maxwell–Boltzmann law is valid *can exchange energy and are in equilibrium, then they must have the same value of* β. Thermodynamically, any two systems which can exchange energy *have the same temperature* when they are in equilibrium. Thus, β in statistical mechanics has the same role as T in thermodynamics; or, mathematically, β is some universal function of T.

(iii) *A decrease in* β *is associated with an increase in temperature.* According to Equation 2.20, the ratio of any two most-probable distribution numbers of one system is:

$$\frac{\overline{n_k}}{\overline{n_l}} = \frac{g_k}{g_l} \mathrm{e}^{-\beta(\epsilon_k - \epsilon_l)}.$$

Therefore, if $\epsilon_k > \epsilon_l$ (i.e., if ϵ_k is a higher energy level than ϵ_l), then, when β decreases, the ratio $\overline{n_k}/\overline{n_l}$ must increase. Hence, a decrease in the value of β causes an increase in the numbers of molecules in the higher energy levels, at the expense of lower levels, which must mean that the total energy of the system is increased. Thermodynamically, such an increase in energy would be associated with a *rise* in temperature.

Before considering the *quantitative* relationship between β and T we must state precisely what is meant by temperature, because there are many ways in which numerical temperature scales can be chosen. Throughout this book we use the term to mean the *absolute thermodynamic temperature* (cf. Fowler and Guggenheim pp. 57 and 662). This is *defined* by the equation

$$dQ = T\,dS \tag{2.29}$$

wherein dS is the increase in entropy of a system when it absorbs reversibly an amount of heat dQ at the temperature T. Experimental measurement of this temperature is accomplished by use of thermodynamical relationships between T and observable properties. For example, it can be shown from Equation 2.29 that if $\tilde{V}$ is the volume of one mole of a perfect gas* and P the pressure, then

$$P\tilde{V} = RT \tag{2.30}$$

where R is the molar gas constant. Therefore, if $P\tilde{V}/R$ is determined by extrapolating PV measurements on real gases to the limit $P \rightarrow 0$, the temperature is obtained; this is the basis of gas thermometry, and, of course, other types of thermometer can be calibrated against the gas thermometer.

In Sections 3.2 and 3.5 we will derive from Equation 2.29 the result

$$\beta = \frac{1}{kT}$$

* The *thermodynamical* definition of a perfect gas (on which the derivation of Equation 2.30 is based) is that it is a system for which the energy and enthalpy derivatives $(\partial E/\partial V)_T$ and $(\partial H/\partial P)_T$ are zero. On the other hand, the *molecular* definition of a perfect gas is that it is one in which the intermolecular forces are negligible. Finally, the *practical* definition of a perfect gas is that it is any real gas at very low pressures. Elementary arguments show that these definitions are in fact equivalent.

wherein k is a positive constant, called the *Boltzmann constant.* We will also prove (in Section 4.5) that

$$P\tilde{V} = \tilde{N}kT$$

($\tilde{N}$ is the Avogadro number) for a perfect gas. Comparing this with Equation 2.30,

$$k = \frac{R}{\tilde{N}} \tag{2.31}$$

That is,

$$k = \frac{8{\cdot}314}{6{\cdot}023 \times 10^{23}} = 1{\cdot}381 \times 10^{-23}\ \mathrm{J\,K^{-1}}.$$

Consequently, at room temperature ($T \approx 300\,\mathrm{K}$), $kT \approx 4 \times 10^{-21}$ J.

Suppose that, in Equation 2.20, l refers to the lowest level (now denoted by the suffix 0), and g_k/g_0 is taken as unity. Then:

$$\frac{\overline{n_k}}{\overline{n_0}} = \mathrm{e}^{-(\epsilon_k - \epsilon_0)/kT}$$

Taking $kT = 4 \times 10^{-21}$ J, we obtain the following results:

$\epsilon_k - \epsilon_0$ (J)	0	10^{-21}	10^{-20}	10^{-19}
$\overline{n_k}/\overline{n_0}$	1	0·78	0·08	4×10^{-6}

Thus, energy levels lying higher than about $10kT$ (or 4×10^{-20} J at room temperature) above the lowest level are not occupied to any important degree.

Problems

2.1 Statistical mechanics can be applied to the theoretical treatment of electrons in metals. We show in Section 13.2 that, because not all the conditions listed in Section 2.1 apply to electrons, the number of complexions associated with a particular distribution is *not* given by Equation 1.6 but by:

$$\Omega_D = \prod \frac{g_i!}{(g_i - n_i)!\, n_i!}$$

Assuming that all the factorial terms are large, evaluate $\ln \Omega_D$ and $\partial \ln \Omega_D / \partial n_i$. Hence show that:

$$n_i = \frac{g_i}{e^{\alpha} e^{\beta \epsilon_i} + 1}$$

the Fermi–Dirac distribution law.

2.2 The partition function of argon at 400 K and 1 atmosphere pressure may be taken as (cf. problem 4.1):

$$f = 12{\cdot}4 \times 10^{30}$$

How many molecules out of one mole of argon at this temperature and pressure will be, on average, in the lowest energy level?

2.3 Two of the energy levels of a molecule X are:

$$\epsilon_1 = 6{\cdot}1 \times 10^{-21}\ \mathrm{J} \quad \text{and} \quad \epsilon_2 = 8{\cdot}4 \times 10^{-21}\ \mathrm{J},$$

the corresponding degeneracies being $g_1 = 3$ and $g_2 = 5$. What is the ratio of the distribution numbers $\overline{n_1}/\overline{n_2}$ in an assembly of molecules of X at (a) 300 K, (b) 3000 K?

2.4 Show that, for systems obeying the conditions listed in Section 2.1:

$$\ln \Omega_{D_0} = N \ln f + \beta E$$

Calculate Ω_{D_0} for one mole of argon at 400 K and 1 atmosphere pressure. [Use the value of f given in problem 2.2 and take (cf. problem 4.2) $\beta E = \frac{3}{2}N$].

2.5 Consider two systems in thermal contact. Show that (in the notation of Section 2.6) the number of complexions of the whole is:

$$\Omega_D = \left(N'! \prod \frac{(g_i')^{n_i'}}{n_i'!}\right) \left(N''! \prod \frac{(g_i'')^{n_i''}}{n_i''!}\right).$$

For the most probable distribution:

$$\delta\Omega_D = \sum \frac{\partial\Omega_D}{\partial n_i'} \delta n_i' + \sum \frac{\partial\Omega_D}{\partial n_i''} \delta n_i'' = 0, \qquad \text{(a)}$$

subject to the restrictions:

$$\sum \delta n_i' = 0 \qquad \text{(b)}$$

$$\sum \delta n_i'' = 0 \qquad \text{(c)}$$

$$\sum \delta n_i' \epsilon_i' + \sum \delta n_i'' \epsilon_i'' = 0. \qquad \text{(d)}$$

Multiply equations (b), (c), and (d) by γ', γ'', and β respectively, subtract from equation (a), and deduce that:

$$\frac{\partial\Omega_D}{\partial n_i'} - \gamma' - \beta\epsilon_i' = 0 \qquad (i = 1, 2, ...)$$

$$\frac{\partial\Omega_D}{\partial n_i''} - \gamma'' - \beta\epsilon_i'' = 0 \qquad (i = 1, 2, ...)$$

Hence obtain Equation 2.28.

CHAPTER THREE

The Calculation of Thermodynamic Properties

Two concepts have now been developed which are of particular importance in the application of statistical mechanics to the calculation of thermodynamic properties. The first is the evaluation of the total energy of a system as the sum of the energies of the separate particles (Equations 1.4 and 2.3); by identifying this total with the thermodynamical *internal energy*, we shall obtain an expression for the latter in terms of the molecular partition function. The second concept is the statistical mechanical interpretation (Section 2.6) of temperature; from this we shall obtain an expression for the entropy of the system. The remaining thermodynamic properties can then be deduced from known thermodynamic equations involving the energy and entropy functions.

The equations derived in this chapter apply, of course, to systems which satisfy the conditions listed in Section 2.1. We shall find, in the following chapter, that the expressions for entropy and work function must be modified slightly in the case of *gaseous* systems.

3.1 Internal energy

The internal energy is identified with the total energy of the system, given as the sum of the energies of the individual particles:

$$E = \sum_i n_i \epsilon_i \tag{3.1}$$

One set of distribution numbers n_i which satisfies Equation 3.1 is:

$$\overline{n_i} = g_i e^{-\alpha} e^{-\beta\epsilon_i}$$

$$= \frac{N}{f} g_i e^{-\beta\epsilon_i}, \tag{3.2}$$

where f is the partition function defined in Section 2.4:

$$f = \sum_i g_i e^{-\beta\epsilon_i}. \tag{3.3}$$

Therefore, substituting for n_i in Equation 3.1:

$$E = \frac{N}{f} \sum_i g_i e^{-\beta\epsilon_i} \epsilon_i \tag{3.4}$$

The energy levels ϵ_i are solutions of the Schrödinger Equation 1.1. They are determined (see Chapters 4 and 5), in the absence of external electric or magnetic fields, by the molecular structure and the *volume* in which the molecule can move. Let us consider the effect upon f of varying the value of β (and hence, according Section 2.6, of the temperature), *keeping the values of* ϵ_i *(and therefore the volume) constant.*

Now:

$$f = g_0 e^{-\beta\epsilon_0} + g_1 e^{-\beta\epsilon_1} + \ldots + g_i e^{-\beta\epsilon_i} + \ldots$$

Hence:

$$\left(\frac{\partial f}{\partial \beta}\right)_V = -\epsilon_0 g_0 e^{-\beta\epsilon_0} - \epsilon_1 g_1 e^{-\beta\epsilon_1} - \ldots - \epsilon_i g_i e^{-\beta\epsilon_i} - \ldots$$

$$= -\sum_i g_i e^{-\beta\epsilon_i} \epsilon_i \tag{3.5}$$

Comparing this with Equation 3.4, we obtain finally:

$$E = -\frac{N}{f}\left(\frac{\partial f}{\partial \beta}\right)_V$$

$$= -N\left(\frac{\partial \ln f}{\partial \beta}\right)_V \tag{3.6}$$

An alternative expression for internal energy can be obtained from Equation 1.9:

$$\ln\Omega_D = N \ln N + \sum_i n_i \ln g_i - \sum_i n_i \ln n_i. \tag{1.9'}$$

If the n_i-values are those of the most probable distribution then, from Equation 3.2:

$$\ln \overline{n_i} = \ln N - \ln f + \ln g_i - \beta\epsilon_i.$$

Substituting this value of $\ln \overline{n_i}$ in Equation 1.9, and remembering that:

$$\sum_i \overline{n_i} = N \quad \text{and} \quad \sum_i \overline{n_i}\,\epsilon_i = E,$$

we obtain:

$$\ln \Omega_{D_0} = N \ln f + \beta E \tag{3.7}$$

or

$$E = \frac{\ln \Omega_{D_0}}{\beta} - \frac{N \ln f}{\beta} \tag{3.8}$$

3.2 Entropy

The *a priori* calculation by statistical mechanics of the entropy of a system is not quite so straightforward as the calculation of the energy, because there is no equation analogous to Equation 3.1 from which to start.

Consider a thermodynamic system, containing a fixed quantity of matter, in equilibrium at temperature T and pressure P. An infinitesimal displacement from one equilibrium state to another with absorption of heat $\mathrm{d}Q$ results in the energy change

$$\mathrm{d}E = \mathrm{d}Q - P\mathrm{d}V$$

if the only work done in the process is the work of expansion. Since an infinitesimal change between two equilibrium states is, by definition, a thermodynamically reversible process, $\mathrm{d}Q$ can be replaced by $T\mathrm{d}S$ according to Equation 2.29, and then

$$\mathrm{d}E = T\mathrm{d}S - P\mathrm{d}V\,. \tag{3.9}$$

We show that this equation has a counterpart in statistical mechanics, and in so doing obtain statistical mechanical equations for temperature and entropy. We here consider a simplified treatment, and postpone the full derivation until Section 3.5. The simplification consists in treating only the case in which a constant volume is maintained, so that Equation 3.9 is replaced by

$$\mathrm{d}E = T\mathrm{d}S\,. \tag{3.10}$$

For an assembly in equilibrium, the most probable distribution numbers are those which lead to Equation 3.7. If the assembly is displaced to a new equilibrium condition, new values of f, β, and E apply. Thus, the values of Ω_{D_0} before and after displacement are given by:

$$\begin{aligned} \ln \Omega_{D_0} &= N \ln f + \beta E \\ \ln \Omega'_{D_0} &= N \ln f' + \beta' E' \end{aligned} \tag{3.11}$$

where the primed values are the new equilibrium values.

If the above change is infinitesimal, then

$$\begin{aligned} \mathrm{d} \ln \Omega_{D_0} &= \ln \Omega'_{D_0} - \ln \Omega_{D_0} \\ &= N(\ln f' - \ln f) + (\beta' E' - \beta E) \\ &= N \mathrm{d} \ln f + (\beta + \mathrm{d}\beta)(E + \mathrm{d}E) - \beta E \\ &= N \mathrm{d} \ln f + \beta \mathrm{d}E + E \mathrm{d}\beta \end{aligned} \tag{3.12}$$

Let us suppose that, in this infinitesimal process, the volume remains constant, so that (cf. Section 3.1) the energy levels ϵ_i do not change. In this case, the only way in which $\ln f$ can vary is through a change in β:

$$\begin{aligned} N \mathrm{d} \ln f &= N \left(\frac{\partial \ln f}{\partial \beta} \right)_V \mathrm{d}\beta \\ &= -E \mathrm{d}\beta \end{aligned} \tag{3.13}$$

by Equation 3.6. Therefore, substituting Equation 3.13 into 3.12:

$$\mathrm{d}E = \frac{\mathrm{d} \ln \Omega_{D_0}}{\beta} \tag{3.14}$$

Equation 3.14 is the statistical mechanical counterpart of the thermodynamical Equation 3.10. In Section 2.6, we saw that β is a universal (i.e., the same for all assemblies) function of temperature, and, moreover, that β decreases as T increases. In view of this, it is tempting* to make Equations 3.10 and 3.14 formally identical, by putting:

$$\beta = \frac{1}{\boldsymbol{k} T} \tag{3.15}$$

$$\mathrm{d}S = \boldsymbol{k}\, \mathrm{d} \ln \Omega_{D_0} \tag{3.16}$$

where $\boldsymbol{k}$ is some constant, which must have the dimensions of (energy)(degree)$^{-1}$, and the value of which must be chosen so as to make the two sides of Equation 3.16 identical.

* A proof of the following equations is given in Section 3.5.

Integrating Equation 3.16, one obtains:

$$\int_{S_1}^{S_2} \mathrm{d}S = k \int_{\ln \Omega_{D_0(1)}}^{\ln \Omega_{D_0(2)}} \mathrm{d} \ln \Omega_{D_0} \tag{3.17}$$

that is:

$$S_2 - S_1 = k \ln \frac{\Omega_{D_0(2)}}{\Omega_{D_0(1)}} \tag{3.18}$$

an equation which relates the *difference* in entropy of two states of a system to the ratio of the values of Ω_{D_0} for these states. Alternatively, indefinite integration of Equation 3.16 gives:

$$S = k \ln \Omega_{D_0} + c \tag{3.19}$$

where c is some constant. Physically, however, only entropy *differences* can be measured experimentally (cf. Chapter 8), and, in the calculation of such differences by means of Equation 3.19, the constant c will cancel. The actual numerical value of c is therefore of no practical importance, and it is conventional to put $c = 0$, so that:

$$S = k \ln \Omega_{D_0} . \tag{3.20}$$

Equation 3.20 was first obtained by Boltzmann, and is called the Boltzmann equation, or Boltzmann hypothesis; as we have already remarked (Section 2.6), the constant k is called the Boltzmann constant, and is equal to $R/\tilde{N}$.

It can be shown (Schrödinger, p. 37) that, within the accuracy of Stirling's theorem, $\ln \Omega_{D_0}$ *has the same value* as $\ln \Omega$.

A rough demonstration of how this arises (but not a proof) is as follows. Suppose that the values of Ω_D are arranged in order of descending magnitude:

$$\Omega = \Omega_{D_0} + \Omega_{D_1} + \Omega_{D_2} + \ldots$$

$$= \Omega_{D_0} \left(1 + \frac{\Omega_{D_1}}{\Omega_{D_0}} + \frac{\Omega_{D_2}}{\Omega_{D_0}} + \ldots \right).$$

Then:

$$\ln \Omega = \ln \Omega_{D_0} + \ln \left(1 + \frac{\Omega_{D_1}}{\Omega_{D_0}} + \frac{\Omega_{D_2}}{\Omega_{D_0}} + \ldots \right)$$

$$\approx \ln \Omega_{D_0} ,$$

since the bracketed term is of a smaller order of magnitude than that of Ω_{D_0}.

An equivalent expression to 3.20 is therefore:

$$S = k \ln \Omega . \tag{3.20a}$$

Equation 3.20a permits the following physical meaning to be assigned to entropy. The entropy is a measure of the number of complexions of an assembly with a given energy, i.e., the number of ways in which the particles can be allocated to the molecular quantum states. For this reason, *it is often said that entropy is a measure of the 'randomness' in the assembly, because the greater the number of complexions, the greater the degree of uncertainty as to the way in which any given molecular quantum state is occupied.*

It might be remarked that Equation 3.20 is often derived in elementary texts by somewhat artless arguments which appear to relate entropy to the 'probability of a thermodynamic state'. Unfortunately, there is invariably much vagueness about the meaning of such probability, and, as Fowler points out, the theory of probability is not really relevant to the discussion. (See Fowler, p.200ff, for detailed criticism, also Mayer and Mayer, Ch. 4.)

3.3 The statistical mechanical values of thermodynamic quantities

(i) *Temperature*

According to Equation 3.15:

$$T = \frac{1}{k\beta} \tag{3.21}$$

Substituting for β in Equation 3.3, we obtain for the molecular partition function:

$$f = \sum_i g_i \mathrm{e}^{-\epsilon_i/kT}. \tag{3.22}$$

(ii) *Energy*

From Equation 3.15:

$$\frac{\partial\beta}{\partial T} = -\frac{1}{kT^2}$$

Therefore, from Equation 3.6:

$$E = \frac{-N\left(\dfrac{\partial \ln f}{\partial T}\right)_V}{\left(\dfrac{\partial\beta}{\partial T}\right)_V}$$

$$= NkT^2 \left(\frac{\partial \ln f}{\partial T}\right)_V \tag{3.23}$$

(iii) *Entropy*

$$S = k \ln \Omega_{D_0} = k \ln \Omega \tag{3.20'}$$

Substituting for $\ln \Omega_{D_0}$ according to Equation 3.7:

$$S = Nk \ln f + \frac{E}{T}. \tag{3.24}$$

Further, substituting for E according to Equation 3.23:

$$S = Nk \left[\ln f - T\left(\frac{\partial \ln f}{\partial T}\right)_V \right]. \tag{3.25}$$

(iv) *Work Function*

The work function, or Helmholtz free energy, A, is defined by the equation:

$$A = E - TS.$$

Substituting for S given by Equation 3.24, one has:

$$A = -NkT \ln f. \tag{3.26}$$

(v) *Pressure*

The relationship between pressure and partition function can be obtained immediately from Equation 3.26, since there is a thermodynamic relationship:

$$P = -\left(\frac{\partial A}{\partial V}\right)_T$$

and therefore:

$$P = NkT\left(\frac{\partial \ln f}{\partial V}\right)_T \tag{3.27}$$

(vi) *Heat Capacity*

The heat capacity C_V is defined as:

$$C_V = \left(\frac{\partial E}{\partial T}\right)_V$$

Hence, differentiating Equation 3.23:

$$C_V = 2NkT\left(\frac{\partial \ln f}{\partial T}\right)_V + NkT^2\left(\frac{\partial^2 \ln f}{\partial T^2}\right)_V \tag{3.28}$$

(vii) *Molar Thermodynamic Properties*

We denote the values of thermodynamic properties *per mole* by $\tilde{E}$, $\tilde{S}$, $\tilde{A}$, $\tilde{C}_V$, etc. For 1 mole, $N = \tilde{N}$, the Avogadro number; moreover, according to Equation 2.29:

$$\tilde{N}k = R$$

Hence, the above equations applied to *molar* quantities become:

$$\tilde{E} = RT^2\left(\frac{\partial \ln f}{\partial T}\right)_V$$

$$\tilde{S} = RT\ln f + \frac{\tilde{E}}{T}$$

$$\tilde{A} = -RT\ln f$$

and so on.

3.4 Factorization of the molecular partition function

The importance of the molecular partition function is now evident; thermodynamic functions for an assembly which obeys Maxwell–Boltzmann statistics can be calculated when the molecular partition function for that assembly is known. At this stage we have therefore obtained, in principle, a method of calculating thermodynamical properties by statistical mechanics; the only requirement is that the permitted energy levels of the molecules be known. Much of the remainder of the present book is concerned with the detailed applications of this method to real systems.

In such applications, one great simplification is the fact that the molecular partition function can usually be expressed as a *product of simpler factors*, each factor being evaluated independently of the others. Since all the equations for thermodynamic functions derived in the preceding section involve only the *logarithm* of f, and since $\ln(ab) = \ln a + \ln b$, this means that the thermodynamic functions themselves can be expressed as the *sums of simpler terms*, and each term is independent of the others. An important example, which we now consider, is the factorization of the partition function for a *gas*.

The energy of a molecule in the gaseous phase can be expressed as the sum of two independent terms:

$$\epsilon = \epsilon_{\mathbf{trans}} + \epsilon_{\mathbf{int}}.$$

$\epsilon_{\mathrm{trans}}$ is the kinetic energy of translation, i.e., the energy associated with the movement of the centre of mass of the molecule in space. The remainder of the energy is referred to as the *internal molecular energy*,* ϵ_{int}. Thus, ϵ_{int} can be defined as the total energy of the molecule when the centre of mass is at rest.

Both $\epsilon_{\mathrm{trans}}$ and ϵ_{int} are quantized. Each value of the translational energy is associated with one or more translational quantum states, i.e., wave functions which depend on ('are functions of') the co-ordinates of the centre of mass of the molecule. The individual translational quantum *states* will be distinguished from one another by labelling them 1,2,3,...,t,..., and the corresponding energies will be denoted by:

$$\epsilon_{\mathrm{trans}(1)}, \epsilon_{\mathrm{trans}(2)}, \ldots, \epsilon_{\mathrm{trans}(t)}, \ldots$$

In a similar way, the internal molecular energy values will be denoted by:

$$\epsilon_{\mathrm{int}(u)}\ , (u\ -\ 1, 2, 3, \ldots)$$

the index u labelling the *internal quantum states* of the molecule, i.e., wave functions of the molecule when the centre of mass is fixed at a definite point in space.

Since a possible value, say ϵ_j, of the total molecular energy is obtained by adding together *any* pair of values of the translational and internal energies, we have:

$$\epsilon_j = \epsilon_{\mathrm{trans}(t)} + \epsilon_{\mathrm{int}(u)} \quad (t = 1, 2, 3, \ldots; u = 1, 2, 3, \ldots)$$

Hence, the molecular partition function is:

$$f = \sum_{t,u} \mathrm{e}^{-(\epsilon_{\mathrm{trans}(t)} + \epsilon_{\mathrm{int}(u)})/kT}$$

the sum being over all the quantum states of the molecule, and therefore over all the values of the indices t and u. Let:

$$x_t = \mathrm{e}^{-\epsilon_{\mathrm{trans}(t)}/kT}$$

$$y_u = \mathrm{e}^{-\epsilon_{\mathrm{int}(u)}/kT}\ .$$

Then:

$$\mathrm{e}^{-(\epsilon_{\mathrm{trans}(t)} + \epsilon_{\mathrm{int}(u)})/kT} = x_t y_u$$

* Not, of course, to be confused with the thermodynamical internal energy *of the system*.

Therefore:

$$\begin{aligned} f &= \sum_{t,u} x_t y_u \\ &= x_1\left(\sum_u y_u\right) + x_2\left(\sum_u y_u\right) + x_3\left(\sum_u y_u\right) + \ldots \end{aligned}$$

when the products with the same x_t are collected together. Hence:

$$\begin{aligned} f &= (x_1 + x_2 + x_3 + \ldots)\left(\sum_u y_u\right) \\ &= \left(\sum_t x_t\right)\left(\sum_u y_u\right) \end{aligned}$$

Let us denote the sum:

$$\sum_t x_t \equiv \sum_t \mathrm{e}^{-\epsilon_{\mathrm{trans}(t)}/kT} \tag{3.29}$$

by the symbol f_{trans}, and:

$$\sum_u y_u \equiv \sum_u \mathrm{e}^{-\epsilon_{\mathrm{int}(u)}/kT}$$

by f_{int}. Then:

$$f = f_{\mathrm{trans}} f_{\mathrm{int}}\,. \tag{3.30}$$

f_{trans} and f_{int} are called, respectively, the *translational partition function* and the *internal partition function*. The method of calculating these quantities is discussed in Chapters 4 and 5; we shall find that f_{int} can be further factorized.

As in the case of the partition function f itself, terms in f_{trans} with the same translational energy can be collected together, and an expression equivalent to Equation 3.29 is:

$$f_{\mathrm{trans}} = \sum_{\substack{\text{translational} \\ \text{level} \\ s}} g_{\mathrm{trans}(s)} \mathrm{e}^{-\epsilon_{\mathrm{trans}(s)}/kT},$$

where $g_{\mathrm{trans}(s)}$ is the degeneracy of the translational energy level $\epsilon_{\mathrm{trans}(s)}$. There are corresponding expressions for the other factors in f.

The individual terms in f_{trans} have a meaning similar to those in f itself. Thus (cf. Section 2.4), it is easily shown that, in the most probable distribution, $\mathrm{e}^{-\epsilon_{\mathrm{trans}(t)}/kT}$ *is proportional to the number of molecules with the translational wave function* ψ_t, *and that* $g_{\mathrm{trans}(s)}\mathrm{e}^{-\epsilon_{\mathrm{trans}(s)}/kT}$ *is proportional to the number of molecules having the particular translational energy* $\epsilon_{\mathrm{trans}(s)}$. *There are corresponding proportionalities for the others in* f.

In case these statements should not appear self-evident, the proof is as follows. The number of molecules in the j-th molecular quantum state is, according to Equation 2.21:

$$\overline{m_j} = \frac{N}{f} e^{-\epsilon_j/kT} = \frac{N}{f} e^{-(\epsilon_{\text{trans}(t)} + \epsilon_{\text{int}(u)})/kT}$$

$$= \frac{N}{f} x_t y_u$$

The number of molecules, $\overline{m}_{\text{trans}(t)}$, in any one translational state t *irrespective of the internal state* is obtained by summing over all the indices u:

$$\overline{m}_{\text{trans}(t)} = \sum_u \frac{N}{f} x_t y_u = \frac{N x_t}{f} \sum_u y_u$$

$$= \frac{N x_t \Sigma y_u}{f_{\text{trans}} f_{\text{int}}}$$

$$= \frac{N x_t}{f_{\text{trans}}},$$

since $\sum_u y_u = f_{\text{int}}$. Thus:

$$\overline{m}_{\text{trans}(t)} = \frac{N}{f_{\text{trans}}} e^{-\epsilon_{\text{trans}(t)}/kT}, \quad \text{or} \quad \overline{m}_{\text{trans}(t)} \propto e^{-\epsilon_{\text{trans}(t)}/kT}$$

The number of molecules, say $\overline{n}_{\text{trans}(s)}$, with a particular translational energy, $\epsilon_{\text{trans}(s)}$, is obtained by adding together $\overline{m}_{\text{trans}(t)}$ values for each of the $g_{\text{trans}(s)}$ states which have the energy $\epsilon_{\text{trans}(s)}$. Thus:

$$\overline{n}_{\text{trans}(s)} = g_{\text{trans}(s)} \times \frac{N}{f_{\text{trans}}} e^{-\epsilon_{\text{trans}(s)}/kT}$$

or

$$\overline{n}_{\text{trans}(s)} \propto g_{\text{trans}(s)} e^{-\epsilon_{\text{trans}(s)}/kT}.$$

3.5 Further discussion on the derivation of the Boltzmann equation

The results established above provide a sufficient basis for the understanding of the following chapters; consequently, if it is desired, the present section and the following one may be omitted in a first reading of this book.

Our purpose is here to investigate more closely the basis of Equations 3.15 and 3.20, and we begin by examining the meaning of the thermodynamical concepts of *work* and *heat*.

If the internal energy of a system is given by:

$$E = \sum m_j\epsilon_j\,,$$

there are two ways in which it may be changed, either by varying the m_j (subject to the restriction that $\Sigma m_j = N$), or by varying the ϵ_j. Thus, if E is changed, in some way, to $E + \Delta E$, one has:

$$\begin{aligned}\Delta E &= \sum_j (m_j + \Delta m_j)(\epsilon_j + \Delta\epsilon_j) - \sum_j m_j\epsilon_j \\ &= \sum_i (\epsilon_j\Delta m_j + m_j\Delta\epsilon_j + \Delta m_j\Delta\epsilon_j) \\ &= \sum_j \epsilon_j\Delta m_j + \sum_j m_j\Delta\epsilon_j + \sum_j \Delta m_j\Delta\epsilon_j \qquad (3.31)\end{aligned}$$

Consider an infinitesimal change, from one equilibrium state to another. Since $\delta m_j\delta\epsilon_j$ is negligible, Equation 3.30 becomes:

$$\delta E = \sum \epsilon_j\delta m_j + \sum m_j\delta\epsilon_j\,. \qquad (3.32)$$

(i) The second term on the right-hand side of Equation 3.32 is the contribution of δE arising from changes in the energy levels. As we have already mentioned, in Section 3.1, the only way in which molecular energy levels can be changed is through such effects as changing externally applied electric or magnetic fields, or changing the volume of the system. For simplicity, we here consider only the effect of changing the volume (although the subsequent generalization to include the other effects will be obvious). When a system is compressed, work is done on it. If the energy ϵ_j is changed by such compression to $\epsilon_j + \delta\epsilon_j$, the work done *on* the m_j molecules is the increase in their energy, namely $m_j\delta\epsilon_j$. For the whole system, the work done is correspondingly:

$$\sum_j m_j\delta\epsilon_j\,.$$

This work is, of course, equal to $-P\mathrm{d}V$ ($\mathrm{d}V$ negative).

(ii) The first term on the right-hand side of Equation 3.32 is the change in energy resulting from a redistribution of molecules among the energy levels. This is obviously equal to the energy absorbed, in the form of heat, when the volume remains constant.

We now derive a more general equation than 3.13, by considering the effect upon f of a change in β *and also* in the energies ϵ_j.

Since:

$$f = \mathrm{e}^{-\beta\epsilon_0} + \mathrm{e}^{-\beta\epsilon_1} + \mathrm{e}^{-\beta\epsilon_2} + \ldots + \mathrm{e}^{-\beta\epsilon_j} + \ldots,$$

$$\mathrm{d}\ln f = \left(\frac{\partial \ln f}{\partial \beta}\right)_{\epsilon_0,\epsilon_1,\epsilon_2,\ldots} \mathrm{d}\beta + \left(\frac{\partial \ln f}{\partial \epsilon_0}\right)_{\beta,\epsilon_1,\epsilon_2,\ldots} \mathrm{d}\epsilon_0 + \left(\frac{\partial \ln f}{\partial \epsilon_1}\right)_{\beta,\epsilon_0,\epsilon_2,\ldots} \mathrm{d}\epsilon_1 + \ldots \tag{3.33}$$

Now:

$$\begin{aligned} \frac{\partial \ln f}{\partial \epsilon_0} &= \frac{1}{f}\frac{\partial}{\partial \epsilon_0}(\mathrm{e}^{-\beta\epsilon_0} + \mathrm{e}^{-\beta\epsilon_1} + \mathrm{e}^{-\beta\epsilon_2} + \ldots) \\ &= -\frac{\beta}{f} g_0 \mathrm{e}^{-\beta\epsilon_0} \\ &= -\frac{\beta \overline{m}_0}{N}, \end{aligned} \tag{3.34}$$

by Equation 2.21.

There are corresponding identities for derivatives of $\ln f$ with respect to the other ϵ_j in Equation 3.33. Furthermore, by Equation 3.6:

$$\left(\frac{\partial \ln f}{\partial \beta}\right)_{\epsilon_0,\epsilon_1,\epsilon_2,\ldots} = \left(\frac{\partial \ln f}{\partial \beta}\right)_V = -\frac{E}{N} \tag{3.35}$$

Substituting Equations 3.34 and 3.35 into 3.33:

$$N \mathrm{d}\ln f = -E\,\mathrm{d}\beta - \beta \sum_j \overline{m}_j \mathrm{d}\epsilon_j$$

From Equation 3.12, therefore:

$$\begin{aligned} \mathrm{d}\ln \Omega_{D_0} &= N\mathrm{d}\ln f + \beta \mathrm{d}E + E\mathrm{d}\beta \\ &= \beta(\mathrm{d}E - \sum_j \overline{m}_j \mathrm{d}\epsilon_j) \end{aligned}$$

If $\mathrm{d}V$ is the *expansion* of the system, then, according to paragraph (i) above:

$$-\sum \overline{m}_j \mathrm{d}\epsilon_j = P\mathrm{d}V$$

and

$$\begin{aligned} \mathrm{d}\ln \Omega_{D_0} &= \beta(\mathrm{d}E + P\mathrm{d}V) \\ &= \beta T \mathrm{d}S, \end{aligned} \tag{3.36}$$

by Equation 3.9.

The value of $\ln \Omega_{D_0}$ is a fixed and definite property of any state of the system. If a given system undergoes a finite change, say from state A to state B, then the value of $\ln \Omega_{D_0}$ is correspondingly changed, but the difference, $\ln \Omega_{D_0(\mathrm{B})} - \ln \Omega_{D_0(\mathrm{A})}$, is independent of the way in which the change is carried out. That is, $\mathrm{d}\ln \Omega_{D_0}$ is an *exact differential*; the value of the integral:

$$\int_{\text{state A}}^{\text{state B}} d\ln\Omega_{D_0} = \ln\Omega_{D_0\,(\text{B})} - \ln\Omega_{D_0(\text{A})}$$

is independent of the path of integration. Consequently, the right-hand side of Equation 3.36 must also be an exact differential, i.e., the value of:

$$\int_{\text{state A}}^{\text{state B}} \beta T \, dS \qquad (3.37)$$

must be independent of the path of integration. This can only be the case if the value of βT depends upon nothing other* than S, i.e., if βT is a function of S.

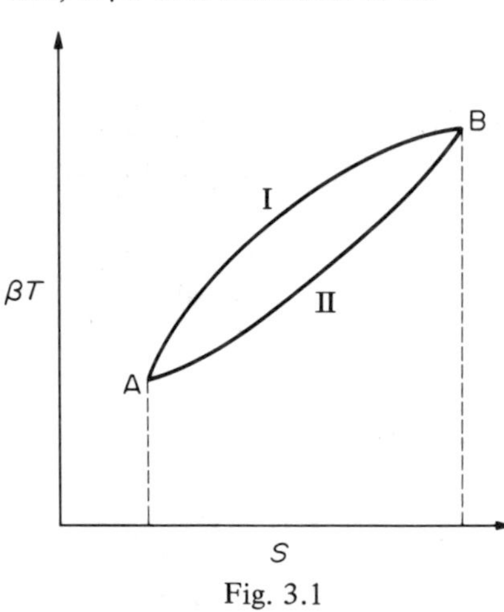

Fig. 3.1

If βT were dependent upon some other property besides S, it would, in general, be possible to go from state A to state B by more than one path, giving different βT *versus* S curves (Fig. 3.1). Since the integral 3.36 is the area under the corresponding curve, this would not have a unique value for a given change.

Suppose that βT is the function $f(S)$. For *any* function $f(S)$ of *any* variable S, it is possible to express $f(S)dS$ in the form $d\varphi(S)$ where $\varphi(S)$ is the function (i.e., indefinite integral):

$$\varphi(S) = \int f(S)\,dS$$

and

$$f(S) = \frac{d\varphi(S)}{dS}$$

Hence, Equation 3.36 becomes:

$$d\ln\Omega_{D_0} = d\varphi(S)$$

Integrating:

$$\varphi(S) = \ln\Omega_{D_0} + \text{constant of integration.}$$

As we have seen (p. 33), $\ln\Omega_{D_0}$ differs negligibly from $\ln\Omega$. Thus,

* We find, in fact, that $\beta T = 1/k$ i.e., βT is *also* independent of S – but we must not anticipate this result.

entropy is some function of $\ln\Omega$.

The nature of this function is easily deduced by the following argument. Consider two systems, A and B, in thermal contact and at equilibrium. Since the systems are in equilibrium, they have the same temperature. Suppose that an amount of heat dQ_{AB} is added reversibly to the combined system, in such a way that A absorbs an amount dQ_A and B an amount dQ_B, where:

$$dQ_{AB} = dQ_A + dQ_B.$$

The entropy changes are correspondingly:

$$\frac{dQ_{AB}}{T} = \frac{dQ_A}{T} + \frac{dQ_B}{T},$$

that is:

$$dS_{AB} = dS_A + dS_B$$

Integrating this expression:

$$S_{AB} = S_A + S_B + \text{constant of integration} \qquad (3.38)$$

i.e., entropy is an *additive function*.*

The number of complexions of the isolated system A is Ω_A and that of B is Ω_B. Assuming that these numbers are unchanged by the presence of the other system,† the number of complexions of the combined system is:

$$\Omega_{AB} = \Omega_A \times \Omega_B$$

since each complexion of A can be taken with each complexion of B to give a complexion of the combined system. Hence:

$$\ln\Omega_{AB} = \ln\Omega_A + \ln\Omega_B$$

or

$$\varphi(S_{AB}) = \varphi(S_A) + \varphi(S_B) + \text{constant.} \qquad (3.39)$$

The *only* function φ which will satisfy Equations 3.38 and 3.39 simultaneously is:

$$\varphi(S) = k'S + c'$$

where c' and k' are constants; that is:

$$\ln\Omega = k'S + c'$$

* It is often stated as a self-evident fact that $S_{AB} = S_A + S_B$. Actually this is not self-evident: neither is it always true; for example, Equation 3.37 has no basis if the temperatures of A and B are different (cf. Fowler, p.202).

† This is equivalent to assuming that there is no nett energy transfer between A and B, which is reasonable since the temperatures of A and B are the same.

or

$$S = k \ln \Omega + c$$

where $\boldsymbol{k} = 1/\boldsymbol{k}'$ and $c = -c'/\boldsymbol{k}'$ are constants. We have thus proved Equation 3.19. To prove Equation 3.15, we first note that (cf. Equation 3.9):

$$T = \left(\frac{\partial E}{\partial S}\right)_V = \left(\frac{\partial E}{\partial \beta}\right)_V \Big/ \left(\frac{\partial S}{\partial \beta}\right)_V \qquad (3.40)$$

From Equation 3.7:

$$S = \boldsymbol{k} \ln \Omega_{D_0} = N\boldsymbol{k} \ln f + \beta \boldsymbol{k} E$$

and therefore:

$$\left(\frac{\partial S}{\partial \beta}\right)_V = N\boldsymbol{k} \left(\frac{\partial \ln f}{\partial \beta}\right)_V + \boldsymbol{k}E + \boldsymbol{k}\beta \left(\frac{\partial E}{\partial \beta}\right)_V$$

According to Equation 3.6:

$$N\left(\frac{\partial \ln f}{\partial \beta}\right)_V = -E$$

Hence:

$$\left(\frac{\partial S}{\partial \beta}\right)_V = \boldsymbol{k}\beta \left(\frac{\partial E}{\partial \beta}\right)_V$$

Substitution of this result in Equation 3.40 now gives:

$$\beta = \frac{1}{\boldsymbol{k}T}$$

3.6 The second law of thermodynamics

Suppose that dQ is an amount of heat absorbed reversibly by a system at temperature T. The *increase in entropy* of the system is then defined to be $dS = dQ/T$, i.e., dQ/T is equated to the differential of some function S, the entropy of the system. The second law of thermodynamics is now contained in the following statements (e.g., Lewis and Randall, Ch. 7).

(i) S is a function of state, i.e., it has a unique value for each thermodynamic state of the system, independent of any previously existing conditions in the system.

(ii) The entropy of an isolated system can never decrease; in particular, for any finite, irreversible (spontaneous) process in an isolated system:

$$\Delta S > 0$$

(By *isolated* it is meant that the system has a fixed amount of matter and energy.)

We now discuss the significance of these statements according to statistical mechanics.

Statement (i) follows from Equation 3.20:

$$S = k \ln \Omega$$

The number Ω is a quantity which has a unique value for each state of the system. It follows that S also has a unique value for each state.

To understand statement (ii), let us consider some examples of changes which may occur in an isolated system. These are illustrated in Fig. 3.2. They all have one property in common, namely, that *the number of complexions* Ω *in the final state is greater than the number in the initial state, if E is the same for both states.* This arises because the number of energy levels in which the molecules can be placed is, in each case, increased; clearly, the greater the number of available energy levels with some range, the greater the number of ways in which they can be occupied, to give the same total, E.

Consider, for example, case (a). The essential difference between the liquid and the vapour phases is that, in the latter, the molecules are free to move, i.e., they have translational energy. Consequently, in the final state there are two sets of energy levels available to the molecules, and Ω is increased.

Energy levels of molecules in vapour

Energy levels of molecules in liquid

Energy levels of molecules in liquid

Case (b) is similar; this time, the extra energy levels are those of the molecules adsorbed upon the surface.

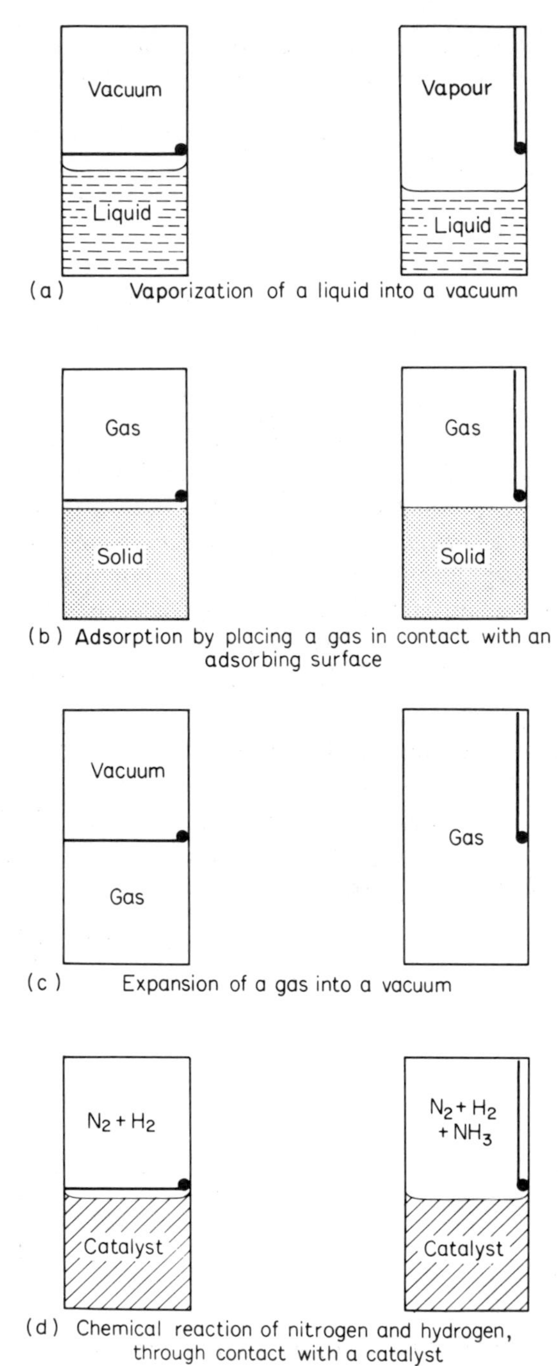

Fig. 3.2 Spontaneous processes

In case (c), as is shown in the following chapter (Section 4.7), the number of energy levels lying within any specified range increases with volume, and therefore Ω increases with expansion.

The final case (d) is a little different from the others, because new molecular species are formed. Suppose that the distribution numbers of nitrogen, hydrogen and ammonia *in the mixture* are:

$$n_{H_2(1)},\ n_{H_2(2)},\ \ldots,\ n_{H_2(i)},\ \ldots$$

$$n_{N_2(1)},\ n_{N_2(2)},\ \ldots,\ n_{N_2(i)},\ \ldots$$

$$n_{NH_3(1)},n_{NH_3(2)},\ldots,n_{NH_3(i)},\ \ldots$$

(We have not yet dealt specifically with mixtures, but here it may be assumed that the Maxwell–Boltzmann law applies to the distribution numbers of each component present.)

In the absence of catalyst (the initial state):

$$n_{NH_3(1)} = n_{NH_3(2)} = \ldots n_{NH_3(i)} = \ldots = 0,$$

and the value of Ω is determined by the number of ways in which the distribution numbers $n_{H_2(i)}$ and $n_{N_2(i)}$ can be chosen such that the total energy is E. In the presence of catalyst (the final state), on the other hand, the $n_{NH_3(i)}$ values are *not* restricted to zero, and the value of Ω is increased over that for the initial state by the number of ways in which the $n_{H_2(i)}$, $n_{N_2(i)}$, *and* the $n_{NH_3(i)}$ can be chosen subject to constant E.

Mayer and Mayer (p.81ff.) describe processes of the above type as the *removal of inhibitions*. In the first three examples, the inhibition is the physical barrier; in example (d), the inhibition is the high activation energy, which prevents the reaction of N_2 and H_2 molecules. In the initial state, certain molecular energy levels, or wave functions, are *inaccessible*. However, when the inhibition is removed, these levels become *accessible* and a spontaneous process occurs. Since more energy levels are available to the molecules, the number of complexions associated with the energy E is increased; that is:

$$\Omega' > \Omega$$

where Ω and Ω' are the numbers of complexions before and after the removal of the inhibition. The change in entropy is, correspondingly:

$$\Delta S = S' - S = k \ln \frac{\Omega'}{\Omega} > 0,$$

which is just statement (ii) above.

Problems

3.1 Show that the enthalpy and Gibbs free energy of a system of the type considered in this chapter are given respectively by:

$$H = NkT\left[\left(\frac{\partial \ln f}{\partial \ln T}\right)_V + \left(\frac{\partial \ln f}{\partial \ln V}\right)_T\right]$$

$$G = -NkT\left[\ln f - \left(\frac{\partial \ln f}{\partial \ln V}\right)_T\right].$$

3.2 Use Equation 2.22 to show that the average energy per molecule is:

$$\bar{\epsilon} = kT^2\left(\frac{\partial \ln f}{\partial T}\right)_V.$$

How is this result related to Equation 3.23?

3.3 Deduce from Equation 2.22 and the preceding problem that the mean value of ϵ^2 is:

$$\overline{\epsilon^2} = kT^2\left(\frac{\partial \bar{\epsilon}}{\partial T}\right)_V + (\bar{\epsilon})^2$$

Hence obtain an expression for the standard deviation σ in the energy (defined by $\sigma = [\overline{\epsilon^2} - (\bar{\epsilon})^2]^{1/2}$) in terms of $\tilde{C}_V$

CHAPTER FOUR

Gases: The Translational Partition Function

An intrinsic property of gases, which distinguishes them from solids or liquids, is that, in a gas, the molecules are perfectly free to move between any points within an overall volume. This has two consequences in the treatment of the statistical thermodynamics of gases. Firstly, the molecular partition function for a gas contains as a factor the partition function f_{trans} for free translation (Sections 3.4 and 4.6). Secondly, we shall find (Sections 4.1–4.3) that, although the Maxwell–Boltzmann law (Equation 2.14) normally holds for gases, the basic Equation 1.6 does not, and therefore certain formulae of Section 3.3 must be modified slightly before they can be applied to gases.

4.1 Indistinguishability of gas molecules

An assumption, upon which our derivation of the Maxwell–Boltzmann law is based, is that the molecules of the system can be distinguished from one another. If this is correct, interchanging the wave functions of any two molecules would give rise to a different complexion, as in Example 1.4, in which the four complexions associated with distribution 1 are represented by (cf. page 4):

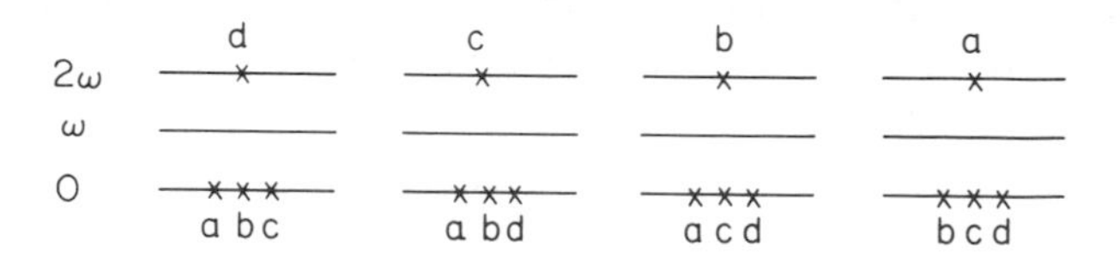

Since there is no way of telling one molecule from another of the same kind, do these diagrams really represent four physically distinct conditions of the system? In other words, is there any property of the first arrangement which would enable it to be distinguished from the other three? Or even more simply: what does labelling the molecules a,b,c,... actually *mean*? The answers to these questions depend upon whether the substance is a solid, liquid or gas, or, more precisely, upon whether the molecules are *localized* or *non-localized*.

Molecules in a solid substance (and, to a large extent, in a liquid) occupy definite *positions*: that is, each is localized within a small region. They can, therefore, be labelled (distinguished) by these positions. It is not, then, the molecules which are different but the positions in which they occur. The first of the above diagrams, for example, really means 'there is a molecule with energy 2ω at position d and there are molecules with energy 0 at positions a,b, and c'. The second diagram denotes a different physical situation: 'there is a molecule with energy 2ω at position c and there are molecules with energy 0 at a,b, and d' – and so on. That is what labelling the molecules really implies.

Now consider the molecules of a gas. One of the first topics encountered in any study of quantum mechanics is the problem of 'a particle in a potential box', in which the Schrödinger equation for a particle undergoing free translation within a given volume is considered. Solution of this problem shows that at any instant, the particle is in one of an infinite number of translational quantum states. Each such state is defined by a translational wave function ψ_{trans} and the value of $|\psi^2_{\text{trans}}|$ at any point measures the probability of finding the particle in the vicinity of that point. Since ψ_{trans} is a periodic (sine) function of position, $|\psi^2_{\text{trans}}|$ has the same value at many points (except for states with the very lowest translational energies). This means, physically, that if the translational quantum state of a particle is known then *its position is not*. Accordingly, *it would be meaningless to label a molecule (in a given translational state) by its position.* Therefore, the four arrangements represented in the above diagrams do not correspond to physically different conditions; instead of four complexions, there is only *one*:

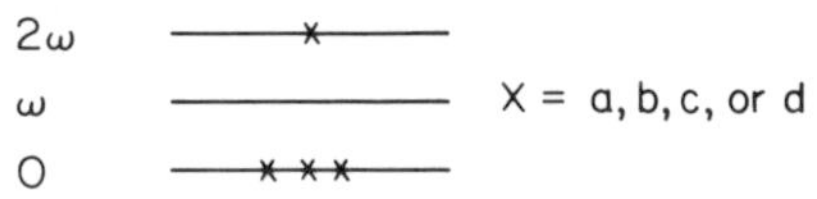

2ω ——X——

ω ———— X = a, b, c, or d

0 ——X X X——

and it is said that molecules of a gas are *indistinguishable* from one another.*

4.2 Enumeration of complexions for gases

Consider, for a gaseous system, the distribution defined by the numbers $n_1, n_2, ..., n_i, ...$. If all the energy levels were *non-degenerate*, there would only be one complexion for this distribution, i.e., that for which *any* n_1 molecules have energy ϵ_1, *any* n_2 molecules have energy ϵ_2, and so on, since the molecules are indistinguishable from one another. Actually, the energy levels will be, respectively, $g_1, g_2, ..., g_i, ...$-fold degenerate; in this case, the number of complexions is found as follows.

In Chapter 13 it will be shown that the number of different ways in which any n_i indistinguishable molecules can be placed in the g_i quantum states belonging to the level ϵ_i is:

$$\omega_i = \frac{(g_i + n_i - 1)!}{n_i!(g_i - 1)!} . \quad (4.1)$$

The number of complexions associated with the distribution is found by taking the product of terms like ω_i over all the energy levels. Hence:

$$\Omega_D = \prod_i \omega_i = \prod_i \frac{(g_i + n_i - 1)!}{n_i!(g_i - 1)!} . \quad (4.2)$$

This formula is identified in Chapter 13 as the basic formula of *Bose–Einstein* statistics. Our present interest in Equation 4.2 is, however, to show that, in most cases, it can be replaced by a much simpler expression for Ω_D.

Writing out in full the first n_i factors in the factorial term $(g_i + n_i - 1)!$, one obtains from Equation 4.1:

$$\omega_i = \frac{[(g_i + n_i - 1)(g_i + n_i - 2)...g_i](g_i - 1)!}{n_i!(g_i - 1)!} .$$

* An argument similar to the above can be given in terms of *classical* mechanics:- As before, labelling the molecules is only permitted if it has some physical meaning. The instantaneous translational *state* of a gas molecule is determined (labelled) in classical mechanics (see Section 11.2) by the momentum and the *position*; therefore, position cannot be used again to label a molecule in a given state. See also Rushbrooke, p.36.

Further discussion of the complexions of a gaseous system is given in Section 13.6, and there is a more detailed account in Chapter 1 of Fowler and Guggenheim.

Cancelling out the terms $(g_i - 1)!$, and taking out g_i as a factor from each of the terms in square brackets:

$$\omega_i = \frac{(g_i)^{n_i}}{n_i!}\left[\left(1+\frac{n_i-1}{g_i}\right)\left(1+\frac{n_i-2}{g_i}\right)\ldots 1\right]$$

We shall prove in Section 4.7 that normally (the exceptions of interest are discussed in Chapter 13), for gases:

$$g_i >> n_i$$

and each of the factors in the square brackets is therefore approximately 1. Hence, to a good degree of approximation:

$$\omega_i = \frac{(g_i)^{n_i}}{n_i!}$$

Thus, *for gases (indistinguishable molecules)*:

$$\Omega_D = \prod_i \frac{(g_i)^{n_i}}{n_i!} \tag{4.3}$$

which may be contrasted with our previous result, *for solids and liquids (distinguishable molecules)*:

$$\Omega_D = \frac{1}{N!}\prod \frac{(g_i)^{n_i}}{n_i!}\,. \tag{4.4}$$

The value of Ω_D is thus reduced by a factor $N!$, if the molecules are indistinguishable. This is understandable, because the $N!$ different ways of ordering N distinguishable particles all become identical when the particles cannot be distinguished from one another. Actually, the simple division of Equation 4.4 by $N!$ *over-compensates* for indistinguishability, because, as a little thought will show, Equation 4.4 already allows for the equivalence of molecules *within a given quantum state.* However, when $g_i >> n_i$, i.e., when the number of available quantum states is much larger than the number of molecules, the probability of more than one molecule being in any one quantum state is very small.*

4.3 The Maxwell–Boltzmann law for gaseous systems

Equation 2.12:

$$\frac{\partial(\ln \Omega_D)}{\partial n_i} - \alpha - \beta\epsilon_i = 0 \tag{2.12$'$}$$

* As in Section 2.3, the average value of m_j is $\overline{n_i}/g_i$. This is less than 1, even though $\overline{n_i}$ itself may be a large number.

is, of course, valid whatever Ω_D happens to be. Since Ω_D (Equation 4.3) for systems of indistinguishable molecules differs from Ω (Equation 4.4) for systems of distinguishable molecules only by the constant $N!$, it is clear that $\partial(\ln\Omega_D)/\partial n_i$ will be the same in both cases. Hence, the Maxwell–Boltzmann law:

$$n_i = g_i \mathrm{e}^{-\alpha}\mathrm{e}^{-\beta\epsilon_i}$$

derived from Equation 2.12, applies equally to systems of distinguishable molecules.

4.4 Thermodynamic functions for gaseous systems

(i) *Energy*

Since the Maxwell–Boltzmann law also applies to gases, Equations 3.15 and 3.21 are again obtained, i.e.:

$$E = NkT^2\left(\frac{\partial \ln f}{\partial T}\right)_V \tag{4.5}$$

(ii) *Temperature and Entropy*

The derivations in Section 3.5 are independent of the form of the expression for Ω_{D_0}, and so are independent of whether the system is, or is not, gaseous. Therefore, the results:

$$\beta = \frac{1}{kT}$$

$$S = k\ln\Omega_{D_0}$$

hold in both cases.* For gaseous systems, however, Ω_{D_0} must be calculated (by substituting the most probable values for the n_i) from Equation 4.3, instead of from 4.4. Thus:

$$S = k\ln\left(\prod_i \frac{(g_i)^{\overline{n}_i}}{\overline{n}_i!}\right)$$

$$= k\ln\left(N!\prod_i \frac{(g_i)^{\overline{n}_i}}{\overline{n}_i!}\right) - \ln N! \tag{4.6}$$

The first term on the right-hand side of Equation 4.6 is just the entropy derived from the Ω_{D_0} given by Equation 4.4, a value which we have already found (cf. Equation 3.24):

* Alternatively, this can be demonstrated by carrying through the simpler derivation of these results, given in Section 3.2, using Ω_{D_0} given by Equation 4.3 instead of 4.4.

$$k \ln\left(N! \prod_i \frac{(g_i)^{\bar{n}_i}}{\bar{n}_i!}\right) = Nk \ln f + \frac{E}{T} \qquad (4.7)$$

Applying Stirling's theorem to the remaining term in Equation 4.6:

$$k \ln N! = k(N \ln N - N) = Nk \ln(N/\mathrm{e}) \qquad (4.8)$$

Now substituting Equations 4.7 and 4.8 into 4.6:

$$S = Nk \ln f + \frac{E}{T} - Nk \ln \frac{N}{\mathrm{e}}$$

$$= Nk \ln\left(\frac{f\mathrm{e}}{N}\right) + \frac{E}{T}. \qquad (4.9)$$

(iii) *Generalization*

Equation 4.9 differs from Equation 3.24 only in that, in the former case, f is multiplied by the factor e/N. Now, this same factor could be introduced into Equation 4.5 without altering the value of E. For:

$$NkT^2 \frac{\partial}{\partial T} \ln\left(\frac{f\mathrm{e}}{N}\right) = NkT^2 \left(\frac{\partial}{\partial T} \ln f + \frac{\partial}{\partial T} \ln \frac{\mathrm{e}}{N}\right)$$

$$= NkT^2 \frac{\partial}{\partial T} \ln f = E.$$

Therefore, because all the other thermodynamic functions can be derived from the equations for E and S, the relationships obtained in the previous chapter are applicable to gases, provided that, *whenever f occurs in our earlier equations, it is now multiplied by the factor* e/N *if the substance is gaseous.*

For example, the work function for a solid or liquid is given by Equation 3.26:

$$A = -NkT \ln f$$

but for a *gas* it is:

$$A = -NkT \ln\left(\frac{f\mathrm{e}}{N}\right). \qquad (4.10)$$

It is only when the e/N factor is included in this way that an internally *consistent* set of equations for the thermodynamic functions of a gas is obtained; historically, the necessity for such a factor was known before the arguments given in Section 4.1 were properly understood. Let us consider, for example, the result of attempting to calculate A for a gas with the omission of e/N from Equation 4.10. It is shown in Section 4.5 that the translational partition function (and, therefore, f, because f_{trans} is the only volume-dependent factor in the partition function for a *gas*) is directly proportional to volume. Suppose that the amount of gas is doubled. If the temperature and pressure are fixed, then

$$N \longrightarrow 2N$$
$$V \longrightarrow 2V,$$

and, because of the latter, $f \longrightarrow 2f$. Therefore, $-NkT\ln f \longrightarrow -2NkT\ln 2f$; that is, A is changed by an amount $-NkT[\ln f + 2\ln 2]$.

Now this is wrong, because, thermodynamically, A is an *extensive* property: doubling the amount of gas at constant pressure and temperature must double A. If, on the other hand, e/N is *not* omitted from Equation 4.10, we obtain correspondingly

$$-NkT\ln\frac{fe}{N} \longrightarrow -2NkT\ln\frac{2fe}{2N} \longrightarrow -2NkT\ln\frac{fe}{N}$$

the expected result.

Another example of the importance of the e/N factor occurs in the following section. If the derivation given there is repeated with the omission of the e/N factor, the absurd result $PV = 0$ is obtained for the equation of state of a perfect gas.

4.5 Equation of state for a perfect gas: evaluation of Boltzmann's constant k

For n moles of a pure substance, the Gibbs free energy and work function are related by:*

$$G = n\left(\frac{\partial A}{\partial n}\right)_{T,V} \tag{4.11}$$

Replacing N in Equation 4.10 by $n\tilde{N}$:

$$A = -n\tilde{N}kT\ln\left(\frac{fe}{n\tilde{N}}\right)$$

Substituting for A in Equation 4.11, we obtain, for the Gibbs free energy of a *gas*:

* For a single substance, $G = n\mu$, where the chemical potential μ is defined by $dE = TdS - PdV + \mu dn$. Whence, $dA = -SdT - PdV + \mu dn$, and so $(\partial A/\partial n)_{T,V} = \mu$.

$$G = -n\tilde{N}kT\ln\left(\frac{fe}{n\tilde{N}}\right) + n\tilde{N}kT$$

$$= A + n\tilde{N}kT \qquad (4.12)$$

But, for *any* system, G and A are also related by:

$$G = A + PV \qquad (4.13)$$

From Equations 4.12 and 4.13, therefore:*

$$PV = n\tilde{N}kT \qquad (4.14)$$

It will be recalled that the equations derived in this chapter apply to any assembly of indistinguishable particles, in which the forces of interaction between the particles are negligible. Physically, such an assembly is a perfect gas. Thus, in deriving Equation 4.14, we have obtained the Boyle–Charles law for a perfect gas:

$$PV = nRT \qquad (4.15)$$

where R is the molar gas constant.

Comparison of Equation 4.14 with 4.15 shows that:

$$k = \frac{R}{\tilde{N}} \qquad (4.16)$$

as stated in Section 3.2.

4.6 The translational partition function

We now consider the other topic mentioned in the opening paragraph of this chapter, namely, the evaluation of the partition function f_{trans} for free translation. In the following derivation, it is supposed that the gas is contained within a rectangular box, of sides a, b, and c, although the result is valid for any shape of container [see Section 11.5(i)].

The translational kinetic energy of a molecule confined to a rectangular box depends upon three quantum numbers, p, q, and s, each of which can be any positive integer. The allowed values of this energy, say ϵ_{pqs} satisfy the equation:

$$\epsilon_{pqs} = \frac{h^2}{8m}\left(\frac{p^2}{a^2} + \frac{q^2}{b^2} + \frac{s^2}{c^2}\right) \qquad (4.17)$$

* A quicker, but less direct way of deriving Equation 4.14 is to substitute Equation 4.23 given below into Equation 3.27.

where h is Planck's constant and m is the mass of the particle. Since each set of p,q, and s values is associated with (i.e., labels) a particular *translational quantum state*, the translational partition function is obtained by summing $\exp(-\epsilon_{pqs}/kT)$ over all positive values of p,q, and s.

Let:

$$\epsilon_{x(p)} = \frac{h^2p^2}{8ma^2}, \quad \epsilon_{y(q)} = \frac{h^2q^2}{8mb^2}, \quad \epsilon_{z(s)} = \frac{h^2s^2}{8mc^2} \tag{4.18}$$

so that:

$$\epsilon_{pqs} = \epsilon_{x(p)} + \epsilon_{y(q)} + \epsilon_{z(s)}$$

We have seen (Section 3.4) that, *whenever the energy can be expressed as the sum of independent terms, the corresponding partition function can be expressed as the product of independent factors.* In this case, therefore:

$$f_{\text{trans}} = f_x f_y f_z \tag{4.19}$$

where:

$$f_x = \sum_p e^{-\epsilon_{x(p)}/kT}; \quad f_y = \sum_q e^{-\epsilon_{y(q)}/kT}; \quad f_z = \sum_s e^{-\epsilon_{z(s)}/kT}$$

Substituting the value given in Equation 4.18 for $\epsilon_{x(p)}$:

$$f_x = \sum_{p=1} e^{-Ap^2} \tag{4.20}$$

where $A = h^2/8ma^2kT$. The sum in Equation 4.20 can be evaluated as an integral* (see Appendix 1):

$$f_x = \int_0^\infty e^{-Ap^2}\,dp = \frac{1}{2}\left(\frac{\pi}{A}\right)^{1/2} \tag{4.21}$$

$$= \frac{(2\pi mkT)^{1/2}a}{h}$$

Similarly:

$$f_y = \frac{(2\pi mkT)^{1/2}b}{h} \quad \text{and} \quad f_z = \frac{(2\pi mkT)^{1/2}c}{h}$$

so that:

$$f_{\text{trans}} = \frac{(2\pi mkT)^{3/2}abc}{h^3}$$

$$= \frac{(2\pi mkT)^{3/2}V}{h^3} \tag{4.22}$$

* Since the lowest value of p is 1, whereas we have effectively summed from zero, Equation 4.21 should, strictly, be:

$$f_x = \int_0^\infty e^{-Ap^2}dp - e^0 = \tfrac{1}{2}\left(\frac{\pi}{A}\right)^{1/2} - 1.$$

Since A is very small, however, the 1 is negligible.

If $\tilde{M}$ is the mass per mole,

$$m = \frac{\tilde{M}}{\tilde{N}}$$

where $\tilde{N}$ is the Avogadro number. Furthermore:

$$k = \frac{R}{\tilde{N}}$$

Hence:

$$f_{\text{trans}} = \frac{(2\pi\tilde{M}RT)^{3/2}\,\tilde{V}}{h^3\tilde{N}^3}, \tag{4.23}$$

where $\tilde{V}$ is the volume *per mole.*

The quantity which it is usually required to calculate in practice (cf. Equations 2.21, 4.9, etc.) is f/N. For a gas at 1 atmosphere ($= 101{\cdot}325\,\text{kN m}^{-2}$) pressure, the volume per mole is:

$$\tilde{V}^\circ = (8{\cdot}206 \times 10^{-5})T\ \text{m}^3\,.$$

From Equation 4.23, therefore:

$$\frac{f^\circ_{\text{trans}}}{\tilde{N}} = 0{\cdot}02559 M^{3/2}\, T^{5/2}$$

where f°_{trans} is the translational partition function *evaluated for the gas at 1 atmosphere pressure* and M is the molecular weight. For example, for hydrogen at 300 K and 1 atmosphere pressure:

$$\frac{f^\circ_{\text{trans}}}{\tilde{N}} \approx 10^5\,.$$

According to Section 3.4, the number of molecules in any one translational *state t* is, for the most probable distribution:

$$\bar{m}_{\text{trans}(t)} = \frac{N}{f_{\text{trans}}}\,\mathrm{e}^{-\epsilon_{\text{trans}(t)}/kT}$$

Since $\epsilon_{\text{trans}(t)}/kT$ is a positive number, $\bar{m}_{\text{trans}(t)}$ cannot be greater than N/f, and, as we have seen, N/f is typically $\sim 10^{-5}$. This is indeed a remarkable result. It means, physically, that, although the number of molecules N is very great, the number of available quantum states is even greater, so that the probability of any one quantum state being occupied by a molecule is very small.

If the translational energy levels were non-degenerate, $\bar{n}_i$ would also be a very small number, and in such a case our derivation of the Maxwell–Boltzmann law (which relies upon Stirling's approximation for $\ln n_i!$, applicable only if n_i is large) could not be valid.

This difficulty can be circumvented in two ways. Firstly, the Maxwell–Boltzmann law can be derived without using Stirling's approximation for $\ln n_i!$, either by the Darwin–Fowler method, referred to in Section 2.5, or by the method of the canonical ensemble, as discussed in Section 14.7. Secondly, since the difference between successive translational energy levels is found to be negligible in comparison to their values, large numbers of levels can be grouped together, and considered as one degenerate level; the value of n_i for this one 'level' is then a large number. We consider this in the following section.

4.7 Degeneracy of translational energy levels

Let:

$$r^2 = \frac{p^2}{a^2} + \frac{q^2}{b^2} + \frac{s^2}{c^2} \tag{4.25}$$

so that:

$$\epsilon_{pqs} = \frac{r^2 h^2}{8m}$$

Suppose that the possible values of p/a, q/b, and s/c are represented as Cartesian coordinates (Fig. 4.1). Each quantum state can then be

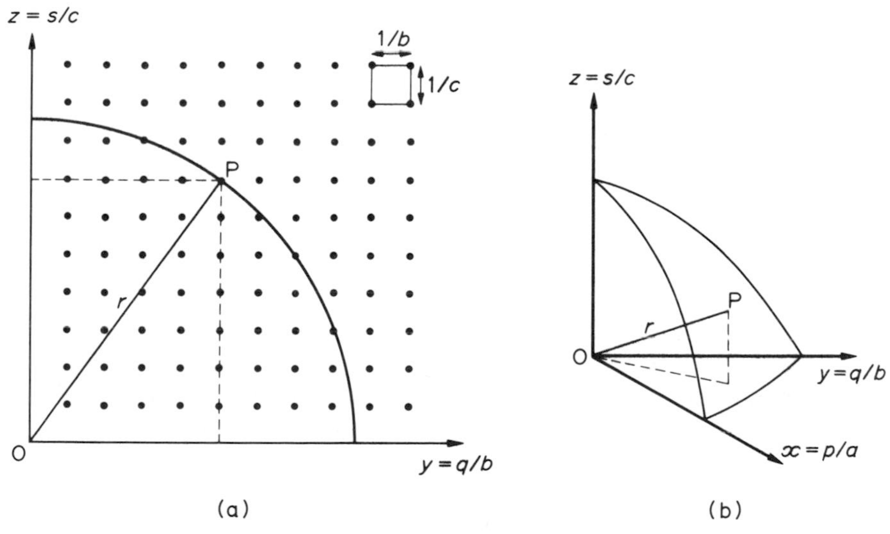

Fig. 4.1 (a) Representation of permitted q/b, s/c values as a two-dimensional array of points. (b) Representation of permitted p/a, q/b, s/c values as a three-dimensional array of points

represented as a point $(p/a, q/b, s/c)$ such as P. If all such points are joined by straight lines to neighbouring points, the Cartesian space is divided up into rectangular boxes ('unit cells'), the sides of which have lengths $1/a$, $1/b$, $1/c$. The volume of each unit cell is then:

$$v = \frac{1}{abc}$$

Since:

$$\text{OP} = \left(\frac{p^2}{a^2} + \frac{q^2}{b^2} + \frac{s^2}{c^2}\right)^{1/2} = r,$$

all quantum states having the same value of r must be represented by points which lie on a sphere of radius r, and (since p, q, and s are positive numbers) on the positive octant of this sphere.

Therefore, all quantum states with smaller values of r than this must be represented by points which *lie inside* the positive octant of the sphere of radius r. The volume of this octant is:

$$\tfrac{1}{8}(\tfrac{4}{3}\pi r^3) = \tfrac{1}{6}\pi r^3.$$

Now, each unit cell has eight corners, each corner being shared with seven other unit cells. Therefore, the number of unit cells within the octant (calculated by dividing the volume of the octant by the volume v of a unit cell) determines the number of quantum states. Thus, the number G_r of quantum states with energies less than $r^2h^2/8m$ is:

$$G_r = \tfrac{1}{6}\pi r^3/v = \frac{abc\pi r^3}{6} = \frac{V\pi r^3}{6} \qquad (4.26)$$

where $V = abc$ is the volume of the gas.

If r is increased by a small amount Δr, this number of quantum states is increased by:

$$\frac{dG_r}{dr}\Delta r = \tfrac{1}{2}V\pi r^2\Delta r$$

$$= g_r, \text{ say.} \qquad (4.27)$$

If Δr is sufficiently small, the translational energy can be treated as constant over the range of r-values, and, in this sense, $\tfrac{1}{2}V\pi r^2\Delta r$ can be regarded as the degeneracy* of the energy $r^2h^2/8m$.

* This 'degeneracy' depends upon the value of Δr chosen. Arbitrariness in this value simply means that there are numerous ways in which the exact translational energy levels can be grouped together to form 'degenerate' levels. The essential thing is that a value of Δr can be chosen which is (a) sufficiently small for the approximation to be valid, and (b) sufficiently large for the 'degenerate' level to include many quantum states.

In the most probable distribution, the number of molecules, say $\overline{n_r}$, with this energy is given, in the usual way, by:

$$\overline{n_r} = \frac{N}{f_{\text{trans}}} \left(\tfrac{1}{2} V \pi r^2 \Delta r\right) \mathrm{e}^{-r^2 h^2 / 8mkT} .$$

Now let us calculate the value of r, say r_{m}, for which $\overline{n_r}$ has the greatest value. For this value:

$$\frac{\mathrm{d}n_r}{\mathrm{d}r} = 0$$

that is:

$$\frac{\mathrm{d}}{\mathrm{d}r}\left(r^2 \mathrm{e}^{-r^2 B}\right) = 0,$$

where $B = h^2/8mkT$. It is thus found that:

$$r_{\mathrm{m}} = \left(\frac{1}{B}\right)^{1/2} = \frac{(8mkT)^{1/2}}{h} = \frac{(8MRT)^{1/2}}{\tilde{N}h}$$

which, for hydrogen at 300 K, is about $10^{10}\,\mathrm{m}^{-1}$. With this value of r, we obtain (taking Δr as $1\,\mathrm{m}^{-1}$):

$$g_r \approx 10^{18}$$

at atmospheric pressure. Now:

$$\frac{r_{\mathrm{m}}^2 h^2}{8mkT} = 1$$

and, as previously:

$$\frac{f^{\circ}_{\text{trans}}}{\tilde{N}} \approx 10^5$$

Therefore:

$$\overline{n_r} \approx \frac{10^{18}}{10^5} = 10^{13},$$

which is a very large number. This is our justification for using Stirling's approximation in deriving the Maxwell–Boltzmann law for gases.

Problems

4.1 Substitute for the values of the constants in Equation 4.23 and so obtain 4.24:

$$\frac{f^{\circ}_{\text{trans}}}{\tilde{N}} = 0{\cdot}02559\, M^{3/2}\, T^{5/2}$$

Calculate $f^{\circ}_{\text{trans}}/\tilde{N}$ for ^{40}Ar at 400 K.

4.2 Show that the translational contribution to the molar internal energy of a gas at temperature T is:

$$\tilde{E}_{\text{trans}} = \tfrac{3}{2}RT$$

and that the contribution to $\tilde{C}_V$ is:

$$\tilde{C}_{V(\text{trans})} = \tfrac{3}{2}R\ .$$

4.3 Calculate the translational contributions to the energy, entropy, and work function of argon at 1 atmosphere pressure and 400 K.

4.4 In classical mechanics, the kinetic energy of a molecule moving with speed c is $\frac{1}{2}mc^2$. As is discussed in Chapter 11, it is permissible, for large quantum numbers, to equate the classical and quantum-mechanical expressions, so that:

$$\tfrac{1}{2}mc^2 = \frac{r^2h^2}{8m}\ .$$

Substitute for r^2 and f_{trans} in Equation 4.28 (and note that $mc\Delta c = rh^2\Delta r/4m$), and so obtain Maxwell's equation for the distribution of molecular speeds:

$$n_c = 4\pi N\left(\frac{m}{2\pi kT}\right)^{3/2} \mathrm{e}^{-mc^2/2kT}\, c^2\Delta c$$

where n_c is the average number of molecules with speeds between c and Δc.

CHAPTER FIVE

The Internal Partition Function

The factor remaining in Equation 3.30, after the evaluation of f_{trans}, is the *internal partition function* f_{int}. It will be recalled that this quantity is defined as:

$$f_{\mathrm{int}} = \sum_{u} \mathrm{e}^{-\epsilon_{\mathrm{int}(u)}/kT}$$

where u labels one of the possible quantum states of a molecule not undergoing translation (i.e., with a fixed centre of mass).

5.1 Factorization of the internal partition function

The internal molecular energy can be regarded as the sum of four approximately independent quantities:

$$\epsilon_{\mathrm{int}} = \epsilon'_{\mathrm{vib}} + \epsilon'_{\mathrm{rot}} + \epsilon'_{\mathrm{el}} + \epsilon'_{\mathrm{nuc}}\,, \tag{5.1}$$

the four subscripts having, respectively, the following meanings.

(a) vib (*vibration*). It is assumed that there is a spatial *configuration* (defined, for example, by a set of bond angles and bond lengths) of the nuclei in the molecule, referred to as the *equilibrium configuration*, for which the potential energy of the molecule has a minimum value. The molecular vibration consists of oscillatory displacements of the bond angles and bond lengths, with respect to the equilibrium values. In such oscillations, the centre of mass remains at a fixed point.

(b) rot (*rotation*). When a molecule rotates freely (i.e., in the absence of external fields; see Section 11.11) about any axis through the centre of mass, without undergoing distortion from its equilibrium configuration, the *potential* energy does not change. The *kinetic* energy is ϵ'_{rot}.

(c) el (*electronic*). ϵ'_{el} is the electronic energy of the molecule in its equilibrium configuration (the equilibrium configuration may be different for different electronic states of the molecule).

(d) nuc (*nuclear*). The energy ϵ'_{nuc} would include, for example, the potential energy associated with the forces which bind a nucleus together. Actually, it is found that detailed specification of ϵ'_{nuc} is unnecessary in applications of statistical mechanics to physical chemistry.

Equation 5.1 can be written in a slightly different form. Let the lowest allowed (i.e., *ground state*) values of the four terms on the right-hand side be $\epsilon'_{vib(0)}$, $\epsilon'_{rot(0)}$, $\epsilon'_{el(0)}$, and $\epsilon'_{nuc(0)}$. The sum:

$$\epsilon_0 = \epsilon'_{vib(0)} + \epsilon'_{rot(0)} + \epsilon'_{el(0)} + \epsilon'_{nuc(0)}$$

is called the *zero-point energy* (or *residual energy*) of the molecule. The molecular vibrational energy *relative to the lowest vibrational level* is denoted by ϵ_{vib}, and is defined as the energy difference:

$$\epsilon_{vib} = \epsilon'_{vib} - \epsilon'_{vib(0)}$$

Correspondingly, we define:

$$\epsilon_{rot} = \epsilon'_{rot} - \epsilon'_{rot(0)}$$

$$\epsilon_{el} = \epsilon'_{el} - \epsilon'_{el(0)}$$

$$\epsilon_{nuc} = \epsilon'_{nuc} - \epsilon'_{nuc(0)}$$

Then Equation 5.1 becomes:

$$\epsilon_{int} = \epsilon_0 + \epsilon_{vib} + \epsilon_{rot} + \epsilon_{el} + \epsilon_{nuc}. \tag{5.2}$$

Each of the quantities on the right-hand side of Equation 5.2 can, in principle, be separately calculated by solving the appropriate version of Schrödinger's equation. Certain discrete values will be allowed for each of these energies, and, associated with any one energy, will be one or more quantum states, i.e., wave functions.

Let the possible vibrational, rotational, electronic, and nuclear states be labelled, respectively, by the indices v,r,e, and n. As is shown below, only nuclear and (in most cases) electronic states of lowest energy need be considered, and it is assumed in the following discussion that the subscripts e and n refer to such states.* In order to avoid complications arising from symmetry restrictions on the allowed wave functions, we also assume that the molecule is unsymmetrical (e.g., HCl but not N_2); the effect of molecular symmetry will be discussed in Section 5.5.

Since a possible value of ϵ_{int} is obtained by adding together one each of the vibrational, rotational, electronic, and nuclear energies, the internal partition function is:

$$f_{int} = \sum_{\substack{\text{quantum} \\ \text{states, } u}} e^{-\epsilon_{int(u)}/kT}$$

$$= \sum_{v,r,e,n} e^{-[\epsilon_0 + \epsilon_{vib(v)} + \epsilon_{rot(r)} + \epsilon_{el(e)} + \epsilon_{nuc(n)}]/kT}$$

By the same arguments which led to Equation 3.30, it follows that:

$$f_{int} = f_0 f_{vib} f_{rot} f_{el} f_{nuc}$$

where:

$$f_{rot} = \sum_{\substack{\text{rotational} \\ \text{states, } r}} e^{-\epsilon_{rot(r)}/kT} \tag{5.3}$$

is called the *rotational partition function*, and the vibrational, electronic, and nuclear partition functions are defined in a corresponding way. The factor f_0 is defined as:

$$f_0 = e^{-\epsilon_0/kT}. \tag{5.4}$$

As in Section 3.4, terms in these partition functions which arise from quantum states having the same energy can be collected together. Thus, if the rotational energy *levels* are labelled by an index J ($J = 0,1,2,3,...,$), Equation 5.3 becomes:

$$f_{rot} = g_{rot(0)}e^{-\epsilon_{rot(0)}/kT} + g_{rot(1)}e^{-\epsilon_{rot(1)}/kT} + ... + g_{rot(J)}e^{-\epsilon_{rot(J)}/kT} + ...$$

$$= \sum_{\substack{\text{rotational} \\ \text{levels, } J}} g_{rot(J)}\, e^{-\epsilon_{rot(J)}/kT},$$

where $g_{rot(J)}$ is the degeneracy in the J-th rotational level. The individual terms in this sum have the usual (cf. Section 3.4) meaning, namely, the J-th term is proportional to $\bar{n}_{rot(J)}$, *the number*

* With this assumption, one avoids complications arising from changes in the geometry of the equilibrium configuration associated with the electronic excitation of molecules. This is discussed in Section 5.3.

of molecules with rotational energy $\epsilon_{\mathrm{rot}(J)}$, *irrespective of their other energies*; similarly for the other factors in f_{int}.

In the remainder of the present chapter, we consider each of the factors of f_{int} in turn.

5.2 The nuclear partition function

The value of $\epsilon_{\mathrm{nuc}(n)}$ for excited nuclear states is extremely large, and the corresponding terms $e^{-\epsilon_{\mathrm{nuc}(n)}/kT}$ are therefore negligible at any terrestial temperatures. Consequently, only the ground state term need be considered:

$$f_{\mathrm{nuc}} = g_{\mathrm{nuc}(0)}e^0 = g_{\mathrm{nuc}(0)}$$

the degeneracy of the lowest nuclear level.

We shall find that, in almost every calculation, f_{nuc} cancels out, and a detailed knowledge of $g_{\mathrm{nuc}(0)}$ is therefore unnecessary. Important exceptions to this are calculations for symmetrical diatomic molecules such as H_2 and D_2. It will be shown in Section 13.7 that such homonuclear diatomic substances can be regarded as mixtures of two species, referred to as *ortho and para molecules*. These differ in the symmetry properties of their nuclear spin wave functions, and it is found that for *ortho* molecules:

$$g^{\mathrm{ortho}}_{\mathrm{nuc}(0)} = (I + 1)(2I + 1) \tag{5.5}$$

and for *para* molecules:

$$g^{\mathrm{para}}_{\mathrm{nuc}(0)} = I(2I + 1) \tag{5.6}$$

In these expressions, I is the spin quantum number of the nucleus, i.e., the number determining the spin angular momentum m of the nucleus according to:

$$m = I(I + 1)(h/2\pi).$$

The number I can be deduced from the hyperfine structure of atomic spectra, and from intensity variations in molecular rotational spectra (Section 13.7).

For the hydrogen atom, $I = \frac{1}{2}$, and so:

$$g^{\mathrm{ortho}}_{\mathrm{nuc}(0)} = 3 \;;\; g^{\mathrm{para}}_{\mathrm{nuc}(0)} = 1 \quad \text{for hydrogen.}$$

For the deuterium atom, $I = 1$, and so:

$$g^{\mathrm{ortho}}_{\mathrm{nuc}(0)} = 6 \;;\; g^{\mathrm{para}}_{\mathrm{nuc}(0)} = 3 \quad \text{for deuterium.}$$

5.3 The electronic partition function

In the term $e^{-\epsilon_{el(e)}/kT}$, $\epsilon_{el(e)}$ is the difference between the energy of the electronic state e and that of the lowest electronic level. Values of $\epsilon_{el(e)}$ can be obtained directly from the electronic spectra of atoms and molecules, and, typically, for an excited level:

$$\epsilon_{el(e)} > 2\text{eV} \approx 3 \times 10^{-19}\text{ J}.$$

In this case (see p. 26), $e^{-\epsilon_{el(e)}/kT}$ is negligible at room temperature. It follows that, in the evaluation of f_{int}, terms arising from excited electronic states can often be neglected. In such cases:

$$f_{el} = g_{el(0)}$$

where $g_{el(0)}$ is the degeneracy of the lowest electronic level.

The lowest electronic levels of *molecules* and *stable ions* are nearly always non-degenerate, i.e., $g_{el(0)} = 1$. An exception is that $g_{el(0)} = 3$ for O_2.

For *free radicals*, $g_{el(0)} \neq 1$. Usually, since one unpaired electron can have either α or β spin, $g_{el(0)} = 2$.

The lowest electronic energy levels of *free atoms* are frequently degenerate, the value of $g_{el(0)}$ depending upon the number of unpaired electrons. In the Russell–Saunders approximation, the degeneracy of an electronic level is $2J + 1$, where J is the quantum number for the total electronic angular momentum. The value of J is shown as the subscript in the Russell–Saunders term symbol. Some examples are shown in Table 5.1.

Table 5.1. Ground State Electronic Degeneracies

Atom	He	Na	Tl	Pb	Cl
Ground state term $^{(2S+1)}L_J$	^1S	$^1S_{1/2}$	$^2P_{1/2}$	$^3P_0$	$^2P_{3/2}$
$g_{el(0)} \equiv 2J + 1$	1	2	2	1	4

In some cases, particularly for free atoms and free radicals, the contributions to f_{el} from excited electronic states are rather important.

For atoms, these can be calculated quite simply. For example, the ground and first excited states of the chlorine atom are $^2P_{3/2}$ and $^2P_{1/2}$, and a good approximation for f_{el} is, therefore:

$$f_{el} = 4 + 2e^{-\epsilon_{el(1)}/kT}.$$

Now, $\epsilon_{el(1)} = 0{\cdot}11$ eV $= 1{\cdot}8 \times 10^{-20}$ J. At 1000 K, therefore:

$$f_{el} = 4 + (2 \times 0{\cdot}28) = 4{\cdot}56.$$

The term from the first excited electronic level thus contributes about 10% to f_{el} in this case.

The problem is more complicated for molecules, because the equilibrium configuration of the molecule may change on electronic excitation. In such cases, the molecular energy is not the sum of *independent* terms (cf. Equation 5.2), and simple factorization of f_{int} is not possible. Clearly, however, f_{int} can be expressed in the following way:

$$f_{int} = f_{int(0)} + f_{int(1)} + f_{int(2)} + \ldots$$

where $f_{int(0)}$ is the sum of those terms in f_{int} which arise from the lowest electronic level, $f_{int(1)}$ is the sum of terms arising from the first excited electronic level, and so on. The individual terms $f_{int(0)}$, $f_{int(1)}$, etc. can *now* be factorized in the usual way, e.g.:

$$f_{int(1)} = f_0 f_{vib(1)} f_{rot(1)} e^{-\epsilon_{el(1)}/kT} f_{nuc}$$

where $f_{rot(1)}$ and $f_{vib(1)}$ are rotational and vibrational partition functions appropriate to the molecular geometry associated with the first excited electronic level.

5.4 The vibrational partition function

(i) *Diatomic Molecules*

Let the equilibrium internuclear distance in a diatomic molecule be r_0. According to Hooke's law, the increase $\mathbf{V}$ in potential energy of the molecule, which results when the internuclear distance is changed to r, is:

$$\mathbf{V} = \tfrac{1}{2}k(r - r_0)^2 \qquad (5.7)$$

where k is the force constant.

Let $\mathbf{V}'$ be the *total* potential energy of the molecule when the internuclear distance is r. Taylor series expansion of $\mathbf{V}'$ about any *arbitrary* point ($\mathbf{V}' = \mathbf{V}'_0$, $r = r_0$) gives:

$$\mathbf{V}' = \mathbf{V}'_0 + \left(\frac{d\mathbf{V}'}{dr}\right)_{r=r_0}(r - r_0) + \frac{1}{2}\left(\frac{d^2\mathbf{V}'}{dr^2}\right)_{r=r_0}(r - r_0)^2 + \ldots$$

Hooke's law, Equation 5.7, is obtained if both of the following hold.

(a) r_0 is the *equilibrium* internuclear distance; in this case, $\mathbf{V}'$ is a minimum at r_0 and:

$$\left(\frac{d\mathbf{V}'}{dr}\right)_{r=r_0} = 0$$

(b) the displacement $(r-r_0)$ is small; in this case, terms in $(r-r_0)^3$, $(r-r_0)^4$, etc., can be neglected. Under these conditions:

$$\mathbf{V}' = \mathbf{V}'_0 + \frac{1}{2}\left(\frac{d^2\mathbf{V}'}{dr^2}\right)_{r=r_0} (r-r_0)^2 .$$

Identification of $(\mathbf{V}'-\mathbf{V}'_0)$ with $\mathbf{V}$ and $(d^2\mathbf{V}'/dr^2)_{r=r_0}$ with k then gives Equation 5.7.

The variation of r with time is, in classical mechanics, simple harmonic motion. This is a vibration of frequency ω given by:

$$\omega = \frac{1}{2\pi}\sqrt{\frac{k}{\mu}} \tag{5.8}$$

where μ is the *reduced mass* of the molecule:

$$\mu = \frac{m_A m_B}{m_A + m_B} \tag{5.9}$$

m_A and m_B being the respective masses of the two nuclei.

The permitted vibrational energy levels are found in quantum mechanics to be:

$$\epsilon'_{vib} = (v + \tfrac{1}{2})h\omega \tag{5.10}$$

where the vibrational quantum number v can have values 0,1,2,3,..., and ω is again given by Equation 5.8. These energy levels are non-degenerate, i.e.:

$$g_{vib(v)} = 1, \quad \text{for all } v.$$

The lowest vibrational energy level is $\epsilon'_{vib(0)} = \frac{1}{2}h\omega$, and, as explained in Section 5.1, we measure vibrational energies from this level. Thus:

$$\begin{aligned}\epsilon_{vib(v)} &= \epsilon'_{vib(v)} - \epsilon'_{vib(0)} \\ &= (v+\tfrac{1}{2})h\omega - \tfrac{1}{2}h\omega \\ &= vh\omega \qquad (v = 0,1,2,3,...).\end{aligned} \tag{5.11}$$

Hence the vibrational partition function for a diatomic molecule is:

$$f_{vib} = \sum_{v=0}^{\infty} e^{-vh\omega/kT}. \tag{5.12}$$

Let:

$$e^{-h\omega/kT} = x$$

Then:

$$e^{-vh\omega/kT} = x^v$$

and Equation 5.9 becomes:

$$f_{vib} = \sum_{v=0}^{\infty} x^v = 1 + x + x^2 + x^3 + \ldots \tag{5.13}$$

The value of this infinite series is simply $1/(1 - x)$. Finally, therefore:

$$f_{vib} = \frac{1}{1 - e^{-h\omega/kT}} \tag{5.14}$$

The quantity $h\omega/k$ has the dimension of temperature. It is called the *characteristic temperature for vibration* of the molecule, and is denoted by Θ_{vib}. Thus:

$$f_{vib} = \frac{1}{1 - e^{-\Theta_{vib}/T}}\,. \tag{5.14$'$}$$

The value of ω, and therefore of Θ_{vib}, is obtained experimentally from the vibration spectrum of the molecule (see Chapter 10). Some results are shown in Table 5.2.

Table 5.2. Characteristic Temperatures for Vibration

Molecule	Θ_{vib} (K)	f_{vib} 300 K	f_{vib} 1000 K
H_2	5987	1·000	1·002
HD	5226	1·000	1·005
D_2	4307	1·000	1·014
N_2	3352	1·000	1·036
O_2	2239	1·000	1·019
Cl_2	798	1·075	1·816
I_2	307	1·556	3·773
CO	3084	1·000	1·046

Values of Θ_{vib} taken from H.J.G. Hayman, *Statistical Thermodynamics*, Elsevier, Amsterdam (1967), p. 26. f_{vib} values are calculated from Equation 5.14.

Equation 5.14 is a valid approximation for the vibrational partition functions of diatomic molecules. It is, however, physically meaningless to take the sum in Equation 5.13 to $v \longrightarrow \infty$, because the vibrational energy would exceed the dissociation energy of the molecule long before such high values of v are reached. More realistically, one could sum over a finite number of vibrational levels (see problem 5.4). Such a calculation is, however, somewhat artificial, because *anharmonicity* (i.e., departure from Equations 5.7 and 5.10) is important in the higher levels. Fortunately, these complications are rather unimportant, because $h\omega/kT$ is so large that only the first few terms in the sum 5.13 contribute significantly to f_{vib}.

A correction for anharmonicity can, in fact, be made but, since it is of the same order as certain rotational correction terms, we will consider this in Section 5.5. However, it must be pointed out here that, because of anharmonicity, the *apparent* value of ω [obtained from the spectroscopic frequency ν for the transition $v' \longrightarrow v''$ by $h\nu = (v'' - v')h\omega$] is not exactly constant. Consequently, Θ_{vib} is usually defined to be $h\omega_0/k$, where ω_0 is the value of ω calculated from the experimentally determined frequency of the $v = 0 \longrightarrow v = 1$ transition.

(ii) *Polyatomic Molecules*

A linear molecule composed of $\mathfrak{N}$ atoms has $3\mathfrak{N} - 5$ *normal modes of vibration.* If the molecule is non-linear, the corresponding number of normal modes is* $3\mathfrak{N} - 6$.

These statements have the following meaning. $3\mathfrak{N}$ spatial coordinates must be specified if the instantaneous positions of all the nuclei in the molecule are to be uniquely defined (see Chapter 11). Three of these are coordinates of the centre of mass, the variation of which with time gives the *translational* motion of the molecule. Two (for linear molecules) or three (for non-linear) angles specify the orientation of the molecule, in its equilibrium configuration, relative to some fixed reference system; the variation of these angles with time gives the *rotational* motion of the molecule. There remain $3\mathfrak{N} - 5$ (linear molecule) or $3\mathfrak{N} - 6$ (non-linear molecule) coordinates, specification of which gives the positions of the nuclei relative to one another. Variation of these with time gives the *vibration* of the molecule.

The positions of the nuclei relative to one another can be expressed in terms of many different coordinates, or variables. It is always possible to choose one set of coordinates, say $q_1, q_2, ..., q_{3\mathfrak{N}-5 \text{ or } 6}$, such that the potential energy of a molecular configuration, relative to that of the equilibrium configuration, can be expressed as:

$$\mathbf{V} = \mathbf{V}_1 + \mathbf{V}_2 + ... + \mathbf{V}_{3\mathfrak{N}-5 \text{ or } 6} = \sum_{l=1}^{3\mathfrak{N}-5 \text{ or } 6} \mathbf{V}_l$$

where

$$\mathbf{V}_l = \tfrac{1}{2}k_l(q_l - q_l^\circ)^2$$

k_l being a constant and q_l° being the equilibrium value of q_l. It is then found that the equation of vibrational motion (involving $3\mathfrak{N} - 5$ or $3\mathfrak{N} - 6$ position variables) can be broken down into $3\mathfrak{N} - 5$ or $3\mathfrak{N} - 6$ simpler equations (Pauling and Wilson, p.289), each of which involves only one of the coordinates q. Coordinates chosen in this way are called *normal coordinates. Each of the* $3\mathfrak{N} - 5$ *or* $3\mathfrak{N} - 6$ *equations has the same form as the equation of motion of a diatomic oscillator*, and corresponds to a vibration of the whole molecule in which every nucleus oscillates with the same frequency ω. Such a molecular vibration is called a *normal mode*. Any general vibration of the molecule can thus be expressed as a superposition of normal modes.

* In some molecules, one or more of these normal modes may be replaced by an internal rotation; see Section 5.6.

It is found (Pauling and Wilson, p.289) that the total vibrational energy of the molecule, ϵ'_{vib} can be expressed as the sum of $3\mathfrak{N} - 5$ (linear molecules) or $3\mathfrak{N} - 6$ (non-linear molecules) independent terms:

$$\epsilon'_{\text{vib}} = \epsilon'_1 + \epsilon'_2 + \ldots + \epsilon'_{3\mathfrak{N} - 5 \text{ or } 6},$$

each of which is given by an equation of the same form as 5.10:

$$\epsilon'_1 = (v_1 + \tfrac{1}{2})h\omega_1 \ , \ (v_1 = 0,1,2,\ldots).$$

$$\epsilon'_2 = (v_2 + \tfrac{1}{2})h\omega_2 \ , \ (v_2 = 0,1,2,\ldots).$$

and so on, ω_i being the vibrational frequency (determined spectroscopically) of the i-th normal mode.

The lowest molecular vibrational energy level is:

$$\epsilon'_{\text{vib}(0)} = \tfrac{1}{2}h\omega_1 + \tfrac{1}{2}h\omega_2 + \ldots + \tfrac{1}{2}h\omega_{3\mathfrak{N} - 5 \text{ or } 6}$$

and the molecular vibrational energy *relative to this* is:

$$\epsilon_{\text{vib}} = \epsilon_1 + \epsilon_2 + \ldots + \epsilon_{3\mathfrak{N} - 5 \text{ or } 6} \tag{5.15}$$

where $\epsilon_1 = v_1 h\omega_1$, $\epsilon_2 = v_2 h\omega_2$, etc. Since ϵ_{vib} is expressible as the sum of *independent* terms, it follows, by the argument used on pages 57 and 65 that the vibrational partition function of a polyatomic molecule can be factorized:

$$f_{\text{vib}} = f_{\text{vib}(1)} f_{\text{vib}(2)} \cdots f_{\text{vib}(3\mathfrak{N} - 5 \text{ or } 6)}$$

where, for example:

$$f_{\text{vib}(1)} = \frac{1}{1 - e^{-h\omega_1/kT}}$$

Finally, therefore:

$$f_{\text{vib}} = \prod_{l=1}^{3\mathfrak{N} - 5 \text{ or } 6} \frac{1}{1 - e^{-h\omega_l/kT}} \tag{5.16}$$

$$= \prod_{l=1}^{3\mathfrak{N} - 5 \text{ or } 6} \frac{1}{1 - e^{-\Theta_{\text{vib}(l)}/kT}} \tag{5.16'}$$

where $\Theta_{\text{vib}(l)} = h\omega_l/kT$. Some value of $\Theta_{\text{vib}(l)}$ are given in Table 5.3.

It is often possible to estimate most of the vibrational frequencies of a molecule, without experimental observation; thus, values of many of the $\Theta_{\text{vib}(l)}$ terms can be deduced from the molecular structure. The basis of this procedure is the theory of Mecke, that to every bond in a molecule there corresponds a vibration in which this bond is stretched, and another, of much smaller frequency, in which it is bent. Actually, a normal mode is a vibration

Table 5.3. Characteristic Temperatures for Vibration in Polyatomic Molecules

Molecule	$\Theta_{vib(l)}$ (K)					
CO_2	1890	3360	954(2)			
NH_3	4780	1360	4880(2)	2330(2)		
$CHCl_3$	523	938	4330	374(2)	1090(2)	1745(2)
H_2O	2290	5160	5360			

Values taken from Fowler and Guggenheim, Ch. 3. The bracketed values are the degeneracies of the corresponding frequencies.

of the *whole* molecule; however, the explanation of Mecke's observation is that, in the molecular vibration, the vibrational energy is essentially concentrated in one bond, i.e., the amplitude of vibration of the remaining atoms is small. Frequencies associated with various bond vibrations have been tabulated (Herzberg II, p.195). Unfortunately, there are limitations to this procedure; in particular, there are often vibrations for which $\Theta_{vib(l)}$ is small (and therefore important), which cannot be regarded as being localized in one bond. It is necessary to refer to books on infrared spectroscopy, for details (Herzberg II Ch. II).

5.5 The rotational partition function

(i) *Heteronuclear Diatomic Molecules*

Initially, to avoid complications which result from molecular symmetry, we consider unsymmetrical diatomic molecules, such as HCl, HD, $^{37}Cl^{35}Cl$. The rotational energies depend upon a quantum number J:

$$\epsilon_{rot(J)} = \frac{J(J+1)h^2}{8\pi^2 I} \quad (J = 0,1,2,...) \tag{5.17}$$

where I is the moment of inertia about an axis perpendicular to the molecular axis and passing through the centre of mass of the molecule. I is related to the *equilibrium* internuclear distance r_0 by:

$$I = \mu r_0^2 \tag{5.18}$$

where μ is the reduced mass of the molecule, given by Equation 5.9.

The energy level $\epsilon_{rot(J)}$ is degenerate, the degeneracy being:

$$g_{rot(J)} = 2J + 1 \tag{5.19}$$

Hence:

$$f_{rot} = \sum_{\substack{\text{rotational} \\ \text{states, } r}} e^{-\epsilon_{rot(r)}/kT} \equiv \sum_{\substack{\text{rotational} \\ \text{levels, } J}} g_{rot(J)}\, e^{-\epsilon_{rot(J)}/kT}$$

$$= \sum_{J=0}^{\infty} (2J + 1) e^{-J(J+1)h^2/8\pi^2 IkT} \quad (5.20)$$

It is found by calculation that, for almost every case of practical interest (the exceptions are the hydrogen isotopes at low temperatures), $h^2/8\pi^2 IkT$ is small, and successive terms in the summation of Equation 5.20 have values sufficiently close together for the summation to be approximated by integration (see Appendix 1). Thus, approximately:

$$f_{rot} = \int_0^{\infty} (2J + 1) e^{-J(J+1)C}\, dJ$$

where:

$$C = h^2/8\pi^2 IkT \quad (5.21)$$

Let:

$$x = J^2 + J$$

Then:

$$dx = (2J + 1) dJ$$

and

$$f_{rot} = \int_0^{\infty} e^{-Cx}\, dx$$

$$= -\frac{1}{C}[e^{-\infty} - e^0] = \frac{1}{C}$$

Thus:

$$f_{rot} = \frac{8\pi^2 IkT}{h^2} \quad (5.22)$$

The quantity $h^2/8\pi^2 Ik$ has the dimension of temperature, and is called the *characteristic temperature for rotation*, denoted by Θ_{rot}; that is:

$$f_{rot} = \frac{T}{\Theta_{rot}} \quad (5.22')$$

Some values of Θ_{rot} are given in Table 5.4.

Equation 5.22 provides a satisfactory approximation in most cases. More generally, however, the following expansions (Mayer and Mayer, p.448) give estimates of the sum 5.20 which are accurate to at least 0·1%.

If $C > 0{\cdot}7$:

$$f_{rot} = 1 - 3e^{-2C} + 5e^{-6C} + 7e^{-12C}$$

(the first terms in Equation 5.20).

Table 5.4. Characteristic Temperatures for Rotation

Molecule	Θ_{rot} (K)	f_{rot} 300 K	f_{rot} 1000 K
H_2	85·38	1·76	5·86
HD	64·27	4·67	15·56
D_2	43·03	3·49	11·62
N_2	2·863	52·41	174·7
O_2	2·069	72·51	241·7
CO	2·766	108·5	361·5
$^{35}Cl_2$	0·3495	429·3	1431
I_2	0·0537	2793	9310

Values of Θ_{rot} taken from H.J.G. Hayman, *Statistical Thermodynamics*, Elsevier, Amsterdam (1967), p. 26. f_{rot} values are calculated from Equation 5.24.

If $C < 0{\cdot}7$:

$$f_{rot} = \frac{1}{C}\left(1 + \frac{C}{3} + \frac{C^2}{15} + \frac{4C^3}{315}\right)$$

(Mulholland's equation; see Appendix 1).

There are also small errors inherent in the use of Equations 5.12 and 5.20. These arise from three effects of significant magnitude.

(a) Because Hooke's law is not obeyed exactly, a more correct expression for the vibrational energy is:

$$\epsilon'_{vib} = (v + \tfrac{1}{2})h\omega - x(v + \tfrac{1}{2})^2 h\omega$$

where x is a small ($\sim 10^{-2}$) positive term called the *anharmonicity constant* of the vibration. The energy relative to the ground state is:

$$\epsilon_{vib} = vh\omega_0 - xh\omega_0 v(v-1)$$

where $h\omega_0 \equiv h(\omega - 2x\omega)$ is the energy of the first ($v = 0 \longrightarrow v = 1$) vibrational transition.

(b) The rotation of the molecule causes centrifugal stretching, which increases the moment of inertia. The rotational levels corrected for this effect are:

$$\epsilon'_{rot(J)} = CJ(J+1) - DJ^2(J+1)^2$$

where C is as defined in Equation 5.21 and D is a parameter for the molecule.

(c) It is more correct to use in Equation 5.18 the value of I calculated by using an internuclear distance $\bar{r}$ which is the value of r averaged over the whole vibration. In simple harmonic motion, $\bar{r} = r_0$. In anharmonic motion, however, $\bar{r}$ varies with ϵ'_{vib}. The term added to the rotational energy to correct for this effect is $\alpha(v + \frac{1}{2})J(J+1)$, where α is a parameter for the molecule.

Mayer and Mayer* give expressions for thermodynamic functions corrected for these effects. However, the corrections are small, being of the order of 0·1% at 500 K (although they increase rather rapidly in importance as T is increased).

(ii) *Homonuclear Diatomic Molecules*

In Section 13.7 we will show that, because of symmetry restrictions on the molecular wave functions, not all integral values of J are permitted for homonuclear diatomic molecules. The following rules governing the allowed values of J will be obtained. Defining the mass number of an atom as the sum of the numbers of protons and neutrons in the nucleus:

(a) for molecules of atoms with odd mass numbers (e.g., H_2),

for *ortho* molecules, $J = 1,3,5,...$

for *para* molecules, $J = 0,2,4,...$

(b) for molecules with even mass numbers (e.g., D_2),

for *ortho* molecules, $J = 0,2,4,...$

for *para* molecules, $J = 1,3,5,...$

For present purposes, however, detailed application of these results is usually unnecessary. In most cases (see Table 5.4), $T >> \Theta_{rot}$ and it can be shown (Appendix 1) that, in such case, a valid approximation is:

$$\sum_{J=0,2,4,...} (2J + 1)e^{-J(J+1)\Theta_{rot}/T} = \sum_{J=1,3,5,...} (2J + 1)e^{-J(J+1)\Theta_{rot}/T}$$

$$= \frac{1}{2}\left(\frac{8\pi^2 IkT}{h^2}\right) \qquad (5.23)$$

The factor $\frac{1}{2}$ appears because, essentially, the sums are evaluated by omitting half the terms from the right-hand side of Equation 5.20.

Equations 5.22 and 5.23 can be combined into one by writing, for any diatomic molecule:

$$f_{rot} = \frac{8\pi^2 IkT}{\sigma h^2} \qquad (5.24)$$

where $\sigma = 1$ for heteronuclear, and 2 for homonuclear diatomic molecules. The factor σ is called the *symmetry number* of the

* Mayer and Mayer, p.60. For a review of accurate methods of calculating partition functions, see Aston pp.560–568.

molecule. For any type of molecule, the symmetry number is defined as the number of equivalent ways in which a given orientation of the molecule *can be obtained by rotation, when like atoms are regarded as indistinguishable*. Thus, for a homonuclear diatomic molecule:

and equivalent, and $\sigma = 2$.

(iii) *Linear Polyatomic Molecules*

The rotational energies are again given by Equation 5.17, in which the moment of inertia is:

$$I = \sum_s m_s r_{0(s)}^2$$

where $r_{0(s)}$ is the equilibrium distance of nucleus s (mass m_s) from the axis of rotation; the degeneracies are again given by Equation 5.19. Consequently, Equation 5.22 also applied to linear unsymmetrical molecules, such as N_2O. If the equilibrium configuration of the molecule has a centre of symmetry, as in the case of O=C=O and H–C≡C–H the arguments leading to Equation 5.23 apply. Hence f_{rot} for the general linear molecule is given by Equation 5.24, with $\sigma = 1$ or 2.

(iv) *Non-linear Polyatomic Molecules*

No *simple* general formula for f_{rot} for non-linear polyatomic molecules has been derived from the quantum theory; an approximation obtained by treating the molecule by classical mechanics is generally used. In the present section, the result is merely stated, the derivation being given in Section 11.6.

Suppose that any three mutually perpendicular (X,Y,Z) axes are drawn through the centre of mass of the molecule. Let the coordinates of the s-th nucleus, referred to these axes, be (x_s,y_s,z_s). By Pythagoras's theorem, the perpendicular distances of the s-th nucleus from the $X, Y,$ and Z axes are, respectively:

$$(y_s^2 + z_s^2)^{1/2}\,; \quad (x_s^2 + z_s^2)^{1/2}\,; \quad (x_s^2 + y_s^2)^{1/2}$$

Hence, the *moments of inertia* about these axes are, respectively:

$$I_{xx} = \sum_s m_s(y_s^2 + z_s^2)$$

$$I_{yy} = \sum_s m_s(x_s^2 + z_s^2)$$

$$I_{zz} = \sum_s m_s(x_s^2 + y_s^2)$$

where the summation is over all the nuclei in the molecule.

The *products of inertia* are defined as:

$$I_{xy} = \sum_s m_s x_s y_s$$

$$I_{xz} = \sum_s m_s x_s z_s$$

$$I_{yz} = \sum_s m_s y_s z_s \, .$$

It is always possible to choose X, Y, and Z axes such that:

$$I_{xy} = I_{xz} = I_{yz} = 0$$

Axes so chosen are called the *principal axes*, and the moments of inertia about them are called the *principal moments of inertia* (denoted by A, B, and C) of the molecule. If $A = B = C$ (as in CH_4), the molecule is a *spherical top*; if $A = B \neq C$ (as in NH_3), the molecule is a *symmetric top*; and if $C = 0$, the molecule must be linear.

In Section 11.6, it will be shown by classical mechanics that:

$$f_{\text{rot}} = \left(\frac{8\pi^2 kT}{h^2}\right)^{3/2} \frac{(\pi ABC)^{1/2}}{\sigma} \qquad (5.25)$$

where σ is the symmetry factor discussed at the end of the present section. In practice, it is unnecessary to find the principal axes of the molecule, the method of calculation being as follows.

(a) Suppose that X, Y, and Z axes *passing through the centre of mass* of the molecule are chosen arbitrarily; they may or may not be principal axes. It can be shown that the product (ABC) is always equal to the value of the determinant:*

* The reader acquainted with matrix algebra will observe that A, B, and C are the eigenvalues of the matrix $[I_{rs}]$, and that the value of the determinant is the product of the eigenvalues, the latter being invariant to rotation of axes. See Jeffreys and Jeffreys, p.95.

$$D = \begin{vmatrix} I_{xx} & -I_{xy} & -I_{xz} \\ -I_{xy} & I_{yz} & -I_{yz} \\ -I_{xz} & -I_{yz} & I_{zz} \end{vmatrix} \tag{5.26}$$

$$= I_{xx} I_{yy} I_{zz} - 2I_{xy} I_{yz} I_{xz} - I_{xx} I_{yz}^2 - I_{yy} I_{xz}^2 - I_{zz} I_{xy}^2 \,.$$

Thus:

$$f_{\text{rot}} = \left(\frac{8\pi^2 kT}{h^2}\right)^{3/2} \frac{(\pi D)^{1/2}}{\sigma} \,. \tag{5.27}$$

(b) Often, the position of the centre of mass is not obvious by inspection. If, however, completely arbitrary Cartesian axes X', Y', Z' are chosen, then the position of the centre of mass is defined by the coordinates $\overline{X}$, $\overline{Y}$, $\overline{Z}$, given by:

$$\overline{X} = \frac{1}{m} \sum_s m_s x_s'$$

$$\overline{Y} = \frac{1}{m} \sum_s m_s y_s'$$

$$\overline{Z} = \frac{1}{m} \sum_s m_s z_s'$$

where m is the mass of the molecule. The coordinates of atom s referred to axes through the centre of mass and parallel to the original axes are then:

$$x_s = x_s' - \overline{X}$$

$$y_s = y_s' - \overline{Y}$$

$$z_s = z_s' - \overline{Z}$$

These values can now be substituted into the expressions for I_{xx}, I_{xy}, etc., for the calculation of D; for example:

$$I_{xy} = \sum_s m_s(x_s' - \overline{X})(y_s' - \overline{Y}) \,.$$

Example 5.5

In the water molecule, the HOH angle is 105° and the H–O bond length is $0{\cdot}9584 \times 10^{-10}$ m. Calculate the product of the principal moments of inertia.

Quite arbitrarily, choose X', Y' axes through the centre of the oxygen atom, as shown in Fig. 5.1, so that the Z' axis is perpendicular

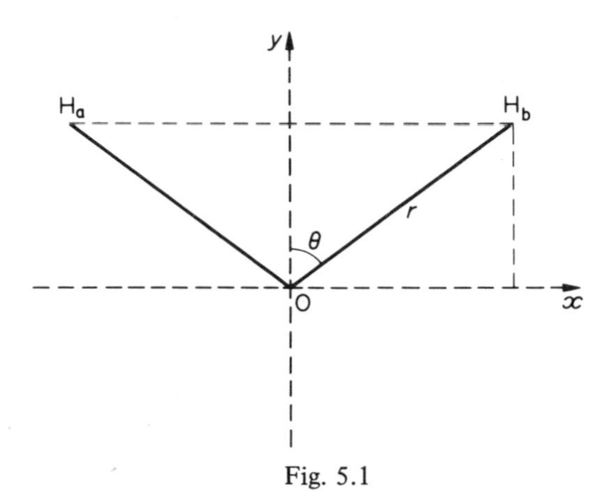

Fig. 5.1

to the plane of the paper. The coordinates of the respective atoms, referred to these axes, are:

$$\begin{aligned}
H_a: &\quad x' = -r\sin\theta \quad & y' &= r\cos\theta \quad & z' &= 0\\
H_b: &\quad x' = r\sin\theta \quad & y' &= r\cos\theta \quad & z' &= 0\\
O: &\quad x' = 0 \quad & y' &= 0 \quad & z' &= 0.
\end{aligned}$$

The atomic masses of H and O are m_H and $m_O = 16m_H$, so that $m = 18m_H$.

Hence the coordinates of the centre of mass are:

$$\bar{X} = \frac{1}{18}(r\sin\theta - r\sin\theta + 0) = 0$$

$$\bar{Y} = \frac{1}{18}(r\cos\theta + r\cos\theta + 0) = \frac{r\cos\theta}{9}$$

$$\bar{Z} = 0.$$

The coordinates of the atoms relative to the centre of mass are then:

$$\begin{aligned}
H_a: &\quad x = -r\sin\theta \quad & y &= \frac{8}{9}r\cos\theta \quad & z &= 0\\
H_b: &\quad x = r\sin\theta \quad & y &= \frac{8}{9}r\cos\theta \quad & z &= 0\\
O: &\quad x = 0 \quad & y &= -\frac{r}{9}\cos\theta \quad & z &= 0.
\end{aligned}$$

Therefore:

$$I_{xx} = m_{\mathrm{H}}r^2\left(\frac{2\times 64}{81}\cos^2\theta + \frac{16}{81}\cos^2\theta\right) = 2m_{\mathrm{H}}r^2 \times \frac{8}{9}\cos^2\theta$$

$$I_{yy} = 2m_{\mathrm{H}}r^2\sin^2\theta$$

$$I_{zz} = 2m_{\mathrm{H}}r^2\left(\sin^2\theta + \frac{8}{9}\cos^2\theta\right) = 2m_{\mathrm{H}}r^2\left(1 - \frac{1}{9}\cos^2\theta\right)$$

$$I_{xy} = I_{xz} = I_{yz} = 0$$

Immediately, therefore:

$$ABC = 8m_{\mathrm{H}}^3r^6\left(\tfrac{8}{9}\cos^2\theta\right)(\sin^2\theta)\left(1 - \tfrac{1}{9}\cos^2\theta\right)$$

$$= 5{\cdot}640 \times 10^{-141}\ \mathrm{kg^3\ m^6}\,.$$

□ □ □

As is the case of linear and diatomic molecules, the symmetry number σ is the number of ways of obtaining a given orientation by rotation of the molecule. Some examples are:

H_2O:

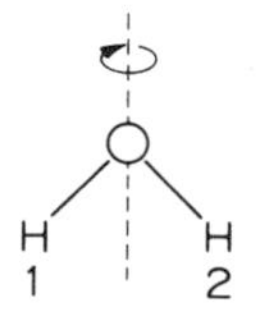

H 2 H 1 $\sigma = 2$

NH_3:

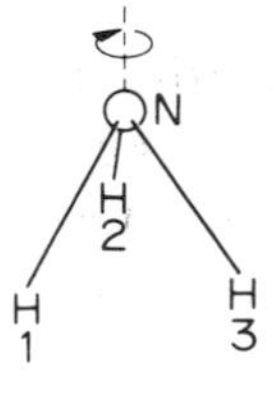

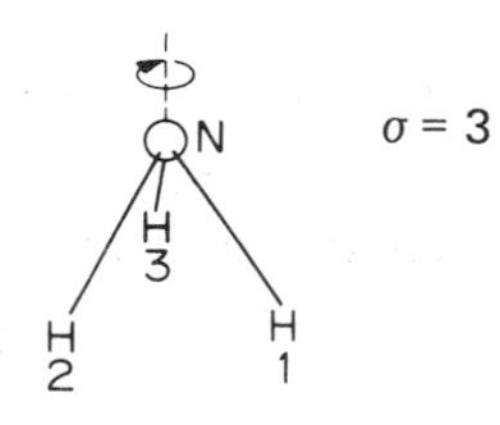

$\sigma = 3$

Note that orientation such as:

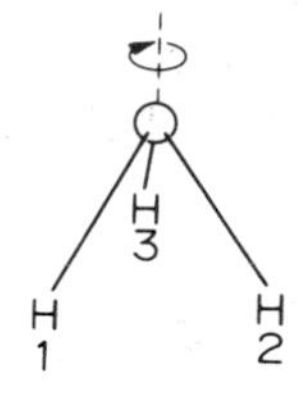

cannot be obtained from the above structures by rotation, and this arrangement is not considered in the evaluation of σ.

CH_4 (tetrahedrally arranged hydrogen atoms):

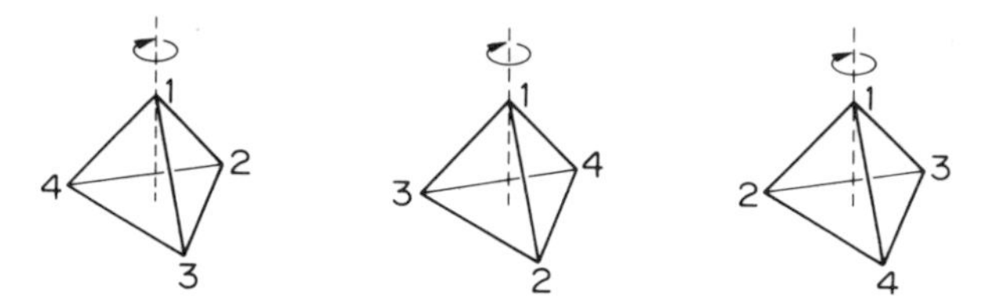

and three other sets of three orientations, obtained by placing (by rotation) in turn atoms 2,3, and 4 at the topmost position. Thus, $\sigma = 4 \times 3 = 12$.

5.6 Internal rotation

Ethylene has $3 \times 6 - 6 = 12$ normal modes of vibration. Eleven of the corresponding vibrational frequencies have been determined spectroscopically. The remaining mode, which is spectroscopically inactive, is the *twisting mode*:

in which the two CH_2 groups are twisted relative to each other about the C=C axis, through an angle φ. If Hooke's law applies, the change in molecular potential energy which results from an angular displacement from the equilibrium position is:

$$\mathrm{V} = \tfrac{1}{2} c \varphi^2 \tag{5.28}$$

where c is the force constant. Classical mechanics shows that the variation of φ with time is then a simple harmonic motion, and equations analogous to 5.8 and 5.10 can be derived.

The corresponding dynamical behaviour of the *ethane* molecule:

is somewhat different, because the C–C bond is now a *single* bond, associated with a (cylindrically symmetrical) σ-electron distribution.

The dependence of **V** upon φ is therefore not very great; what dependence there is can be regarded as arising from the repulsion between the C–H bonds of different CH_3 groups. In such a case, the twisting mode is replaced by an internal rotation, i.e., complete revolution of one CH_3 group relative to the other, around the C–C axis.

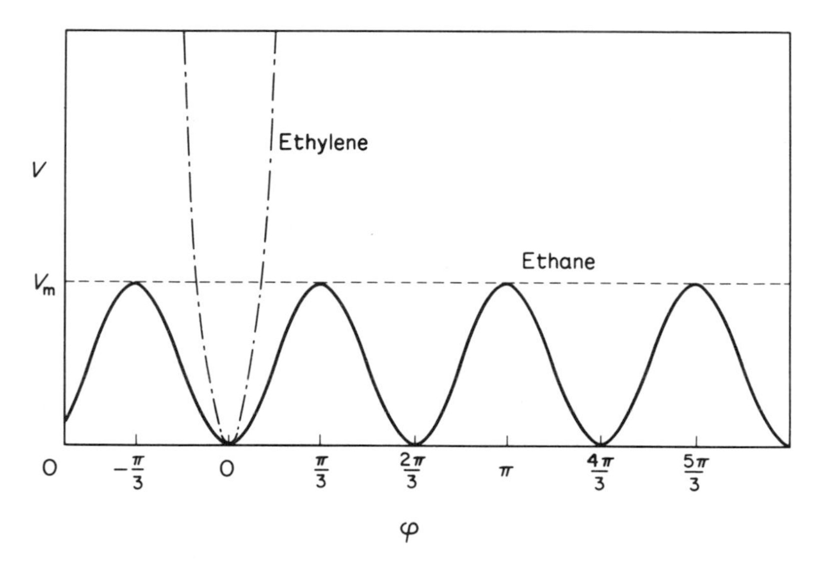

Fig. 5.2 Potential functions for ethane and ethylene

Figure 5.2 represents the difference between the potential functions **V** for ethylene and ethane. Positions of minimum potential energy (i.e., minimum repulsion) occur when the two CH_3 groups have a staggered conformation; it is convenient to measure **V** from this minimum value. The positions of maximum potential energy ($\mathbf{V} = \mathbf{V}_m$) are the eclipsed conformations. $\mathbf{V}_m$ is called the *potential barrier to internal rotation.*

$V = 0$ $\qquad\qquad$ $V = V_m$

In general, in one complete revolution ($0 \leqslant \varphi < 2\pi$) each *relative* configuration will occur σ_{int} times, where σ_{int} is the *symmetry number for internal rotation*. For example, for ethane, $\sigma_{int} = 3$ and the conformation of minimum potential energy occurs at 0, $2\pi/3$, and $4\pi/3$.

The following expression for $\mathbf{V}$ is consistent with* the form of the curve in Fig. 5.2:

$$\mathbf{V} = \tfrac{1}{2}\mathbf{V}_m(1 - \cos \sigma_{int}\varphi) \tag{5.29}$$

with $\sigma_{int} = 3$ for ethane.

Three types of behaviour can be distinguished, depending on the magnitude of $\mathbf{V}_m$.

(i) $\mathbf{V}_m$ *is large*

If $\mathbf{V}_m$ is greater than about $10\boldsymbol{k}T$, relatively few molecules have sufficient rotational energy to surmount the potential barrier (see page 26). The predominant motion is then an oscillation about the value of φ for which $\mathbf{V} = 0$; only small values of φ can be achieved, and expansion of the cosine term in Equation 5.29 leads to an expression of the type 5.28 (see problem 5.14). This is just the behaviour which we have described in the case of ethylene.

(ii) $\mathbf{V}_m$ *is essentially zero*

In a few cases, $\mathbf{V}_m$ is sufficiently small for the dependence of $\mathbf{V}$ on φ to be neglected; in such cases, there is essentially *free* internal rotation. For example, the experimentally observed entropies of cadmium dimethyl, $CH_3{-}Cd{-}CH_3$, and dimethyl acetylene, $CH_3{-}C{\equiv}C{-}CH_3$, agree with the entropies calculated from a partition function derived for free internal rotation.

The derivation of the partition function for free internal rotation is quite easy. In classical mechanics, the *kinetic* energy of internal rotation is:

$$\epsilon = \frac{1}{2} I_{red} \left(\frac{d\varphi}{dt}\right)^2 \tag{5.30}$$

where I_{red} is the reduced moment of inertia about the axis around which φ is measured. If $\mathbf{V}_m = 0$, this is the *total* energy of internal rotation. If the molecule consists of two coaxial symmetric tops:

* The terms on the right-hand side of Equation 5.29 are actually the initial members of a Fourier series, and the periodic function $\mathbf{V}$ could probably be represented more accurately if higher terms in this series were included. However, in view of the uncertain nature of the function $\mathbf{V}$, it is usually assumed that Equation 5.29 is adequate as it stands.

$$I_{\text{red}} = \frac{I_1 I_2}{I_1 + I_2} \tag{5.31}$$

where I_1 and I_2 are the moments of inertia of the two groups about the axis of internal rotation.

Taking the axis of internal rotation as the z-axis, a group is a symmetric top if its moments of inertia about the x- and y-axes are the same; this is the case for CH_3. Groups which are symmetric tops have the property that their rotation around the z-axis does not alter the principal moments of inertia of the molecule. Consider the kinetic energy of rotation of the groups around the z-axis (which, in the present case is a principal axis of the molecule; see page 78). This energy is:

$$\epsilon = \tfrac{1}{2}I_1(\dot{\alpha}_1)^2 + \tfrac{1}{2}I_2(\dot{\alpha}_2)^2 \tag{5.32}$$

where α_1 and α_2 are the angles of rotation of the CH_3 groups referred to a common zero, and $\dot{\alpha}$ means the angular velocity, $d\alpha/dt$.

Let $I = I_1 + I_2$, the total moment of inertia of the molecule about the z-axis. Noting that $\alpha_1 - \alpha_2 = \varphi$, the angle of twist, one can rearrange Equation 5.32 to obtain:

$$\epsilon = \tfrac{1}{2}I(\dot{\theta})^2 + \tfrac{1}{2}I_{\text{red}}(\dot{\varphi})^2 \tag{5.33}$$

where $\dot{\theta} = (I_1\dot{\alpha}_1 + I_2\dot{\alpha}_2)/I$. The first term in Equation 5.33 is the total energy of rotation about the z-axis when φ is constant, i.e., when there is no internal rotation; the contribution of this term is included in Equation 5.25 for the rotation of the molecule as a whole. The second term in Equation 5.33 is the kinetic energy of internal rotation, ϵ in Equation 5.30.

The quantum mechanical energy levels corresponding to Equation 5.30 are:

$$\epsilon_r = \frac{r^2h^2}{8\pi^2 I_{\text{red}}} \qquad (r = 0, \pm 1, \pm 2, \ldots) \tag{5.34}$$

(there is one value of the quantum number r for each quantum *state*).

The partition function f_{free} for free internal rotation can now be derived by the method used above, in the derivation of f_{rot}. Thus:

$$\begin{aligned} f_{\text{free}} &= \frac{1}{\sigma_{\text{int}}} \sum_{r=-\infty}^{\infty} e^{-r^2h^2/8\pi^2 I_{\text{red}}kT} \\ &\approx \frac{1}{\sigma_{\text{int}}} \int_{-\infty}^{\infty} e^{-r^2h^2/8\pi^2 I_{\text{red}}kT}\, dr \\ &= \frac{1}{\sigma_{\text{int}}} \left(\frac{8\pi^3 I_{\text{red}}kT}{h^2}\right)^{1/2} \end{aligned} \tag{5.35}$$

the symmetry number being introduced in the usual way.

The derivation of Equation 5.35 refers specifically to molecules consisting only of two symmetric tops. The calculation of f_{free} for the general case, in which there may be several internal rotations, is a more difficult problem (Aston, p.590). Simplifications are, however, possible if the only internal rotations are those of rigid groups attached to a single atom or rigid framework of atoms (e.g., in toluene, acetone, etc.). The following are particularly useful:

(a) Equation 5.35 is approximately true, even if the groups are not symmetric tops, provided that one of them has a moment of inertia which is small compared with the principal moments of inertia of the molecule.

(b) Alternatively, for a symmetric top attached to a rigid unsymmetrical framework Equation 5.35 is true, provided that I_{red} is defined as follows:
Let I_1 be the moment of inertia of the symmetric top and let λ_x, λ_y, λ_z respectively be the direction cosines between the axis of internal rotation and the principal X, Y, Z axes of the molecule. Then:

$$I_{\mathrm{red}} = I_1 - I_1^2\left(\frac{\lambda_x^2}{A} + \frac{\lambda_y^2}{B} + \frac{\lambda_z^2}{C}\right) \tag{5.36}$$

where A,B, and C are the principal moments of inertia of the molecule.

(c) If there are, in all, h independently rotating groups attached to a rigid framework:

$$f_{\mathrm{free}} \approx f_{\mathrm{free}(1)} f_{\mathrm{free}(2)} \cdots f_{\mathrm{free}(i)} \tag{5.37}$$

where $f_{\mathrm{free}(i)}$ is the value obtained as above for the i-th rotating group. This is an approximation, valid when the moments of inertia of the internally rotating groups are small compared with A,B, and C.

Example 5.6(a)

Assuming that the C–H bond length is $1{\cdot}10 \times 10^{-10}$ m, and the HCC angle is 109°, calculate f_{free} for ethane at 298 K.

Let I_1 be the moment of inertia of one CH_3 group about the C–C axis. From Equation 5.31:

$$I_{\mathrm{red}} = \frac{I_1^2}{2I_1} = \frac{I_1}{2} = \frac{3}{2}mr^2$$

where r is the distance shown in Fig. 5.3, and m is the mass of the

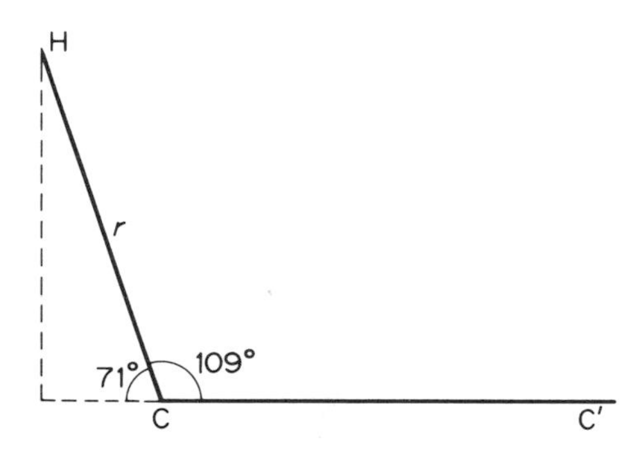

Fig. 5.3

hydrogen atom. Now:

$$m = 1{\cdot}67 \times 10^{-27}\,\text{kg}$$

$$r = 1{\cdot}1 \times 10^{-10} \sin 71$$

$$= 1{\cdot}04 \times 10^{-10}\,\text{m}$$

$$\sigma_{\text{int}} = 3.$$

It therefore follows, from Equation 5.35, that:

$$f_{\text{free}} = \frac{1}{3}\left(\frac{8\pi^3 \times \frac{3}{2} mr^2kT}{h^2}\right)^{1/2}$$

$$= 2{\cdot}64 \text{ at } 298\,\text{K}.$$

Example 5.6(b)

Calculate f_{free} for propane at 231 K.

By symmetry, the centre of mass is at a point G in the plane of the three carbon atoms, which lies on the line bisecting the $C_1C_2C_3$ angle. Take x- and y-axes as shown. The z-axis is then perpendicular to the plane of the paper.

It is easily seen that, because of the symmetry of the molecule, the products of inertia such as:

$$I_{xy} = \sum_s m_s x_s y_s$$

vanish. Thus, the chosen axes are principal axes.

Consider the internal rotation of the methyl group at C_1. From the figure, the direction cosines are:

$$\lambda_x = \sin\alpha \; ; \; \lambda_y = \cos\alpha \; ; \; \lambda_z = \cos 90° = 0.$$

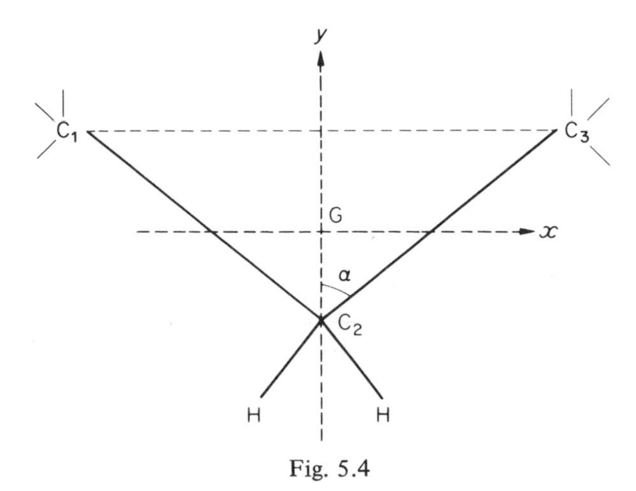

Fig. 5.4

Since both rotating groups are attached to the same atom, C_2, Equation 5.36 can be applied, and, for each of the methyl group rotations:

$$I_{red} = I_1\left(1 - \frac{I_1}{A}\sin^2\alpha - \frac{I_1}{B}\cos^2\alpha\right)$$

where I_1 is the moment of inertia of the methyl group, calculated as in the previous example. Using the geometry of propane assumed by Kemp and Egan (*J. Amer. Chem. Soc.*, 1938, **60**, 1521), we obtain, by straightforward calculation:

$$I_1 = 5{\cdot}18 \times 10^{-47}\,\mathrm{kg\,m^2}$$

$$A = 26{\cdot}9 \times 10^{-47}\,\mathrm{kg\,m^2}$$

$$B = 96{\cdot}8 \times 10^{-47}\,\mathrm{kg\,m^2}$$

Hence:

$$I_{red} = 4{\cdot}40 \times 10^{-47}\,\mathrm{kg\,m^2}.$$

For a single methyl group rotation (note that $\sigma_{int} = 3$), therefore, $f_{free(1)} = 2{\cdot}97$ at 231 K and, for the whole molecule, the approximation 5.37 gives:

$$f_{free} = (2{\cdot}97)^2 = 8{\cdot}81.$$

□ □ □

(iii) $\mathbf{V}_m$ *is neither very large nor very small*

For intermediate values of $\mathbf{V}_m$, the variation of the angle φ with time is described as a *restricted internal rotation*. In such a case, neither of the models considered above (vibration or free rotation) is adequate. For example, the entropy derived from f_{free} is then typically too high by about 5 $JK^{-1}mol^{-1}$ for each internal rotation. It is necessary, in such cases, to calculate the partition function for internal rotation from the energy levels derived for the potential 5.29.

Substitution of **V** given by Equation 5.29 into the wave equation for internal rotation leads to a differential equation of a type known as Mathieu's equation. The energy levels for internal rotation are the eigenvalues W_i of this equation. Thus, the partition function $f_{\text{i.r.}(1)}$ for a single internal rotation is:

$$f_{\text{i.r.}(1)} = \sum_{i=0}^{\infty} e^{-(W_i - W_0)/kT} \tag{5.38}$$

where W_0 is the lowest eigenvalue. Unfortunately, no simple expression can be given, either for the W_i or for $f_{\text{i.r.}(1)}$. K.S. Pitzer has, however, solved the equations numerically, and given tables which permit the calculation of thermodynamic functions for any value of $\mathbf{V}_m$. One first evaluates f_{free}, as above. Pitzer's tables then give the entropies, free energies, heat capacities, etc. corresponding to various values of f_{free} and $\mathbf{V}_m$. The tables can also be used to determine $\mathbf{V}_m$ from the measured entropy or heat capacity. Some values of $\mathbf{V}_m$ so obtained are given in Table 5.5. Numerical details and examples are discussed in Chapter 10.

Table 5.5 Potential Barriers to Internal Rotation

Molecule	$\mathbf{V}_m$ ($kJmol^{-1}$)	Molecule	$\mathbf{V}_m$ ($kJmol^{-1}$)
MeC≡CMe	0	Me_4C	18·8
Toluene	4	$MeCCl_3$	12·42
$MeCH{=}CH_2$	8·86	MeOH	4
MeCH=CHMe	8·36	$MeCH_2Cl$	15·5
Ethane	11·5	Me_2CO	4
$MeCH_2Me$	14·6	Me_4Si	5·36

A more extensive compilation is given in Herzberg II, p. 520. For a more recent review, see Dale, *Tetrahedron*, **22**, 3373 (1966).

5.7 Equipartition of energy: classical mechanics

In the preceding sections, discussion has been confined almost entirely to the results obtained from the quantum theory. Often, however, it is a valid approximation to use *classical*, instead of quantum mechanics; we have encountered such a case in Section 5.5.

The classical expression for the translational kinetic energy of a molecule of mass m is:

$$\epsilon_{\text{trans}} = \tfrac{1}{2}mv_x^2 + \tfrac{1}{2}mv_y^2 + \tfrac{1}{2}mv_z^2$$

where v_x, v_y, and v_z are the components of velocity in the x, y, and z directions. Each term on the right-hand side of this equation is an example of a *squared term*, i.e., each is of the form cw^2, where c is a constant and w is a variable.

A corresponding expression is found for rotation. The rotational kinetic energy is the sum of two (for linear molecules) or three (for non-linear molecules) squared terms; see Chapter 11. Similarly, the vibrational energy of a diatomic molecule is the sum of two squared terms, there being one term for the kinetic energy and one for the potential energy of the vibration.

A general theorem, which we derive in Section 11.8 from *classical* mechanics, is the *principle of equipartition of energy*:

For each squared term in the expression for the energy of the molecule, there is a contribution $RT/2$ to the molar internal energy.

Thus, for translation, there are three squared terms in the expression for the molecular energy, and hence the translational contribution to the *molar* internal energy is $3RT/2$. Similarly, there are the contributions:

rotation:	RT	(linear or diatomic molecules)
	$(3/2)RT$	(non-linear)
vibration:	RT	(diatomic molecules)

The nuclear and electronic energies are not expressible as squared terms, so the equipartition principle, even in classical mechanics, would not apply to these energies.

In classical mechanics, the equipartition principle is rigorously true; we now show that, in contrast, in quantum mechanics the equipartition principle is valid only as an *approximation*, the approximation becoming better as the temperature increases.

It was pointed out in Section 3.4 that, when f can be factorized, the thermodynamic functions can be expressed as the *sums* of simpler quantities. Thus, if:

$$f = f_{\text{trans}} f_{\text{vib}} f_{\text{rot}} f_{\text{nuc}} f_{\text{el}} f_0 ,$$

then the molar internal energy is:

$$\tilde{E} = \tilde{E}_0 + \tilde{E}_{\text{trans}} + \tilde{E}_{\text{vib}} + \tilde{E}_{\text{rot}} + \tilde{E}_{\text{nuc}} + \tilde{E}_{\text{el}} \tag{5.39}$$

(plus a term for internal rotation, if this is present), where:

$$\tilde{E}_{\text{vib}} = RT^2\left(\frac{\partial \ln f_{\text{vib}}}{\partial T}\right)_V, \quad \text{etc.}$$

The term:

$$\tilde{E}_0 = RT^2\left(\frac{\partial \ln f_0}{\partial T}\right)_V \equiv N\epsilon_0 \tag{5.40}$$

is the *molar zero-point energy*.

According to Equation 4.22:

$$f_{\text{trans}} = \text{constant} \times T^{3/2} \quad \text{at constant volume;}$$

that is:

$$\ln f_{\text{trans}} = \ln(\text{constant}) + \tfrac{3}{2}\ln T$$

Hence:

$$\tilde{E}_{\text{trans}} = \tfrac{3}{2}RT$$

which is just the answer given by the equipartition principle.

Similarly, Equation 5.22 leads to:

$$\tilde{E}_{\text{rot}} = RT \quad \text{(linear and diatomic molecules)}$$

and 5.26 to:

$$\tilde{E}_{\text{rot}} = \tfrac{3}{2}RT \quad \text{(non-linear molecules)}$$

These are again the values predicted from the equipartition principle.

On the other hand, the equipartition value, RT, is *not* obtained for E_{vib} when f_{vib} given by Equation 5.14 is used; a considerably more complicated expression results (see Section 6.1). If, however, the temperature is sufficiently high, the expression for f_{vib} can be simplified.

According to Equation 5.12:

$$f_{\text{vib}} = \sum_{v=0}^{\infty} e^{-v\Theta_{\text{vib}}/T}$$

If $T \gg \Theta_{\text{vib}}$, the terms in the sum are sufficiently close together for the approximation (already used for the translational and rotational partition functions) of replacing summation by integration to be applied (see Appendix 1):

$$f_{\text{vib}} = \int_0^{\infty} e^{-v\Theta_{\text{vib}}/T}\,dv = \frac{T}{\Theta_{\text{vib}}} \tag{5.41}$$

With this *high temperature approximation* to f_{vib} it follows immediately that:

$$E_{vib} = RT.$$

The above results demonstrate a general rule, which we shall examine more closely in Chapter 11, namely, *classical mechanics gives a valid approximation* only *if values of successive terms in the partition function are sufficiently close together for the summation to be replaced by integration*. It is easily shown that this condition holds if the value of kT is *large in comparison with the spacing between consecutive, appreciably occupied energy levels*.

The difference between two consecutive terms in f is (omitting degeneracy factors, for simplicity):

$$\begin{aligned} e^{-\epsilon_2/kT} - e^{-\epsilon_1/kT} &= e^{-\epsilon_2/kT}[1 - e^{-(\epsilon_1-\epsilon_2)/kT}] \\ &= e^{-\epsilon_2/kT}\left[1 - \left(1 - \frac{\epsilon_1 - \epsilon_2}{kT} + \ldots\right)\right] \\ &\approx e^{-\epsilon_2/kT}\left(\frac{\epsilon_1 - \epsilon_2}{kT}\right) \end{aligned}$$

This difference is numerically small if $|\epsilon_1 - \epsilon_2| \ll kT$. Note that if $\epsilon_2 \gg kT$, $e^{-\epsilon_2/kT}$ is small, and the difference between the two terms in f may be small even though $|\epsilon_1 - \epsilon_2| > kT$. The Maxwell–Boltzmann law shows, however, that such levels are not *appreciably* occupied.

It is usual, for brevity, to refer to a partition function calculated by classical mechanics as a *quasi-classical partition function*, (quasi = 'as though') and, if this is a valid approximation at a given temperature, the corresponding motion is said to be *classical* at that temperature. Thus:
(i) Translation is classical at all temperatures of practical interest.
(ii) Rotation is nearly always classical. Exceptions are for small molecules (H_2, D_2, etc.), at low temperatures.
(iii) Vibration is classical only if $T \gg \Theta_{vib}$ and (see Table 5.2) Θ_{vib} is of the order of 1000 K.

One anomaly may be mentioned. Electrons in a metal which are responsible for the electrical and thermal conductivity of the metal are essentially mobile, and the difference between successive translational energy levels can be calculated to be considerably less than kT. It might be expected, therefore, that these 'conduction' electrons would contribute the term $(3/2)RT$ to the internal energy. In fact, as we show in the following chapter, the observed specific heats of metals can only be interpreted if it is assumed that the energy of the electrons is essentially *independent* of T. However, as we show in Chapter 13, another statistical method, which differs in an important way from the Maxwell–Boltzmann method, must be used for electrons.

Problems

5.1 The following data are for Si(g) at 5000 K:

Level	$^3P_0$	$^3P_1$	$^3P_2$	$^1D_2$	$^1S_0$
ϵ_i/kT	0·0	0·022	0·064	1·812	4·430

Determine the g_i values and hence the electronic partition function for Si(g) at 5000 K. What fraction of the atoms are in the $^1D_2$ level, in the most probable distribution at 5000 K?

5.2 The lowest energy level of O_2 is three-fold degenerate. The next ($^1\Delta$) is doubly degenerate and lies 0·97 eV ($= 1{\cdot}55 \times 10^{-19}$ J) above the lowest level. Calculate the percentage contribution to f_{el} from the $^1\Delta$ level at (a) 1000 K (b) 3000 K.

5.3 The vibrational frequency of the $^{35}Cl_2$ molecule is $1{\cdot}663 \times 10^{13}\ s^{-1}$. Calculate Θ_{vib} for $^{35}Cl_2$. Evaluate $e^{-\epsilon_{vib(v)}/kT} = e^{-v\Theta_{vib}/T}$ for $v = 0,1,2,3$, with $T = 300$ K. Plot a graph showing how the number of molecules in a vibrational level varies with v.

5.4 Estimate the vibrational partition function of $^{35}Cl_2$ by adding together the four terms calculated in the previous problem, and compare with the result obtained from Equation 5.14′.

Show that if summation is taken to $v = p$ instead of $v = \infty$, then Equation 5.14 is replaced by:

$$\frac{1 - e^{-p\Theta_{vib}/T}}{1 - e^{-\Theta_{vib}/T}}\ .$$

Deduce that f_{vib} is evaluated to an accuracy of 0·1% by taking the sum up to $p = 7(T/\Theta_{vib})$. Refer to Table 5.2 and list the values of p found in this way for $T = 300$ K and $T = 1000$ K for H_2 and I_2.

5.5 Show that the contributions of vibration to the molar internal energy and entropy of a diatomic molecule are respectively:

$$\tilde{E}_{\text{vib}} = \frac{RTu}{e^u - 1}$$

and

$$\tilde{S}_{\text{vib}} = R\left[\frac{u}{e^u - 1} - \ln(1 - e^{-u})\right]$$

where:

$$u = \frac{h\omega}{kT} = \frac{\Theta_{\text{vib}}}{T}.$$

5.6 Values of the expression for $\tilde{S}_{\text{vib}}$ in the previous example are tabulated, for various values of u, on page 198. Using this table, evaluate $\tilde{S}_{\text{vib}}$ for I_2 at 500K, given that $\Theta_{\text{vib}} = 307\,\text{K}$.

5.7 Proceeding as in problem 5.6, and using the data in Table 5.3, calculate $\tilde{S}_{\text{vib}}$ for CO_2 at 1000K.

5.8 Show that the contribution from rotation to the molar entropy of a diatomic molecule is:

$$\tilde{S}_{\text{rot}} = R\left(\ln\frac{T}{\sigma\Theta_{\text{rot}}} + 1\right).$$

From the data in Table 5.4, determine $\tilde{S}_{\text{rot}}$ and $\tilde{A}_{\text{rot}}$ for CO at 500K.

5.9 The equilibrium internuclear distance in $^{127}I_2$ is $2{\cdot}666 \times 10^{-10}$ m. Calculate (a) the moment of inertia, (b) the characteristic temperature for rotation, (c) the rotational partition function of I_2 at 300K, and (d) $\tilde{S}_{\text{rot}}$ for I_2 at 300K.

5.10 Determine the symmetry number σ for the following molecules: CH_3Cl, CH_3D, CH_2Cl_2, benzene, toluene, *cis*- and *trans*-butadiene, SF_6.

5.11 (a) Calculate the product ABC of the principal moments of inertia of ethylene from the following data. The molecule is planar, with the HCC angle equal to 120°; the lengths of the C=C and C–C bonds are respectively $1{\cdot}35 \times 10^{-10}$ m and $1{\cdot}07 \times 10^{-10}$ m.
(b) Calculate the rotational partition function of ethylene.

5.12 By means of Equation 5.36, or otherwise, show that the reduced moment of inertia for internal rotation in dimethylacetylene has the same value as that for ethane (assuming that corresponding bond lengths and angles are the same in the two molecules).

5.13 Calculate I_{red} and f_{free} for the internal rotation in methanol. Assume that all bond angles are 109°, and that the C–H C–O and O–H bond lengths are, respectively, $1{\cdot}10 \times 10^{-10}$ m, $1{\cdot}43 \times 10^{-10}$ m, and $0{\cdot}96 \times 10^{-10}$ m.

5.14 By expanding the cosine term in Equation 5.29, show that, if $\sigma_{int}\varphi$ is small:

$$\mathbf{V} \approx \tfrac{1}{4}\mathbf{V}_m(\sigma_{int}\varphi)^2 .$$

Show that this is of the form 5.28; hence express the force constant c and the corresponding vibrational frequency in terms of $\mathbf{V}_m$. (Note: the equation of motion for a torsional oscillation has the same form as that for a linear vibration, except that reduced moment of inertia occurs in the former where reduced mass occurs in the latter).

CHAPTER SIX

Heat Capacities

Our summary of the general principles of the calculation of the thermodynamic properties of systems of independent molecules is now complete. At this point, therefore, we turn to the application of the theory to specific problems, beginning in the present chapter with the theoretical evaluation of the heat capacities of gases and solids; this was one of the earliest contributions of statistical mechanics to thermodynamics.

6.1 Molar heat capacity of gases

The molar heat capacity at constant volume is given by the equation (cf. Section 3.3):

$$\tilde{C}_V = 2RT\left(\frac{\partial \ln f}{\partial T}\right)_V + RT^2\left(\frac{\partial^2 \ln f}{\partial T}\right)_V \qquad (6.1)$$

As explained in Section 5.1, f can, in most cases, be factorized:

$$f = f_0 f_{\text{nuc}} f_{\text{el}} f_{\text{vib}} f_{\text{rot}} f_{\text{trans}}$$

the only important exceptions (which we consider later) being the hydrogen isotopes at low temperatures.

Now, f_{nuc} and f_{el} have, respectively, the values $g_{\text{nuc}(0)}$ and (usually) $g_{\text{el}(0)}$, so that they are independent of temperature. Thus:

$$\ln f = \ln(\text{constant}) + \ln f_0 + \ln f_{\text{vib}} + \ln f_{\text{rot}} + \ln f_{\text{trans}}. \qquad (6.2)$$

Moreover:

$$2RT\left(\frac{\partial \ln f_0}{\partial T}\right)_V + RT^2\left(\frac{\partial^2 \ln f_0}{\partial T^2}\right)_V = 2RT\left(-\frac{\epsilon_0}{kT^2}\right) + RT^2\left(\frac{2\epsilon_0}{kT^3}\right) = 0.$$

Consequently, Equation 6.1 can be written in the form:

$$\tilde{C}_V = \tilde{C}_{V(\text{vib})} + \tilde{C}_{V(\text{rot})} + \tilde{C}_{V(\text{trans})} \tag{6.3}$$

where $\tilde{C}_{V(\text{rot})}$ is the rotational contribution to $\tilde{C}_V$, and is defined as:

$$\tilde{C}_{V(\text{rot})} = 2RT\left(\frac{\partial \ln f_{\text{rot}}}{\partial T}\right)_V + RT^2\left(\frac{\partial^2 \ln f_{\text{rot}}}{\partial T^2}\right)_V. \tag{6.4}$$

$\tilde{C}_{V(\text{trans})}$ and $\tilde{C}_{V(\text{vib})}$ are, respectively, the translational and vibrational contributions to $\tilde{C}_V$, and are defined in terms of f_{trans} and f_{vib} by equations corresponding to 6.4.

The terms on the right-hand side of Equation 6.3 are now evaluated by substituting the expressions for the partition functions which were obtained in the previous chapter.

(i) *Diatomic molecules*

Translation

$$f_{\text{trans}} = \frac{(2\pi mkT)^{3/2}V}{h^2};$$

$$\frac{\partial \ln f_{\text{trans}}}{\partial T} = \frac{3}{2T}; \quad \frac{\partial^2 \ln f_{\text{trans}}}{\partial T^2} = -\frac{3}{2T^2}$$

Hence:

$$\tilde{C}_{V(\text{trans})} = \tfrac{3}{2}R \tag{6.5}$$

Rotation

If rotation is classical in the sense defined on page 92, then:

$$f_{\text{rot}} = \frac{T}{\sigma\Theta_{\text{rot}}}$$

where σ is 1 for heteronuclear and 2 for homonuclear diatomic molecules. Now:

$$\left(\frac{\partial \ln f_{\text{rot}}}{\partial T}\right)_V = \frac{1}{T} \quad \text{and} \quad \left(\frac{\partial^2 \ln f_{\text{rot}}}{\partial T^2}\right)_V = -\frac{1}{T^2}$$

and so:

$$C_{V(\text{rot})} = R \tag{6.6}$$

Vibration

The simplest method of evaluating $C_{V(\text{vib})}$ starts with the expression for $\tilde{E}_{vib}$ (cf. page 91):

$$\tilde{E}_{vib} = RT^2\left(\frac{\partial \ln f_{vib}}{\partial T}\right)_V = \frac{RT^2}{f_{vib}}\left(\frac{\partial f_{vib}}{\partial u}\right)_V \frac{du}{dT},$$

where (cf. Equation 5.14) $f_{vib} = 1(1 - e^{-u})$ with $u = \Theta_{vib}/T$. Then:

$$\tilde{E}_{vib} = RT^2 \frac{e^{-u}}{1 - e^{-u}} \frac{\Theta_{vib}}{T^2}$$

$$= R\Theta_{vib} \frac{1}{e^u - 1}. \tag{6.7}$$

Therefore:

$$\tilde{C}_{V(vib)} = \left(\frac{\partial \tilde{E}_{vib}}{\partial T}\right)_V$$

$$= R\left(\frac{\Theta_{vib}}{T}\right)^2 \frac{e^{\Theta_{vib}/T}}{(e^{\Theta_{vib}/T} - 1)^2} \tag{6.8}$$

$$= Ru^2 \frac{e^u}{(e^u - 1)^2} \tag{6.9}$$

Values of this function are tabulated in Chapter 10 (Table 10.8). The limiting values of the right-hand side of Equation 6.9 can be shown to be:

$$R \text{ when } u \longrightarrow 0 \quad \text{and} \quad 0 \text{ when } u \longrightarrow \infty.$$

Let:

$$\tau = \frac{u}{e^u - 1} e^{u/2}$$

so that:

$$C_{V(vib)} = R\tau^2$$

Then:

$$\tau = \frac{u}{e^{u/2} - e^{-u/2}}$$

$$= \frac{u}{\left[1 + \frac{u}{2} + \frac{1}{2!}\left(\frac{u}{2}\right)^2 + \dots\right] - \left[1 - \frac{u}{2} + \frac{1}{2!}\left(\frac{u}{2}\right)^2 - \dots\right]}$$

$$= \frac{u}{2\left[\frac{u}{2} + \frac{1}{3!}\left(\frac{u}{2}\right)^3 + \dots\right]} = \frac{1}{1 + \frac{u^2}{24} + \dots}$$

When $u \longrightarrow 0$ this simplifies to $\tau \approx 1$, and when $u \longrightarrow \infty$ to $\tau \approx 0$.

It follows that if $T >> \Theta_{vib}$ (i.e. u is small), Equation 6.9 simplifies to:

$$\tilde{C}_{V(vib)} \approx R$$

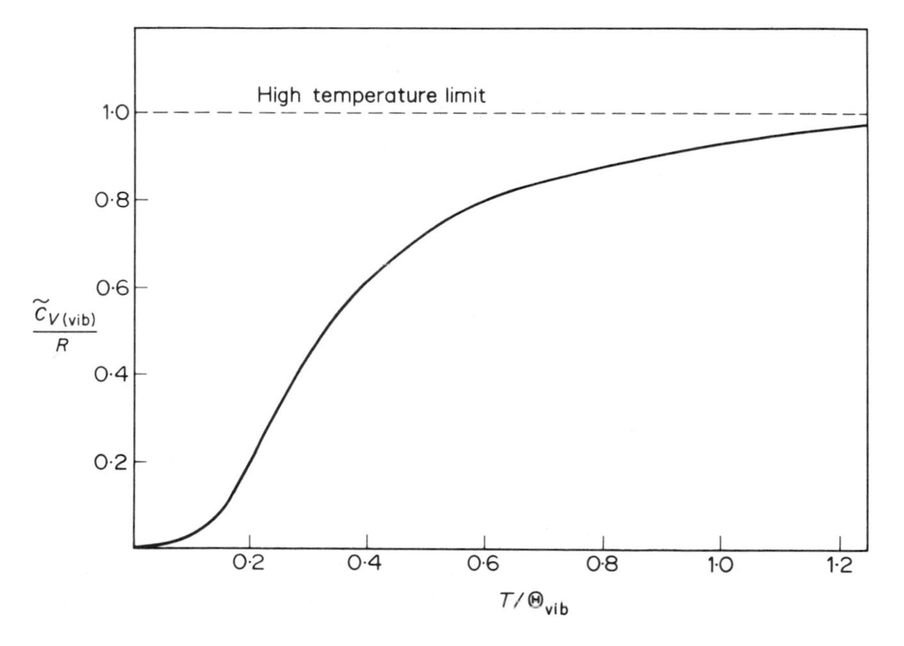

Fig. 6.1

In contrast, if $T << \Theta_{\text{vib}}$, u is large, and:

$$\tilde{C}_{V(\text{vib})} \approx 0$$

Thus, if the temperature is much below the characteristic temperature for vibration, there is a negligible contribution to $\tilde{C}_V$ from vibration. Correspondingly, from Equation 6.7, $\tilde{E}_{\text{vib}} \approx 0$ if u is large, which means, physically, that nearly all the molecules are in the lowest ($v = \frac{1}{2}$) vibrational level, i.e., the *vibration is unexcited* if $T << \Theta_{\text{vib}}$.

The variation of $\tilde{C}_{V(\text{vib})}$ with T is shown in Fig. 6.1. Note that, in consequence of Equation 6.8, if T is plotted in units of Θ_{vib}, a *curve common to all diatomic gases is obtained*, even though Θ_{vib} has different values for different gases.

Adding together the three contributions to $\tilde{C}_V$, we obtain:

$$\tilde{C}_V = \begin{cases} \dfrac{5}{2}R & (T << \Theta_{\text{vib}}) \\[2ex] \dfrac{7}{2}R & (T >> \Theta_{\text{vib}}) \\[2ex] R\left(\dfrac{5}{2} + u^2 \dfrac{e^u}{(e^u - 1)^2}\right) & (\text{intermediate temperatures}), \end{cases}$$

for the molar heat capacity of a diatomic gas.

It may be noted that Equations 6.5 and 6.6 are identical with the results obtained from the classical principle of equipartition of energy, discussed in Section 5.7. According to the latter, the translational contribution to the molar internal energy of a gas is $(3/2)RT$. Hence:

$$\tilde{C}_{V(\text{trans})} = \left(\frac{\partial \tilde{E}_{\text{trans}}}{\partial T}\right)_V = \frac{3}{2}R$$

in agreement with Equation 6.5. Similarly, the classical value of $\tilde{C}_{V(\text{rot})}$ is R, in agreement with Equation 6.6. The classical value of $\tilde{C}_{V(\text{vib})}$ is also R, but the quantum theory predicts that this value is reached only when $T >> \Theta_{\text{vib}}$.

(ii) *Polyatomic molecules*

The *translational* contribution to $\tilde{C}_V$ has, of course, the same value $(3/2)R$ whether the molecule is diatomic or polyatomic. Furthermore, from Equations 5.25 and 5.26, the *rotational* contribution is:

$$\tilde{C}_{V(\text{rot})} = R \qquad \text{if the molecule is linear,}$$

$$\tilde{C}_{V(\text{rot})} = \tfrac{3}{2}R \qquad \text{if the molecule is non-linear.}$$

Since the *vibrational* partition function can be factorized into $3\mathfrak{N} - 5$ or $3\mathfrak{N} - 6$ separate terms, one for each normal mode, $\tilde{C}_{V(\text{vib})}$ is the sum of $3\mathfrak{N} - 5$ or $3\mathfrak{N} - 6$ terms of the type in Equation 6.9:

$$\tilde{C}_{V(\text{vib})} = R \sum_{l=1}^{3\mathfrak{N}-5\text{ or }6} C_{\text{vib}(l)} \tag{6.10}$$

where:

$$C_{\text{vib}(l)} = u_l^2 \frac{e^{u_l}}{(e^{u_l} - 1)^2} \quad \text{and} \quad u_l = \frac{h\omega_l}{kT} \tag{6.11}$$

Example 6.1

Calculate $\tilde{C}_V$ for H_2O at 500K.

From Table 5.3, the values of $\Theta_{\text{vib}(l)}$ for H_2O are 2290, 5160, and 5360K. The corresponding values of $u_l = \Theta_{\text{vib}(l)}/T$ are 4·58, 10·32, and 10·72. From Table 10.8, these values of u_l give the following contributions to $\tilde{C}_V$: 0·219R, 0·003R, 0·002R. Therefore:

$$\tilde{C}_{V(\text{vib})} = 0{\cdot}224R$$

Now:

$$\tilde{C}_{V(\text{trans})} = \tfrac{3}{2}R$$

and

$$\tilde{C}_{V(\text{rot})} = \tfrac{3}{2}R \quad \text{(non-linear molecule).}$$

Finally, therefore:

$$\tilde{C}_V \text{ for } H_2O \text{ at } 500\,\text{K} = 3{\cdot}224R$$
$$= 26{\cdot}81\ \text{J K}^{-1}\,\text{mol}^{-1}$$

□ □ □

A modification of the calculation is necessary if there is internal rotation; this is discussed in Section 10.5.

6.2 The heat capacity of hydrogen at low temperatures

The theory developed in the preceding section is applicable to hydrogen at temperatures greater than about 0°C. At lower temperatures than this, the approximation to f_{rot} given by Equation 5.24 is unsatisfactory, and f_{rot} must be evaluated by a direct summation, of the type in Equation 5.20. As we have seen (Section 5.7), the classical expression 5.24 is only applicable at those temperatures for which $T >> \Theta_{\text{rot}}$, and (cf. Table 5.4) Θ_{rot} for hydrogen has the unusually high value, 85·38 K.

The first calculations of $\tilde{C}_V$ for hydrogen were based upon the evaluation of f_{rot} by summing over *all* the terms (odd J and even J) in Equation 5.20; the dependence of $\tilde{C}_V$ upon T, calculated in this way, is shown as curve I in Fig. 6.2.

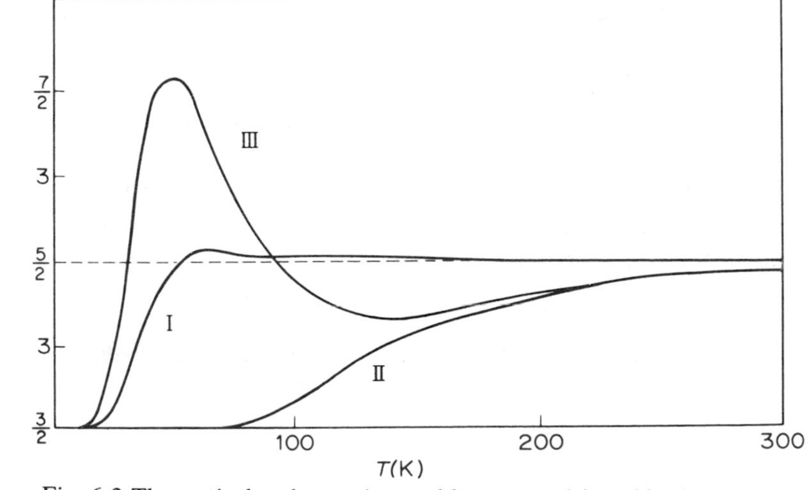

Fig. 6.2 Theoretical and experimental heat capacities of hydrogen

Table 6.1 Rotational Heat Capacities of Symmetrical Diatomic Molecules

T/Θ_{rot}	$\tilde{C}_{\mathrm{rot}}/R$ odd J values*	$\tilde{C}_{\mathrm{rot}}/R$ even J values†
0·3	0·887	1·117
0·4	0·635	1·343
0·5	0·382	1·465
0·6	0·206	1·377
0·7	0·104	1·146
0·8	0·050	0·875
0·9	0·023	0·630
1·0	0·011	0·435
1·2	—	0·192
1·4	—	0·079

* Rotational heat capacity calculated by summing over odd values of J only (e.g., for orthohydrogen).

† Rotational heat capacity calculated by summing over even values of J only (e.g., for parahydrogen).
(Values from Mayer and Mayer pp. 451–2)

At 300 K, $\tilde{C}_V = \frac{5}{2}R$, the classical value of the (rotational + translational) contributions with vibration unexcited. At 50 K $\tilde{C}_V = \frac{3}{2}R$, the classical value of the translational contribution, with rotation unexcited. These predictions agree with the experimentally determined values. However, at temperatures intermediate between 50 K and 300 K, curve I does *not* agree with experiment. The experimentally determined dependence of $\tilde{C}_V$ upon T is shown as curve II in Fig. 6.2.

The discrepancy arises because hydrogen must be regarded as a mixture of *ortho* and *para* molecules (Sections 5.2, 5.5, and 13.7); the calculation of $\tilde{C}_V$ is then a little less straightforward than previously.

Let x^{ortho} and x^{para} be the mole fractions of ortho and para molecules present at the temperature under consideration. The molar internal energy of the mixture is:

$$\tilde{E} = x^{\mathrm{ortho}}\tilde{E}^{\mathrm{ortho}} + x^{\mathrm{para}}\tilde{E}^{\mathrm{para}}$$

where E^{ortho} and E^{para} are the internal energies per mole of ortho and para molecules. The heat capacity is then:

$$\tilde{C}_V = \left(\frac{\partial \tilde{E}}{\partial T}\right)_V$$

$$= x^{\text{ortho}}\tilde{C}_V^{\text{ortho}} + x^{\text{para}}\tilde{C}_V^{\text{para}} + \left[\tilde{E}^{\text{ortho}}\left(\frac{\partial x^{\text{ortho}}}{\partial T}\right)_V + \tilde{E}^{\text{para}}\left(\frac{\partial x^{\text{para}}}{\partial T}\right)_V\right] \tag{6.12}$$

where: $\tilde{C}_V^{\text{ortho}} = \left(\dfrac{\partial \tilde{E}^{\text{ortho}}}{\partial T}\right)_V$ and $\tilde{C}_V^{\text{para}} = \left(\dfrac{\partial \tilde{E}^{\text{para}}}{\partial T}\right)_V$

the molar heat capacities of ortho- and para-hydrogen, respectively. The derivatives in the square brackets can be expressed in terms of f (see problem 6.3).

$\tilde{C}_V^{\text{ortho}}$ and $\tilde{E}^{\text{ortho}}$ are evaluated in the usual way from the partition function, f^{ortho}, for ortho molecules. Since, at low temperatures:

$$f_{\text{el}} = f_{\text{vib}} = 1,$$

$$f^{\text{ortho}} = f_{\text{nuc}}^{\text{ortho}} f_{\text{rot}}^{\text{ortho}} f_{\text{trans}}$$

(the superscript ortho being omitted from f_{trans} because this is the same for ortho and para molecules). According to Sections 5.2 and 5.5, therefore:

$$f^{\text{ortho}} = f_{\text{trans}} \times 3 \times \sum_{J=1,3,5,\ldots} (2J+1)\,\mathrm{e}^{-J(J+1)\Theta_{\text{rot}}/T}$$

Similarly $\tilde{C}_V^{\text{para}}$ and $\tilde{E}^{\text{para}}$ are evaluated from:

$$f^{\text{para}} = f_{\text{trans}} \times 1 \times \sum_{J=0,2,4,\ldots} (2J+1)\,\mathrm{e}^{-J(J+1)\Theta_{\text{rot}}/T}$$

It remains to calculate x^{ortho} and x^{para}. If n^{ortho} and n^{para} are the numbers of ortho and para molecules:

$$x^{\text{ortho}} = \frac{n^{\text{ortho}}}{n^{\text{ortho}} + n^{\text{para}}}$$

The equilibrium number of molecules in any one energy level is (Equation 2.14):

$$\overline{n_i} = g_i \mathrm{e}^{-\alpha}\mathrm{e}^{-\epsilon_i/kT}$$

The total number of molecules in ortho states will be given by adding together all the $\overline{n_i}$ values corresponding to such states. Thus:

$$n^{\text{ortho}} = \mathrm{e}^{-\alpha} \sum_{\substack{\text{ortho}\\\text{energy}\\\text{levels}}} g_i \mathrm{e}^{-\epsilon_i/kT} = \mathrm{e}^{-\alpha} f^{\text{ortho}}.$$

Similarly:

$$n^{\text{para}} = \mathrm{e}^{-\alpha} \sum_{\substack{\text{para}\\\text{energy}\\\text{levels}}} g_i \mathrm{e}^{-\epsilon_i/kT} = \mathrm{e}^{-\alpha} f^{\text{para}}$$

Hence:

$$\frac{\bar{n}^{\text{ortho}}}{\bar{n}^{\text{para}}} = \frac{f^{\text{ortho}}}{f^{\text{para}}} = \frac{3 \sum_{J=1,3,5,\ldots} (2J+1)\mathrm{e}^{-J(J+1)\Theta_{\text{rot}}/T}}{\sum_{J=0,2,4,\ldots} (2J+1)\mathrm{e}^{-J(J+1)\Theta_{\text{rot}}/T}} \tag{6.13}$$

from which x^{ortho} and x^{para} are obtained.

The values of $\tilde{C}_V$ calculated* from Equations 6.12 and 6.13 are shown as curve III in Fig. 6.2. *This still does not coincide with the experimentally determined curve.*

The explanation for this totally unexpected result was given, in 1927, by D.M. Dennison. We see from Equation 6.12 that the variation of $\tilde{C}_V$ with temperature arises from two things:

(a) the dependence of energy terms, $\tilde{E}^{\text{ortho}}$, $(\partial\tilde{E}^{\text{ortho}}/\partial T)_V$, etc., upon temperature;

(b) the dependence of composition terms, x^{ortho}, $(\partial x^{\text{ortho}}/\partial T)_V$, etc., upon temperature.

Dennison pointed out that transitions between ortho and para states are unlikely to occur readily. Hence, during the determination of the experimental $\tilde{C}_V$ *versus* T curve, most of the molecules originally in *ortho* states *remain throughout* in *ortho* states, and those in *para* states remain in *para* states. There is, therefore, *metastable equilibrium*, and *the composition remains constant at all temperatures*, instead of varying according to Equation 6.13. Accordingly, Equation 6.12 simplifies to:

$$\tilde{C}_V = x_0^{\text{ortho}}\tilde{C}_V^{\text{ortho}} + (1 - x_0^{\text{ortho}})\tilde{C}_V^{\text{para}} \tag{6.14}$$

where x_0^{ortho} is the mole fraction of ortho molecules throughout the experiment. The value of x_0^{ortho} is found from Equation 6.13. At temperatures greater than about 300 K (i.e., those temperatures at which the hydrogen is prepared), the sums in the numerator and denominator in Equation 6.13 have nearly the same value (see Appendix 1). Thus, at such temperatures:

$$\frac{\bar{n}^{\text{ortho}}}{\bar{n}^{\text{para}}} = 3$$

and

$$x_0^{\text{ortho}} = \frac{3}{3+1} = \frac{3}{4}.$$

* The above method of evaluating $\tilde{C}_V$ has been given to show how $\tilde{C}_V$ depends upon the ortho/para composition. More usually, the computation of $\tilde{C}_V$ proceeds as follows. It is easily shown that $\tilde{E} = (RT^2/f)(\partial f/\partial T)_V$ where $f = f^{\text{ortho}} + f^{\text{para}}$, and $\tilde{C}_V$ is now found as $(\partial E/\partial T)_V$.

Equation 6.14 then yields:

$$\tilde{C}_V = \tfrac{3}{4}\tilde{C}_V^{\text{ortho}} + \tfrac{1}{4}\tilde{C}_V^{\text{para}}$$

The C_V *versus* T curve calculated from this equation is found to be almost exactly coincident with the experimentally determined curve II.

Further confirmation of Dennison's theory was provided by the work of Bonhoeffer and Harteck, who showed, in 1929, that active charcoal acts as a catalyst for ortho–para hydrogen conversion. If a sample of hydrogen is allowed to remain in contact with this catalyst at a given temperature T K, true equilibrium is achieved. The value of the ratio $\bar{n}^{\text{ortho}}/\bar{n}^{\text{para}}$ is therefore calculated from Equation 6.13. Hence the value of x_0^{ortho} corresponding to the sample in equilibrium at T K can be calculated. The heat capacities of this sample at various temperatures should now be given by Equation 6.14. This has been found, experimentally, to be the case.

6.3 Heat capacities of monatomic crystals: the Einstein model

The experimentally determined values of the heat capacity of most solid elements at room temperature obey the empirical law of Dulong and Petit:

$$\begin{aligned}\tilde{C}_P &\approx 6{\cdot}4\ \text{cal K}^{-1}\,\text{mol}^{-1}\\ &\approx 27\ \text{J K}^{-1}\,\text{mol}^{-1}\end{aligned}$$

where $\tilde{C}_P$ is the molar heat capacity measured at constant (atmospheric) pressure. The difference $(\tilde{C}_P - \tilde{C}_V)$ can be calculated* by thermodynamics, and, typically:

$$\begin{aligned}C_P - C_V &\approx 0{\cdot}4\ \text{cal K}^{-1}\,\text{mol}^{-1}\\ &\approx 2\ \text{J K}^{-1}\,\text{mol}^{-1}\end{aligned}$$

Thus, the law of Dulong and Petit implies that:

$$\begin{aligned}C_V &\approx 6\ \text{cal K}^{-1}\,\text{mol}^{-1}\\ &\approx 25\ \text{J K}^{-1}\,\text{mol}^{-1}\end{aligned}$$

* In general, $\tilde{C}_P - \tilde{C}_V = \alpha^2\tilde{V}T/\beta$, where α and β are respectively the coefficient of cubical expansion and the compressibility coefficient (Lewis and Randall, p.107). Nernst and Lindemann showed that, for solids, an approximate form of this equation is:

$$\tilde{C}_P - \tilde{C}_V = 0{\cdot}109\tilde{C}_V^2 T/RT_{\text{m}}$$

where T_{m} is the melting point.

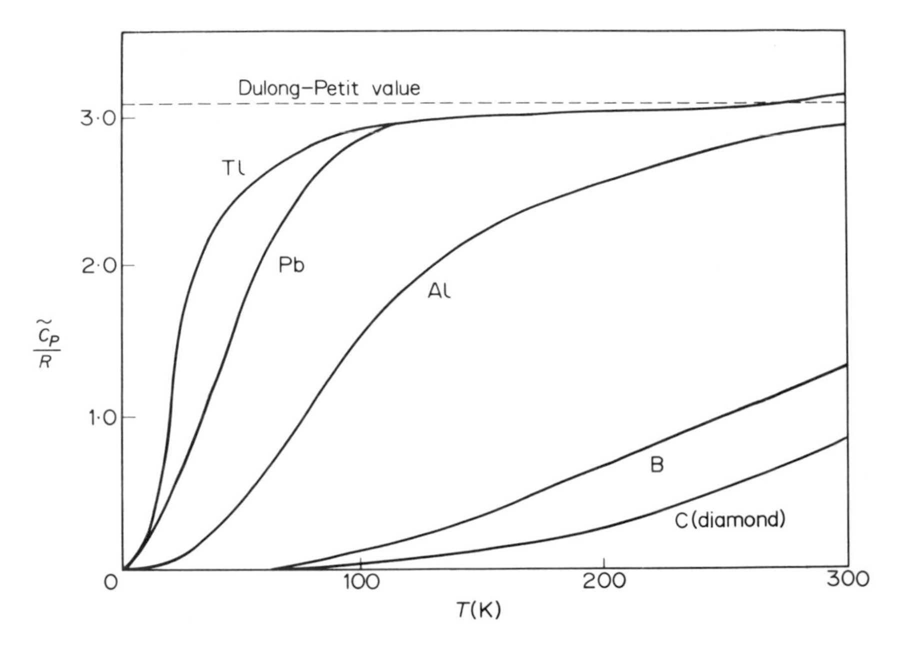

Fig. 6.3 Experimental heat capacities of crystals

for elements at room temperature.

At lower temperatures, $\tilde{C}_V$ decreases, extrapolation of the experimental values indicating (but see Section 8.1) that $\tilde{C}_V = 0$ at 0K. Some examples are shown in Fig. 6.3. At any given temperature, greatest departures from the Dulong and Petit value of $\tilde{C}_V$ are generally exhibited by elements of low atomic weight and high melting point. In the case of the diamond crystal, the deviations are so large that Dulong and Petit's law is not even approximately obeyed at room temperatures.

The first more or less satisfactory explanation of these observations was given by A. Einstein, and is based upon the following assumptions.

(a) the factors* f_{el} and f_{nuc} do not contribute to $\tilde{C}_V$. Moreover, there are no translational or rotational† factors in the molecular partition function for a monatomic crystal. It is, however, possible for the atoms to undergo small oscillations about their equilibrium

* A further discussion of the 'free' electrons in metals is given in Chapter 13.

† The 'rotation' of a single atom consists of the spin of its nucleus and the orbital motion of its electrons. The associated energies are, of course, contained in ϵ'_{nuc} and ϵ'_{el}.

positions, and the observed heat capacity of the crystal arises from the temperature dependence of the vibrational energy.

(b) Each atom vibrates independently of the others. The vibration of an atom can be resolved into three normal modes (i.e., oscillations of the atom in the x, y, and z directions).

(c) Each of these normal modes is a simple harmonic vibration. The corresponding vibrational frequency ω is the same for the three modes, and the same for all atoms.

With these assumptions, our treatment (Section 5.2) of the diatomic oscillator is also applicable to the vibrations of the atoms in the crystal. Thus, the vibrational partition function for a monatomic crystal is:

$$f_{\text{vib}} = f_x f_y f_z$$

where:

$$f_x = f_y = f_z = \frac{1}{1 - e^{-h\omega/kT}},$$

and hence the molar capacity is:

$$\tilde{C}_V = \tilde{C}_{V(\text{vib})} = 3c_{\text{vib}(x)}$$

where:

$$c_{\text{vib}(x)} = 2RT\left(\frac{\partial \ln f_x}{\partial T}\right)_V + RT^2\left(\frac{\partial^2 \ln f_x}{\partial T^2}\right)_V$$

This is given by Equation 6.9; hence:

$$\tilde{C}_{V(\text{vib})} = 3Ru^2 \frac{e^u}{(e^u - 1)^2}$$

$$= 3R\left(\frac{\Theta_E}{T}\right)^2 \frac{e^{\Theta_E/T}}{(e^{\Theta_E/T} - 1)^2} \qquad (6.15)$$

where

$$\Theta_E = \frac{h\omega}{k}$$

is now called the *Einstein characteristic temperature* of the crystal.

As in the case of Equation 6.9, at temperatures for which $T >> \Theta_E$, Equation 6.15 reduces to:

$$\tilde{C}_V = \tilde{C}_{V(\text{vib})} \approx 3R$$

$$\approx 6 \text{ cal K}^{-1} \text{ mol}^{-1}$$

$$\approx 25 \text{ J K}^{-1} \text{ mol}^{-1}$$

which is essentially the law of Dulong and Petit.

The variation of $\tilde{C}_V$ with T at lower temperatures is (apart from the factor 3 in Equation 6.15) just the same as that of $\tilde{C}_{V(\mathrm{vib})}$ for a diatomic gas (see Fig. 6.1), and if C_V is plotted against T/Θ_E a curve common to all monatomic crystals should be obtained. These predictions are confirmed by the experimental results.

By substituting the measured value of $\tilde{C}_V$ at any temperature T into Equation 6.15, the value of Θ_E can be calculated. The numerical values so obtained (see Table 6.2, and the closing sentences of this section) are interpreted as follows.*

In classical mechanics, the vibrational frequency ω is given by an equation analogous to 5.8:

$$\omega = \frac{1}{2\pi}\sqrt{\frac{k}{m}}$$

where m is the mass of the atom and k is the force constant. Thus, Θ_E which is proportional to ω decreases with (a) increasing atomic mass, and (b) decreasing k. If k is regarded as a measure of the forces which bind the crystal together, condition (b) can be replaced by the statement: Θ_E decreases with decreasing binding forces. These considerations are the basis of an equation proposed by Lindemann:

$$\Theta_E = C(T_m/M\tilde{V}^{2/3})^{\frac{1}{2}}$$

where T_m is the melting point, M is the molecular weight, $\tilde{V}$ is the molar volume, and C is an empirical constant.

Lead, mercury, and thallium are heavy metals with low melting points, and they have low values of Θ_E. In contrast, the diamond crystal is composed of light atoms, held together by strong, covalent bonds; Θ_E for diamond is therefore exceptionally large. The departures from the law of Dulong and Petit are thus satisfactorily accounted for.

As might be expected, in view of the sweeping assumptions which underlie the theory, *perfect* agreement between theory and experiment is not found. It is found that Θ_E calculated from high-temperature (say $T \approx 300\,\mathrm{K}$) values of $\tilde{C}_V$ does not have exactly the same value as Θ_E obtained from low temperature measurements on the same substance, whereas, of course, Θ_E should be a constant. An improved theory, due to Debye, is discussed in the following section. In the latter, a new characteristic temperature (the Debye

* The value of the characteristic temperature is also related to the elastic constants and other properties of the crystal (see page 114).

Table 6.2 Debye Characteristic Temperatures

Substance	Θ_D
Pb	88
Tl	96
Hg	97
I	106
Na	172
Al	398
Fe	453
diamond	1860
NaCl	281

temperature) Θ_D is introduced. This is more nearly constant, and is the quantity tabulated in Table 6.2. The value of Θ_E derived from *high temperature* $\tilde{C}_V$ measurements can be shown* to be related to the value Θ_D by:

$$\Theta_E = \Theta_D \sqrt{\frac{3}{5}} .$$

6.4 The Debye theory of crystal heat capacity

For low temperatures, the 1 in the denominator on the right hand side of Equation 6.15 can be neglected in comparison with the exponential term. Then, Equation 6.15 simplifies to:

$$\tilde{C}_V = 3R\left(\frac{\Theta_E}{T}\right)^2 e^{-\Theta_E/T} \tag{6.16}$$

This relation is *not* found experimentally; the measured values of $\tilde{C}_V$ *at very low temperatures* are, in fact, found to be proportional to T^3. Thus, the Einstein theory predicts a more rapid (exponential) decrease in $\tilde{C}_V$ at low temperatures than is observed.

* It follows from the expansion on page 98 that:

$$C_V = 3R\tau^2 \approx 3R(1 - u^2/12 - ...)$$

if $u(\equiv \Theta_E/T)$ is small. Mayer and Mayer (p.253) give the corresponding expansion of C_V in terms of Θ_D/T as:

$$C_V = 3R[1 - (\Theta_D/T)^2/20 + ...].$$

Comparing these expressions:

$$u^2 = \frac{12}{20}\left(\frac{\Theta_D}{T}\right)^2 \quad \text{or} \quad \Theta_E = \Theta_D\sqrt{\frac{3}{5}}$$

Better agreement between theoretical and measured values of $\tilde{C}_V$ is obtained when the treatment due to P.P. Debye is used. Debye's method avoids the most serious assumption of the Einstein theory, namely, that the vibrations of the atoms can be regarded as mutually independent. In a crystal, the motion of any one atom inevitably affects the motion of the neighbouring atoms, and the consequent *coupling of vibrations* is taken into account in Debye's treatment.

The vibrations of the atoms can be studied by regarding the crystal as one *macro-molecule* containing $\tilde{N}$ atoms. Thus, there are, in the usual way, $3\tilde{N}$ vibrational modes (strictly speaking, $3\tilde{N} - 6$, but 6 is, of course, negligible here), and, as in the case of a polyatomic molecule, each of these modes will have some characteristic frequency. As we have seen in simple cases (Table 5.3), several modes may have the same frequency. Let us suppose therefore that, in general, g_l modes of the crystal have the same frequency ω_l. Since the number of modes is finite, one value of ω_l must be greater than all the others; let us denote this value by ω_{max}.

Debye's method consists of replacing $C_{V(vib)}$ given by Equation 6.15 by *a value of* $C_{V(vib)}$ *averaged over all the values of* ω_l. A rigorous justification for this averaging is given later (Section 14.4). Thus:

$$\tilde{C}_{V(vib)} = \frac{1}{3\tilde{N}} \sum_l g_l C_{V(vib)(l)} \tag{6.17}$$

where:

$$C_{V(vib)(l)} = 3Ru_l^2 \frac{e^{u_l}}{(e^{u_l} - 1)^2}$$

with $u_l = h\omega_l/\boldsymbol{k}T$, and the summation is over all different values of ω_l. This summation is approximated by integrating over a continuous variable ω (just as, for example, the summation over different values of J in the rotational partition function was approximated in Section 5.5 by integration over the variable J; see also Appendix 1). Thus:

$$\tilde{C}_{V(vib)} = \frac{1}{3\tilde{N}} \int_0^{\omega_{max}} g(\omega) C_{vib}(\omega) d\omega \tag{6.18}$$

where $C_{vib}(\omega)$ is calculated from Equation 6.15, replacing ω_l by the variable ω, and $g(\omega)d\omega$ is the number of modes which have frequencies between ω and $\omega + d\omega$.

It can be shown (see below) that, for *small* values of ω:

$$g(\omega) = AV\omega^2 \tag{6.19}$$

where V is the volume of the crystal and A is a constant for the crystal. Debye assumed that Equation 6.19 can be used for *all* the frequencies up to ω_{max}.

The total number of modes is obtained by integrating $g(\omega)$ with respect to ω, between 0 and ω_{max}. This total must also be $3\tilde{N}$. Thus:

$$3\tilde{N} = AV\int_0^{\omega_{max}} \omega^2 \, d\omega = \frac{AV\omega_{max}^3}{3} \tag{6.20}$$

Eliminating AV between Equations 6.19 and 6.20, therefore:

$$g(\omega)d\omega = \frac{9\tilde{N}}{\omega_{max}} \omega^2 \, d\omega \tag{6.21}$$

Substituting Equation 6.21 into 6.18:

$$\begin{aligned} \tilde{C}_{V(vib)} &= \frac{9R}{\omega_{max}^3} \int_0^{\omega_{max}} \omega^2 C_{vib}(\omega) \, d\omega \\ &= \frac{9R}{\omega_{max}^3} \int_0^{\omega_{max}} u^4 \left(\frac{kT}{h}\right)^2 \frac{e^u}{(e^u - 1)^2} \, d\omega \end{aligned} \tag{6.22}$$

where $u = h\omega/kT$. Let $h\omega_{max}/k = \Theta_D$, where Θ_D is called the *Debye characteristic temperature*. Noting that $du = (h/kT)d\omega$, and that the value of u corresponding to ω_{max} is $h\omega_{max}/kT = \Theta_D/T$, we obtain from Equation 6.21:

$$\tilde{C}_{V(vib)} = 9R\left(\frac{T}{\Theta_D}\right)^3 \int_0^{\Theta_D/T} \frac{u^4 e^u}{(e^u - 1)^2} \, du \tag{6.23}$$

The integral is evaluated numerically, and tables of $\tilde{C}_{V(vib)}$ values corresponding to the various values of Θ_D/T are available (see Table 6.3). Mayer and Mayer (p.253) give series-expansion approximations to Equation 6.23.

Table 6.3 Debye Heat Capacity Calculated According to Equation 6.23

Θ_D/T	$\tilde{C}_V/3R$	Θ_D/T	$\tilde{C}_V/3R$	Θ_D/T	$\tilde{C}_V/3R$
0·0	1·000	6·0	0·266	14·0	0·028
1·0	0·952	7·0	0·191	16·0	0·019
2·0	0·825	8·0	0·138	18·0	0·013
3·0	0·663	9·0	0·102	20·0	0·010
4·0	0·503	10·0	0·076	22·0	0·007
5·0	0·369	12·0	0·045	24·0	0·006

Data from Fowler and Guggenheim, p. 144.

The high and low temperature limits of $\tilde{C}_{V(\text{vib})}$ are easily obtained. Firstly, expanding the integrand in Equation 6.23:

$$\frac{u^4 e^u}{(e^u - 1)^2} = \frac{u^4(1 + u + u^2/2! + ...)}{(u + u^2/2! + ...)^2}$$

$$\approx u^2$$

if u is small enough for the series to be approximated by their first terms. Hence, if $T >> \Theta_D$:

$$C_{V(\text{vib})} \approx 9R\left(\frac{T}{\Theta_D}\right)^3 \int_0^{\Theta_D/T} u^2\,du$$

$$= 9R\left(\frac{T}{\Theta_D}\right)^3 \left[\frac{1}{3}\left(\frac{\Theta_D}{T}\right)^3\right]$$

$$= 3R\,.$$

The Debye formula has therefore the same high-temperature limit as the Einstein equation. For low temperatures, Θ_D/T is large, and a good approximation is obtained by replacing the integral in Equation 6.23 by:

$$\int_0^\infty \frac{u^4 e^u}{(e^u - 1)^2} = \frac{4}{15}\pi^4\,.$$

Thus, for low temperatures (contrast with Equation 6.16):

$$\tilde{C}_{V(\text{vib})} = \frac{36}{15}\pi^4 R\left(\frac{T}{\Theta_D}\right)^3 \qquad (6.24)$$

or

$$C_{V(\text{vib})}/R = 239{\cdot}3(T/\Theta_D)^3$$

i.e., $C_{V(\text{vib})}$ is proportional to T^3, as required by the experimental data.

Note on the determination of the function $g(\omega)$ (Fowler, Ch. 4)

Figure 6.4 shows some examples of the normal modes of vibration of a *one-dimensional crystal*, in which neighbouring atoms are distance d apart. The diagrams represent the displacement from equilibrium of each atom at some instant in time. Solution of the equation of motion shows that, for such normal modes, the instantaneous displacements of the atoms are related sinusoidally to the distance from the end of the crystal, i.e., they vary in a periodic way along the crystal. The distance after which the pattern is called the *wavelength* λ *of the normal mode*.

Each atom can vibrate in the x-, y-, and z-directions. Vibrations in the x-direction (the direction in which λ is measured) are called *longitudinal*, and those in the y- and z-directions are called *transverse waves*. Boundary conditions restrict the values of λ to those which satisfy the condition:

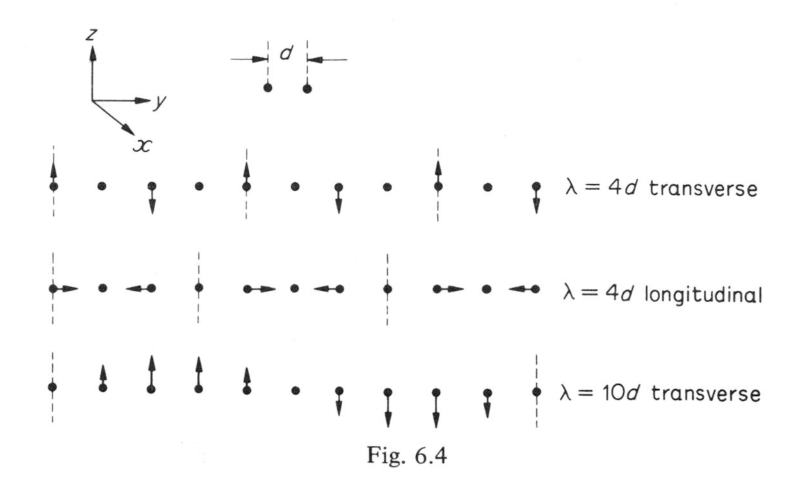

Fig. 6.4

$$\frac{\lambda}{2} = \frac{p}{a} \quad (p = 1,2,3,...)$$

for a one-dimensional crystal of length a. Correspondingly, for a rectangular prism with sides a, b, and c:

$$\left(\frac{2}{\lambda}\right)^2 = \left(\frac{p}{a}\right)^2 + \left(\frac{q}{b}\right)^2 + \left(\frac{s}{c}\right)^2 \tag{6.25}$$

where p, q, and s can have any integer values.

Equation 6.25 closely resembles 4.17 for the translational energy of a particle in a rectangular box (the mathematical basis is essentially the same for both equations). Just as we used Equation 4.25 to obtain the number of translational states with energies in a given range, so we use Equation 6.25 to calculate the number, denoted by $g(\omega)\mathrm{d}\omega$, of modes with frequencies in the range from ω to $\omega + \mathrm{d}\omega$. Thus, let:

$$r^2 = \left(\frac{p}{a}\right)^2 + \left(\frac{q}{b}\right)^2 + \left(\frac{s}{c}\right)^2$$

Then, the number of possible wavelengths with values of $(2/\lambda)^2$ lying between r and $r + \mathrm{d}r$ is (cf. Equation 4.27):

$$g_r = \frac{V\pi r^2\,\mathrm{d}r}{2} \tag{6.26}$$

with:

$$r = 2/\lambda \tag{6.27}$$

In fact, we require the *frequency* distribution $g(\omega)$ rather than the distribution of $1/\lambda$ values given by Equation 6.26. The frequency and wavelength of sound waves passing through a medium are related by

$$\omega = \frac{c}{\lambda} \tag{6.28}$$

where c is the velocity of the wave. For waves with $\lambda \gg d$, it is a good approximation to regard the crystal as a continuous medium (as an extreme example, one can discuss the vibration of a violin string without reference to the molecular structure of the latter!). The basis of the Debye theory is the assumption that Equation 6.28 holds for all the vibrational modes of the crystal.

From Equations 6.27 and 6.28:

$$r = \frac{2\omega}{c} \quad \text{and} \quad dr = \frac{2d\omega}{c}$$

so that:

$$g(\omega)d\omega = \frac{4\pi V\omega^2 d\omega}{c_t^3} \quad \text{for transverse waves,}$$

$$g(\omega)d\omega = \frac{4\pi V\omega^2 d\omega}{c_l^3} \quad \text{for longitudinal waves,}$$

where c_t and c_l are the velocities of the two kinds of wave. In general, $c_t \neq c_l$.

Since a wave travelling in any direction can be analysed into one longitudinal and two transverse waves, we obtain, finally:

$$g(\omega)d\omega = \left[4\pi V\left(\frac{2}{c_t^3} + \frac{1}{c_l^3}\right)\right]\omega^2 \tag{6.29}$$

which has the same form as Equation 6.19, with:

$$A = 4\pi V\left(\frac{2}{c_t^3} + \frac{1}{c_l^3}\right).$$

As explained above, Equation 6.29 is a good approximation for low frequencies (long wavelengths). However, the most serious defect of the Debye theory is the use of this equation for higher frequencies, but in recent years considerable progress has been made in the proper calculation of $g(\omega)d\omega$ from first principles.

The velocities c_t and c_l can be expressed in terms of two elastic constants (Poisson's ratio σ and the compressibility κ) of the (isotropic) solid:

$$c_l^2 = \frac{3(1-\sigma)}{(1+\sigma)\kappa\rho}\ ; \ c_t^2 = \frac{3(1-2\sigma)}{2(1+\sigma)\kappa\rho}$$

where ρ is the density of the solid. The constant A can thus be calculated from σ and κ. From Equation 6.20:

$$\Theta_D = \frac{h\omega_{max}}{k} = \left(\frac{9N}{AV}\right)^{1/3}$$

and hence Θ_D can be calculated from the elastic constants of the solid. Satisfactory agreement is found between the value of Θ_D so calculated and that derived from the measured heat capacity, via Equation 6.23, for such substances as Al, Cu, and Pb.

6.5 More complicated crystal structures

The derivations of the Einstein and Debye equations, outlined in the previous paragraphs, apply specifically to *isotropic* solids, i.e., those belonging to the cubic system. However, it has been found experimentally that the Debye equation represents the values of $\tilde{C}_V$ for certain other monatomic solids, such as zinc (which crystallizes in the hexagonal system).

With a small modification, the theories can also be applied to *molecular crystals*. Suppose that the crystal contains $\tilde{N}$ molecules, each composed of s atoms. Since there are $\tilde{N}s$ atoms, the crystal as a whole has $3\tilde{N}s$ vibrational modes. A reasonable approximation is obtained by classifying the vibrations into

(a) $3\tilde{N}$ lattice vibrations, which are the normal modes discussed in the Debye treatment, they are often referred to as the *acoustical modes*. The bond lengths and angles will be unchanged in these vibrations.

(b) Independent vibrations of the individual molecules, in which bond angles and lengths may vary. There must be $3\tilde{N}s - 3\tilde{N} = 3\tilde{N}(s - 1)$ of these. They are referred to as the *optical modes*, because (just as in the case of free molecules), some of the corresponding frequencies ω_l can be detected spectroscopically.

The contribution of type (a) modes to $\tilde{C}_V$ is given by the Debye formula, and that of the type (b) modes by Equation 6.11. Thus:

$$\tilde{C}_{V(\text{vib})} = 9R\left(\frac{T}{\Theta_D}\right)^3 \int_0^{\Theta_D/T} \frac{u^4 e^u}{(e^u - 1)^2}\, du + R\sum_{l=1}^{3s-3} u_l^2 \frac{e^{u_l}}{(e^{u_l} - 1)^2} \quad (6.30)$$

where $u_l = h\omega_l/kT$. The two expressions on the right-hand side of Equation 6.30 are called, respectively, the Debye term and the Einstein term.

An equation similar to 6.30 also results from a generalization of the Debye procedure, due to M. Born. This is also applicable to *ionic crystals*. In the latter case, the number s can be deduced from the crystal structure. For example, the value $s = 2$ is used for the alkali halide crystals, so that there are $3(2 - 1) = 3$ terms in the Einstein sum in Equation 6.30. By symmetry, ω_l is the same for all three.

An accurate test of the applicability of Equation 6.30 to simple ionic substances was made by Försterling, as follows. Firstly, Θ_D

was calculated from the elastic constants, by the method described at the end of the previous section. The Debye term in Equation 6.30 was thus evaluated for various temperatures. Subtraction of these from the measured C_V values gave the Einstein term at these temperatures. The values of ω_l were calculated from the latter. For NaCl, the result is $\omega_l = 4{\cdot}50 \times 10^{12}\,s^{-1}$. The frequencies ω_l have also been determined directly by infrared absorption spectroscopy and by the selective reflection (Reststrahlen) method for various crystals. For the NaCl crystal, the value so found is $4{\cdot}65 \times 10^{12}\,s^{-1}$, which is in very satisfactory agreement with that calculated from the heat capacities.

Problems

Note that the values of the function:

$$\frac{\tilde{C}_{V(\mathrm{vib})}}{R} = \frac{u^2 e^u}{(e^u - 1)^2}$$

where $u = h\omega/kT = \Theta_{\mathrm{vib}}/T$, are given in Table 10.8.

6.1 From the data in Table 5.2, calculate $\tilde{C}_V$ at 400 K for H_2 and Cl_2.

6.2 From the data in Table 5.3, calculate C_V for $CHCl_3$ at 300 K. Compare the result with the experimentally determined $\tilde{C}_P = 7{\cdot}96R$.

6.3 Show that (using the notation of Section 6.2):

$$x^{\mathrm{ortho}} = f^{\mathrm{ortho}}/f$$

where $f = f^{\mathrm{ortho}} + f^{\mathrm{para}}$. Deduce that:

$$\left(\frac{\partial x^{\mathrm{ortho}}}{\partial T}\right)_V = \frac{x^{\mathrm{ortho}}}{NkT^2}(E^{\mathrm{ortho}} - E)$$

6.4 Calculate the high-temperature value of the ratio $n^{\mathrm{ortho}}/n^{\mathrm{para}}$ for D_2 (note the value of $g_{\mathrm{nuc}(0)}$ given in Section 5.2). From the values given in Table 6.1, plot a graph of C_V against T/Θ_{rot} for temperatures up to 200 K for a metastable mixture of ortho and para D_2 having the high-temperature equilibrium composition. (This graph accurately represents the experimentally observed behaviour of D_2.)

6.5 Taking the first two terms in the electronic partition function:

$$f_{el} = g_0 + g_1 e^{-x}$$

where $x = \epsilon_{el(1)}/kT$, show that the electronic contribution to the heat capacity is:

$$\tilde{C}_{V(el)} = \frac{Rx^2}{\left(1 + \frac{g_0}{g_1} e^{x}\right)\left(1 + \frac{g_0}{g_1} e^{-x}\right)}$$

If $g_0/g_1 = 1$, this simplifies to:

$$\tilde{C}_{V(el)} = R\left(\frac{x}{e^{x/2} + e^{-x/2}}\right)^2 .$$

Proceeding as on page 98 , show that $\tilde{C}_{V(el)} \approx 0$ when $\epsilon_{el(1)} >> kT$ and $\tilde{C}_{V(el)} \approx R$ when $\epsilon_{el(1)} << kT$. Noting that:

$$\tfrac{1}{2}(e^{x} + e^{-x}) = \cosh x ,$$

use mathematical tables to plot a graph of $C_{V(el)}/R$ against $kT/\epsilon_{el(1)}$.

6.6 Plot a graph of Einstein heat capacity against T/Θ_E (multiply the values in Table 10.8 by 3).

The following values of $\tilde{C}_P$ have been found for copper:

T(K)	10	25	50	100	150	200	298
C_P/R	0·05	0·12	0·75	1·95	2·48	2·74	2·96

Correct, in so far as it is necessary to do so, for the difference between $\tilde{C}_P$ and $\tilde{C}_V$ (see the footnote on page 105; the melting point of copper is 1083°C). Determine the values of T/Θ_E corresponding to each of the above pairs of data; hence calculate Θ_E for the various temperatures.

6.7 Using the data in Table 6.3 and the values given in the preceding problem, determine Θ_D for copper at the various temperatures. Comment on the constancy of Θ_E and Θ_D.

CHAPTER SEVEN

Mixtures

We have, so far, discussed the calculation of the thermodynamical properties of pure substances. However, in physical chemistry, one is often concerned with mixtures of several components, and in the present chapter we extend the discussion to embrace such systems.

The simplest subjects of study are the perfect gas mixtures and ideal (liquid and solid) solutions, treated in the first two sections of this chapter. The remainder of the chapter is devoted to an elementary account of non-ideal mixtures (regular solutions).

It might be remarked that Sections 7.3–7.5 can be omitted in a first reading of this book.

7.1 Entropy of mixing of perfect solid and liquid solutions

Consider two substances A and B which are sufficiently similar to be able to form a mixed crystal. Let the energy levels of A be $\epsilon_{a(1)}, \epsilon_{a(2)}, \ldots, \epsilon_{a(i)}$, etc., and the corresponding degeneracies, $g_{a(1)}, g_{a(2)}, \ldots, g_{a(i)}$, etc. Analogously, for B let the energy levels and degeneracies be $\epsilon_{b(1)}, \epsilon_{b(2)}, \ldots, \epsilon_{b(i)}$, etc., and $g_{b(1)}, g_{b(2)}, \ldots, g_{b(i)}$, etc. Let the numbers of molecules of A and B be, respectively, N_A and N_B.

An essential feature of Einstein's theory of the heat capacities of monatomic crystals (Section 6.3) is that the motion of any one atom is independent of the motion of neighbouring atoms. If this *Einstein model* is also adopted for the molecular crystal A, then *each individual molecule of* A *can occupy any of the allowed energy levels, independently of the other molecules present.* The Maxwell–Boltzmann theory then leads to a set of distribution numbers:

$$\bar{n}_{a(1)},\ \bar{n}_{a(2)},\ \ldots,\ \bar{n}_{a(i)},\ \ldots$$

such that the number:

$$\Omega_{D(A)} = N_A! \prod \frac{(g_{a(i)})^{\bar{n}_{a(i)}}}{(\bar{n}_{a(i)})!} \tag{7.1a}$$

is the number of complexions associated with the most probable distribution of molecules among the energy levels. The corresponding number for B is:

$$\Omega_{D(B)} = N_B! \prod \frac{(g_{b(i)})^{\bar{n}_{b(i)}}}{(\bar{n}_{b(i)})!} \tag{7.1b}$$

where the most probable distribution numbers are:

$$n_{b(1)},\ n_{b(2)},\ \ldots,\ n_{b(i)},\ \ldots$$

Consider, now the mixed crystal (A + B). The molecules of A will be distributed among the levels $\epsilon_{a(1)}$, $\epsilon_{a(2)}$, ..., $\epsilon_{a(i)}$, etc., and those of B among $\epsilon_{b(1)}$, $\epsilon_{b(2)}$, ..., $\epsilon_{b(i)}$, etc. If the Einstein model of the crystal is applicable, the most probable distribution numbers $\bar{n}_{a(1)}$, $\bar{n}_{a(2)}$, ..., $\bar{n}_{a(i)}$ and $n_{b(1)}$, $n_{b(2)}$, ..., $n_{b(i)}$, ... have just the same values for the molecules in the mixture as for the pure substances. Since each of the $\Omega_{D(A)}$ complexions can be combined with each of the $\Omega_{D(B)}$ complexions, the number of ways of achieving the most probable distribution of molecules amongst the energy levels is:

$$\Omega_{D(A)} \times \Omega_{D(B)}$$

for the mixed crystal.

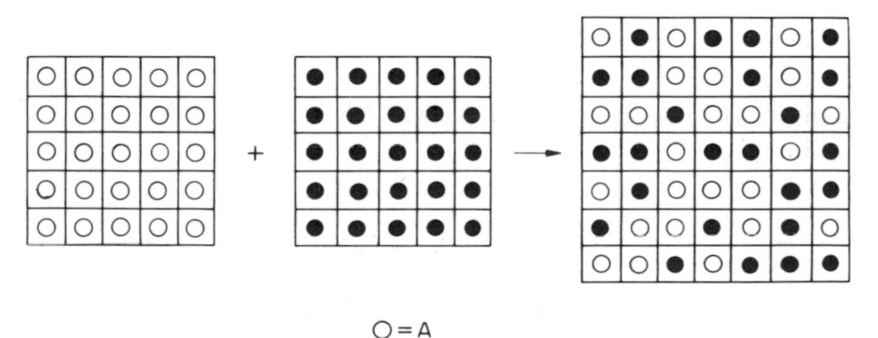

Fig. 7.1 Lattice model of a mixture

The *number of complexions* of the mixed crystal is, however, greater than this. Any one position ('lattice site') in the mixed crystal may be occupied by one molecule of *either* A *or* B, and, since the mixing is perfectly random, there are:

$$\frac{(N_A + N_B)!}{N_A!N_B!} \tag{7.2}$$

distinguishable ways in which N_A molecules of A and N_B molecules of B can occupy the total $(N_A + N_B)$ lattice sites. This follows immediately from the combinatory rule (Equation 1.3); thus, the $(N_A + N_B)$ lattice sites can be grouped into N_A which contain A molecules, and N_B which do not. For *each* of these arrangements there are $\Omega_{D(A)} \times \Omega_{D(B)}$ ways of placing the molecules in the energy levels. Hence, the number of complexions of the mixed crystal corresponding to the most probable distribution of energy is:

$$\Omega_{D_0} = \frac{(N_A + N_B)!}{N_A!N_B!} \times \Omega_{D(A)} \times \Omega_{D(B)} \tag{7.3}$$

The entropy of the mixed crystal is, from Equation 6.1:

$$S = k \ln \Omega_{D_0} = k \ln \frac{(N_A + N_B)!}{N_A!N_B!} + k \ln \Omega_{D(A)} + k \ln \Omega_{D(B)}$$

Before being mixed together, A and B have, respectively, the entropies:

$$S_A = k \ln \Omega_{D(A)} \quad \text{and} \quad S_B = k \ln \Omega_{D(B)}$$

Hence, the change in entropy when A and B are mixed, i.e., the *entropy of mixing*, is:

$$\begin{aligned} \Delta S_{mix} &= S - (S_A + S_B) \\ &= k \ln \frac{(N_A + N_B)!}{N_A!N_B!} \end{aligned}$$

Applying Stirling's theorem (Equation 1.8) to $\ln(N_A + N_B)!$, $\ln N_A!$, and $\ln N_B!$:

$$\begin{aligned} \Delta S_{mix} &= k\,[(N_A + N_B) \ln(N_A + N_B) - (N_A + N_B) \\ &\quad - (N_A \ln N_A - N_A + N_B \ln N_B - N_B)] \\ &= -k \left(N_A \ln \frac{N_A}{N_A + N_B} + N_B \ln \frac{N_B}{N_A + N_B} \right) \end{aligned}$$

Now, $N_A/(N_A + N_B) = x_A$, the mole fraction of A, and:

$$kN_A = kn_A\tilde{N} = n_AR$$

where n_A is the number of moles of A. Therefore:

$$\Delta S_{mix} = -R(n_A \ln x_A + n_B \ln x_B) \qquad (7.4a)$$

The treatment can obviously be generalized for a mixture of any number of components, and:

$$\Delta S_{mix} = -R \sum_r n_r \ln x_r \qquad (7.4b)$$

where the summation is over all the components.

The total energy of the mixed crystal is the sum of the energies of the individual molecules of A and B:

$$E = \sum_i \bar{n}_{a(i)}\epsilon_{a(i)} + \sum_i n_{b(i)}\epsilon_{b(i)}$$

The two sums on the right-hand side of this equation are, however, just the energies E_A and E_B of the separate crystals A and B, since the numbers $\bar{n}_{a(i)}$, $\epsilon_{a(i)}$, $\bar{n}_{b(i)}$, and $\epsilon_{b(i)}$ are unaltered by mixing. Thus, the energy change on mixing A and B is:

$$\Delta E_{mix} = 0 \qquad (7.5)$$

We have so far referred specifically to mixed crystals. Since, however, the molecules in a liquid mixture occupy more or less definite positions, it is assumed that the preceding discussion also applies to liquid solutions.

Liquid or solid mixtures for which Equations 7.4 and 7.5 are valid may be called *perfect solutions*. The properties of perfect solutions (for example, that the vapour pressures obey Raoult's law), which follow as consequences of these equations, can be discussed by classical thermodynamics.* The importance of the derivation of Equations 7.4 and 7.5 by statistical mechanics lies in the fact that one can obtain thereby some insight into the conditions under which such solutions may be formed.

Apart from the simplification of using the Einstein model, we have made, in deriving Equations 7.4 and 7.5, two tacit assumptions concerning the nature of perfect solutions:

(a) The molecules A and B are assumed to be similar in size and

* Alternatively, perfect solutions may be defined as those for which Raoult's law holds at all temperatures. Equation 7.4 and the result $\Delta H_{mix} = 0$ (equivalent to Equation 7.5 since V is implicitly assumed to be constant, in the above discussion) follow. See Lewis and Randall, Ch. 18.

shape, so that they can be arranged in the crystal lattice or liquid in a completely random manner. Thus, the positions of any two molecules can be interchanged without necessitating changes in the final positions of other molecules. Only* under this condition can Equation 7.4 apply.

(b) The values of $\Omega_{D(\mathrm{A})}$ and $\Omega_{D(\mathrm{B})}$ are unaffected by any rearrangement of the molecules amongst themselves. This, in turn, implies that the distribution numbers, and therefore the energy levels $\epsilon_{\mathrm{a}(i)}$ and $\epsilon_{\mathrm{b}(i)}$, are unaffected by such rearrangement. This can only be the case if intermolecular forces between pairs of neighbouring molecules are essentially the same, whether these pairs be A–A, B–B, or A–B. This will be true if A and B have chemically similar structures.

Thus, it is expected that if molecules of A have similar size and chemical structure to those of B, then A and B will form a perfect solution. This is confirmed experimentally; for example (Lewis and Randall, p. 215), ethylene bromide and propylene bromide form a perfect mixture, but acetone and carbon disulphide do not.

7.2 Entropy of mixing of perfect gases

Suppose that two perfect gases A and B are mixed at constant pressure and temperature (and therefore at constant total volume):

P,T	P,T	→	P,T
V_A	V_B		$(V_A + V_B)$

Using the notation of Section 7.1, but remembering that A and B are gaseous, we have (cf. Section 4.4):

$$\Omega_{D(\mathrm{A})} = \prod \frac{(g_{\mathrm{a}(i)})^{\bar{n}_{\mathrm{a}(i)}}}{(n_{\mathrm{a}(i)})!}, \tag{7.6}$$

$$\begin{aligned} S_{\mathrm{A}} &= N_{\mathrm{A}} k \ln \Omega_{D(\mathrm{A})} \\ &= N_{\mathrm{A}} k \ln \left(f_{\mathrm{A}} \frac{\mathrm{e}}{N_{\mathrm{A}}} \right) + \frac{E_{\mathrm{A}}}{T} \end{aligned} \tag{7.7}$$

(f_{A} is the partition function of A), with corresponding equations

* An exception to this rule is that mixtures of rod-like molecules, of the same thickness but different lengths, may be perfect; see Rushbrooke, p. 225.

for $\Omega_{D(B)}$ and S_B.

As in Section 7.1, Equations 7.6 and 7.7 apply to A whether or not B is present, and *vice versa*. Hence, the number of complexions for the most probable distribution of energies among the molecules in the gas mixture is:

$$\Omega = \Omega_{D(A)} \times \Omega_{D(B)}$$

In contrast with Equation 7.3, there is no factor corresponding to 7.2, because *the molecules do not occupy definite (fixed) positions in the volume containing the mixture.*

The entropy of the mixture is therefore:

$$\begin{aligned} S &= k \ln \Omega \\ &= N_A k \ln \left(f'_A \frac{e}{N_A} \right) + \frac{E_A}{T} + N_B k \ln \left(f'_B \frac{e}{N_B} \right) + \frac{E_B}{T} \end{aligned}$$

where f'_A and f'_B are the partition functions of A and B *in the mixture.*

The partition function of a gas contains the translational factor (Equation 4.22) and is therefore proportional to V, the volume occupied by the gas. The volume of A before mixing is V_A; after mixing, A occupies the volume $(V_A + V_B)$. Consequently, f'_A is not equal to f_A, the partition function of A before mixing, but:

$$\frac{f'_A}{f_A} = \frac{V_A + V_B}{V_A} \tag{7.8}$$

and similarly:

$$\frac{f'_B}{f_B} = \frac{V_A + V_B}{V_B}$$

The change in entropy on mixing A and B is:

$$\begin{aligned} \Delta S_{mix} &= S - (S_A + S_B) \\ &= N_A k \ln \left(f'_A \frac{e}{N_A} \right) + N_B k \ln \left(f'_B \frac{e}{N_B} \right) \\ &\quad - N_A k \ln \left(f_A \frac{e}{N_A} \right) - N_B k \ln \left(f_B \frac{e}{N_B} \right) \\ &= N_A k \ln \frac{f'_A}{f_A} + N_B k \ln \frac{f'_B}{f_B} \\ &= N_A k \ln \left(\frac{V_A + V_B}{V_A} \right) + N_B k \ln \left(\frac{V_A + V_B}{V_B} \right), \end{aligned} \tag{7.9}$$

from Equation 7.8. Since:

$$V_A = \frac{N_A kT}{P} \quad \text{and} \quad V_B = \frac{N_B kT}{P}$$

Equation 7.10 simplifies to:

$$\Delta S_{mix} = N_A k \ln\left(\frac{1}{x_A}\right) + N_B k \ln\left(\frac{1}{x_B}\right)$$

$$= -R(n_A \ln x_A + n_B \ln x_B) \tag{7.10a}$$

where n and x are the number of moles and the mole fraction of a component. The generalization of Equation 7.10a for a mixture of several gases is:

$$\Delta S_{mix} = -R \sum_r n_r \ln x_r \tag{7.10b}$$

It is remarkable that this expression for the entropy of mixing of perfect gases at constant pressure is identical with the corresponding expression 7.4 for the entropy of perfect mixing of solids or liquids. Note, however, that the origins of Equation 7.4 and 7.10 are quite different. Equation 7.4 arises because the molecules in the solid or liquid are located in fixed positions, and able to occupy $(N_A + N_B)$ lattice positions in $(N_A + N_B)!/N_A!N_B!$ different ways. In contrast, Equation 7.10 arises because the molecules in the gas mixture are *free to move* within the total volume $(V_A + V_B)$, instead of the original volume V_A or V_B. Thus, the entropy change 7.10 is produced by the expansion of each gas [or, equivalently, by the lowering of the partial pressures, to which (see Section 8.3) the entropies are related by the Sackur–Tetrode equation].

7.3 Regular solutions: the Bragg–Williams model

The term *regular solution* was introduced by J.H. Hildebrand in 1929, to describe those solutions which can be regarded as essentially random mixtures, i.e., solutions in which specific interactions, such as hydrogen bonding and solvation, are absent. The difference between regular solutions and perfect solutions is that, in the former case, the forces of interaction between equivalently placed A–A, A–B, and B–B pairs *may* not be equal, and therefore (cf. Section 7.1) the energy of mixing, ΔE_{mix}, may not be zero. On the other hand, the entropy of mixing might still have, essentially, the ideal-solution value given by Equation 7.4.

In the present discussion of regular solutions, the following simplifying assumptions are made.

(i) The lattice model (Section 7.1) is adopted for the pure components A and B, and for the mixture (A + B); thus, in the liquids, the molecules are packed together in a way which resembles the close packing in a crystal.

(ii) It is assumed that A and B are sufficiently similar in molecular structure and size (a) to allow perfectly random mixing in the lattice, (b) so that the volume change on mixing A and B is negligible.

(iii) It is assumed that intermolecular forces are only appreciable when they are between *nearest neighbours* in the lattice.

Other tacit assumptions will actually be made in our treatment, but the discussion of these is deferred until Section 7.4.

Let us suppose that, in the lattice, each molecule has z nearest neighbours (e.g., for closely packed spheres in a face-centred cubic lattice, $z = 12$). Consider the pure liquid A, and suppose that in this liquid there are N_A molecules. Since each of the latter has z neighbours, the number of *pairs* of nearest neighbours is $\frac{1}{2}N_A z$.

Let w_{AA} be the average potential energy of interaction between pairs of nearest-neighbouring molecules at their equilibrium positions in the liquid. Then the total *intermolecular potential energy* of liquid A, when all the molecules are at their equilibrium positions in the lattice, is:

$$U_{AA} = \tfrac{1}{2}N_A z w_{AA} \tag{7.11a}$$

Similarly, using a corresponding notation, for liquid B:

$$U_{BB} = \tfrac{1}{2}N_B z w_{BB} \tag{7.11b}$$

Now consider the mixture of N_A molecules of A and N_B molecules of B. In the mixture, neighbouring *pairs* will be of three kinds, namely, A–A, B–B, and A–B. Let the numbers of such neighbours be, respectively, N_{AA}, N_{BB}, and N_{AB}. The total intermolecular potential energy, U_{AB}, when all the molecules are in their equilibrium positions, is:

$$U_{AB} = N_{AA}w_{AA} + N_{BB}w_{BB} + N_{AB}w_{AB} \tag{7.12}$$

where w_{AB} is the average potential energy of interaction between neighbouring A and B molecules at their equilibrium positions.

The number of neighbours of *each* A molecule is z, and the sum of this number over the whole crystal is zN_A. This total is also:

$$N_{AB} + 2N_{AA}$$

(note that, in calculating the sum zN_A, we shall have considered each A–A pair twice, but each A–B pair once only). Therefore:

$$N_{AA} = \tfrac{1}{2}(zN_A - N_{AB}) \tag{7.13a}$$

and, similarly:

$$N_{BB} = \tfrac{1}{2}(zN_B - N_{AB}) \tag{7.13b}$$

Substituting for N_{AA} and N_{BB} in Equation 7.12:

$$U_{AB} = \tfrac{1}{2}zN_A w_{AA} + \tfrac{1}{2}zN_B w_{BB} + N_{AB}(w_{AB} - \tfrac{1}{2}w_{AA} - \tfrac{1}{2}w_{BB}) \tag{7.14}$$

The change in intermolecular potential energy, when the mixture is formed, is:

$$\Delta U = U_{AB} - U_{AA} - U_{BB}$$

and it is assumed that this is also the change in internal energy on mixing, i.e.:

$$\Delta E_{mix} = \Delta U$$

From Equations 7.11 and 7.14, therefore:

$$\Delta E_{mix} = N_{AB}(w_{AB} - \tfrac{1}{2}w_{AA} - \tfrac{1}{2}w_{BB})$$

$$= N_{AB}w \tag{7.15}$$

where:

$$w = w_{AB} - \tfrac{1}{2}w_{AA} - \tfrac{1}{2}w_{BB} \tag{7.16}$$

The next problem is to evaluate N_{AB}. Let us choose, at random, one molecule from the mixture; the choice can be made in N ways, where $N = N_A + N_B$, the total number of molecules in the mixture. A second molecule can now be chosen in $(N - 1)$ ways, and there are therefore $N(N-1)/2 \approx \tfrac{1}{2}N^2$ ways of picking out any two molecules in this manner (the factor $\tfrac{1}{2}$ arises because it does not matter which of the two molecules is chosen first). On the other hand, if the first choice is restricted to an A molecule, and the second to a B molecule, there are $N_A N_B$ ways of making the choice. Consequently:

$$\frac{\text{number of A–B pairs}}{\text{number of all pairs}} = \frac{2N_A N_B}{N^2} \tag{7.17}$$

Note that this refers to all pairs in the crystal, *not* just pairs of *neighbouring* molecules.

Since each of the N molecules has z neighbours, the number of neighbouring pairs is $Nz/2$. If the molecules are randomly distributed, the fraction of these pairs which are of type A–B should also be given by Equation 7.17. Hence:

$$N_{AB} = \frac{Nz}{2} \times \frac{2N_A N_B}{N^2}$$

$$= \frac{zN_A N_B}{N}$$

Substituting this value into Equation 7.15:

$$\Delta E_{mix} = \frac{N_A N_B wz}{N} \quad (7.18a)$$

$$= \tilde{N}wz \frac{n_A n_B}{n_A + n_B} \quad (7.18b)$$

where n_A and n_B are the numbers of moles of A and B in the mixture.

Since it was assumed that mixing is perfectly random, the calculated entropy of mixing is the same as for an ideal mixture, i.e., from Equation 7.4:

$$\Delta S_{mix} = -R(n_A \ln x_A + n_B \ln x_B) \quad (7.19)$$

According to assumption (ii) above, the change in volume on mixing is:

$$\Delta V_{mix} = 0$$

Hence the *Gibbs free energy change of mixing* at temperature T is:

$$\Delta G_{mix} = (\Delta E_{mix} + P\Delta V_{mix}) - T\Delta S_{mix}$$

$$= \Delta E_{mix} - T\Delta S_{mix}$$

$$= \tilde{N}wz \frac{n_A n_B}{n_A + n_B} + RT(n_A \ln x_A + n_B \ln x_B) \quad (7.20)$$

The Gibbs free energy, G_m, of the mixture is:

$$G_m = G_A + G_B + \Delta G_{mix}$$

where G_A and G_B are the Gibbs free energies of the separate components A and B. If the mixing is at standard pressure (1 atmosphere), then these are the standard free energies G_A° and G_B°. By definition (Lewis and Randall, p. 202) the *chemical potential* of A in the mixture is:

$$\mu_A = \left(\frac{\partial G_m}{\partial n_A}\right)_{T,P,n_B}$$

and therefore:

$$\mu_A = \left(\frac{\partial G^\circ}{\partial n_A}\right)_{T,P,n_B} + \left(\frac{\partial G_B^\circ}{\partial n_A}\right)_{T,P,n_B} + \left(\frac{\partial \Delta G_{mix}}{\partial n_A}\right)_{T,P,n_B}$$

$$= \mu_A^\circ + \left(\frac{\partial \Delta G_{mix}}{\partial n_A}\right)_{T,P,n_B}$$

since G_B° is independent of n_A, and $(\partial G_A^\circ/\partial n_A)_{T,P,n_B}$ is, by definition, the standard chemical potential μ_A°. Therefore, by differentiating Equation 7.20, we obtain:

$$\mu_A = \mu_A^\circ + RT(\ln x_A + \alpha x_B^2) \qquad (7.21a)$$

where:

$$\alpha = \frac{\tilde{N}zw}{RT} = \frac{zw}{kT}$$

Similarly:

$$\mu_B = \mu_B^\circ + RT(\ln x_B + \alpha x_B^2) \qquad (7.21b)$$

The activities (a_A and a_B) of A and B in the mixture are defined by the equations:

$$\mu_A = \mu_A^\circ + RT \ln a_A \qquad (7.22a)$$

$$\mu_B = \mu_B^\circ + RT \ln a_B \qquad (7.22b)$$

Comparing Equations 7.21 and 7.22:

$$\ln a_A = \ln x_A + \alpha x_B^2$$

$$\ln a_B = \ln x_B + \alpha x_A^2$$

Hence, the activity coefficients $\gamma (\ln \gamma_A = \ln a_A - \ln x_A)$ are given by:

$$\ln \gamma_A = x_B^2 \alpha$$

$$\ln \gamma_B = x_A^2 \alpha$$

Some values of α determined from measured activity coefficients are shown in Table 7.1.

The partial vapour pressures of A and B over the solution are (see Lewis and Randall, p. 246):

$$P_A = P_A^\circ a_A = P_A^\circ \gamma_A x_A$$

$$P_B = P_B^\circ a_B = P_B^\circ \gamma_B x_B$$

where P_A° and P_B° are the vapours pressures of pure A and pure B respectively, at the temperature under consideration, i.e.:

Table 7.1 ΔS^{E}_{mix} = observed ΔS_{mix} − ideal ΔS_{mix}. The values of α and ΔS^{E}_{mix} are those obtained for $x_A = x_B = 0{\cdot}5$ mixtures. Data from Lewis and Randall, p.286.

	T (K)	α	$\Delta S^{E}_{mix}/R$
CCl_4–benzene	298	0·13	0·011
CCl_4–cyclohexane	313	0·10	0·037
CS_2–acetone	308	1·6	0·16
benzene–1,2-$C_2H_4Cl_2$	298	0·036	0·015
benzene–cyclohexane	293	0·52	0·21
benzene–cyclohexane	343	0·36	0·15

$$P_A = P_A^{\circ} x_A\, e^{\alpha x_B^2} \qquad P_B = P_B^{\circ} x_B\, e^{\alpha x_A^2} \tag{7.23}$$

The Equations 7.23 account, in a simple way, for the observed deviations from Raoult's law. Figure 7.2 shows the theoretical dependence of vapour pressure upon concentration, for positive and negative values of α. Curves having these general forms have been observed experimentally (Lewis and Randall, p. 215). When $\alpha > 0$, $P_A > P_A^{\circ} x_A$ and the system exhibits positive deviations from Raoult's law. Now, $\alpha > 0$ implies a positive value of w, i.e., according to Equation 7.16 that:

$$w_{AB} > \tfrac{1}{2}(w_{AA} + w_{BB})$$

and thus that the potential energy of interaction of an A–B pair is greater than the mean of the potential energies of interaction of A–A and B–B pairs.

Another property of mixtures which can be deduced from the above treatment is *phase separation*. Since:

$$\frac{n_A}{n_A + n_B} = x_A \quad \text{and} \quad \frac{n_B}{n_A + n_B} = x_B$$

the mole fractions of A and B in the mixture, Equation 7.20 can be written as:

$$\frac{\Delta G_{mix}}{n_A + n_B} = \tilde{N} w z x_A x_B + RT(x_A \ln x_A + x_B \ln x_B),$$

that is:

$$\frac{\Delta G_{mix}}{nRT} = \alpha x_A x_B + x_A \ln x_A + x_B \ln x_B \tag{7.24}$$

where $n = n_A + n_B$. Figure 7.3 shows the theoretical dependence

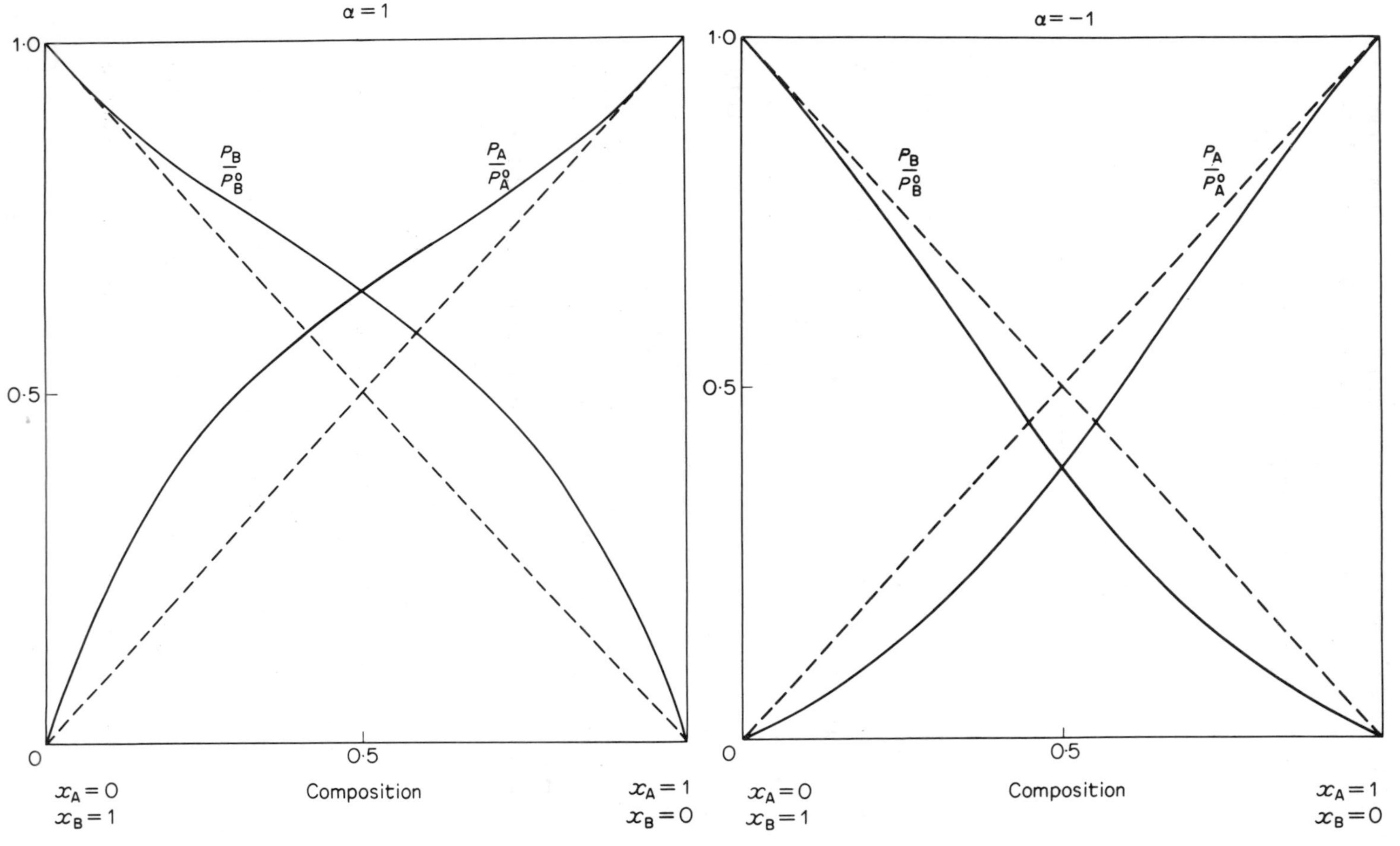

Fig. 7.2 Vapour pressures of regular solutions

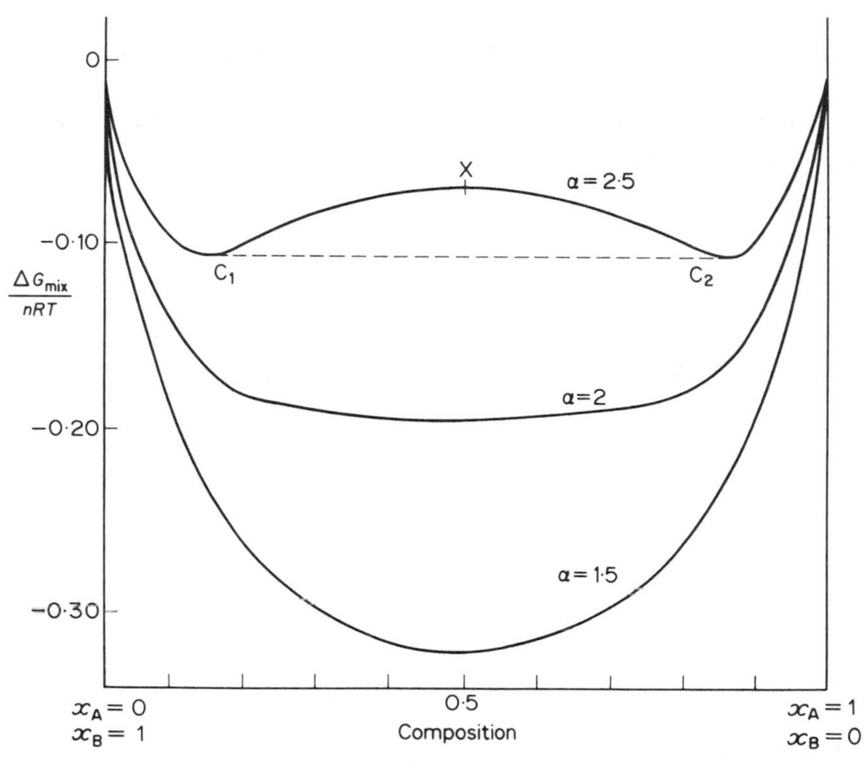

Fig. 7.3

of $\Delta G_{\text{mix}}/nRT$ on x_A and x_B for various values of α. The curves are of two types: those which have a maximum at $x_A = x_B = 0{\cdot}5$, and those which have not.

Consider the $\alpha = 5$ curve. At point X, the value of ΔG_{mix} is greater than the value at C_1 and C_2. Thermodynamics shows that the system will tend to reach equilibrium by minimizing the free energy, and so a liquid of composition X will break up into two phases having the respective compositions C_1 and C_2, the amount of each phase being determined by the overall composition of the mixture. In these circumstances, therefore, A and B are *not completely miscible*. It is shown below that:

$$\begin{aligned} &\text{when } \alpha > 2, \text{ there is a } \textit{maximum} \text{ at } x_A = x_B = 0{\cdot}5\,; \\ &\text{when } \alpha < 2, \text{ there is a } \textit{minimum} \text{ at } x_A = x_B = 0{\cdot}5\,. \end{aligned} \qquad (7.25)$$

Thus:

$$\text{when } \frac{zw}{kT} > 2, \text{ A and B are incompletely miscible,}$$

but when $\dfrac{zw}{kT} < 2$, A and B are miscible over the whole range of concentration.

The critical solution (consolute) temperature T_c above which the liquids are completely miscible, is therefore:

$$T_c = \frac{zw}{2k} \tag{7.26}$$

Proof of the result 7.25

Let:

$$y = \Delta G_{mix}/nRT$$

Since $x_A = 1 - x_B$:

$$\frac{d}{dx_A} = -\frac{d}{dx_B}$$

Hence, differentiating Equation 7.24:

$$\frac{dy}{dx_A} = \alpha(x_B - x_A) + (\ln x_A + 1) - (\ln x_B + 1)$$

$$= \alpha(x_B - x_A) + \ln x_A - \ln x_B \tag{7.27}$$

and

$$\frac{d^2y}{dx_A^2} = -2\alpha + \frac{1}{x_A} + \frac{1}{x_B}$$

When $x_A = x_B = 0{\cdot}5$:

$$\frac{dy}{dx_A} = 0 \quad \text{and} \quad \frac{d^2y}{dx_A^2} = 4 - 2\alpha$$

At this point, therefore:

$$\frac{d^2y}{dx_A^2} > 0 \text{ if } \alpha < 2 \text{ (a minimum)}$$

$$\frac{d^2y}{dx_A^2} < 0 \text{ if } \alpha > 2 \text{ (a maximum)}$$

To complete the proof, we have to show that, when $\alpha \leqslant 2$, there is only one turning point; i.e., that $dy/dx = 0$ *only* when $x_A = x_B = 0{\cdot}5$. Firstly, we note that, according to Equations 7.24, the curve is symmetrical in x_A and x_B; hence we consider only that portion of the curve with $x_A \geqslant 0{\cdot}5 \geqslant x_B$.

Let $\frac{1}{2}(1 + r) = x_A$, so that $\frac{1}{2}(1 - r) = x_B$ and $0 \leqslant r \leqslant 1$. Then Equation 7.27 becomes:

$$\frac{dy}{dx_A} = -r\alpha + \ln\left(\frac{1 + r}{1 - r}\right)$$

Now:

$$\ln\left(\frac{1 + r}{1 - r}\right) = \left(2 \;\; r + \frac{r^3}{3} + \frac{r^5}{5} + \ldots\right)$$

if $|r| < 1 \geqslant 2r$, with equality only if $r = 0$.

$$\text{Hence, if } r \neq 0: \quad \frac{dy}{dx_A} > 2r - r\alpha$$
$$> 0 \text{ if } \alpha < 2$$

i.e., dy/dx cannot be zero if $\alpha < 2$ and $r \neq 0$. When $\alpha < 2$, therefore, the only turning point is at $r = 0$, i.e., $x_A = x_B = 0{\cdot}5$.

7.4 Further discussion of the Bragg–Williams theory

The theoretical model, described in the preceding section, was proposed by Bragg and Williams in 1934. Its applicability is not restricted to liquid mixtures. Bragg and Williams used it to study the order–disorder transition in alloys; it was thus shown, for example, that there is a critical temperature T_c given by an equation of the type 7.26, above which order in the alloy crystal disappears completely (Fowler and Guggenheim, p. 520). A closely related application is to the theory of ferromagnetism; in this case, the 'critical temperature' given by Equation 7.26 is the so-called *Curie temperature*, above which the residual magnetism in ferromagnetic materials disappears.

There are, however, three points of criticism of the Bragg–Williams theory. We shall consider these briefly in turn.

(i) *Non-randomness in the mixture*

The Bragg–Williams theory contains an element of inconsistency. It assumes that mixing is completely random; if, however, the forces between A–A, B–B, and A–B pairs are not equal, then some arrangements in the lattice will be energetically more favoured than others, and therefore mixing will tend to be non-random. The effect is probably small because thermal agitation tends to destroy any order; nevertheless, an improvement in the theory was obtained by Guggenheim by taking it into account.

Calculation of ΔE_{mix} by Equation 7.15:

$$\Delta E_{mix} = N_{AB} w$$

depends upon the evaluation of N_{AB}. In the Bragg–Williams theory, N_{AB} is obtained by assuming a perfectly random arrangement of A and B. Guggenheim suggested that a better estimate of N_{AB} can be obtained by considering the equilibrium of the process:

$$AA + BB \longrightarrow 2AB$$

in which one A of a neighbouring A–A pair, and one B of a neighbouring B–B pair change places. Such a process is, in many ways, analogous to a chemical reaction, and Guggenheim referred to the equilibrium as *quasi-chemical* equilibrium.

In Chapter 9, we discuss the evaluation of equilibrium constants of chemical reactions in terms of the partition functions of reactants and products. Anticipating Equation 9.16a:

$$\frac{N_{AB}^2}{N_{AA}N_{BB}} = \frac{f_{AB}^2}{f_{AA}f_{BB}} e^{-\Delta\epsilon_0/kT}$$

Here, $\Delta\epsilon_0$ is the potential energy increase when an A–A and a B–B pair are replaced by two A–B pairs; clearly, $\Delta\epsilon_0 = 2w$. f_{AA}, f_{BB}, and f_{AB} are the respective partition functions of pairs of molecules in the lattice. It is assumed* that f_{AB} differs from the product $f_{AA}f_{BB}$ only in that the partition functions of A–A and B–B have a symmetry factor 2 compared with that of A–B, i.e.:

$$\frac{f_{AB}^2}{f_{AA}f_{BB}} = 4.$$

The equilibrium constant is therefore:

$$\frac{N_{AB}^2}{N_{AA}N_{BB}} = 4e^{-2w/kT}$$

or

$$N_{AB}^2 = 4e^{-2w/kT}N_{AA}N_{BB}$$

Substituting for N_{AA} and N_{BB} given by the Equations 7.13:

$$N_{AB} = e^{-2w/kT}(zN_A - N_{AB})(zN_B - N_{AB}) \tag{7.28}$$

The resulting quadratic equation can now be solved for N_{AB}, and substitution into Equation 7.15 gives, finally:

$$\Delta E_{mix} = \tilde{N}wz\frac{n_A n_B}{n_A + n_B}\left(\frac{2}{1+\beta}\right) \tag{7.29}$$

where:

$$\beta = [1 - 4x_A x_B(1 - e^{-2w/kT})]^{1/2}$$

Note that, if w/kT is small, $\beta \approx 1$ and Equation 7.29 reduces to 7.18. The other thermodynamic functions for mixing can be derived by elementary, but somewhat tedious, algebra (e.g., Rushbrooke, p.302).

* More complete accounts are given in Hill, Ch. 14, and Fowler and Guggenheim, p.361

The Guggenheim theory predicts a somewhat smaller entropy of mixing than that given by Equation 7.19; this, of course, is to be expected, since the mixing of the solution is slightly less random than that in the ideal case. And here arises the most serious difficulty of the theory: ΔS_{mix} *is sometimes larger than the ideal value*; this fact is quite inexplicable in terms of the Bragg–Williams or Guggenheim models.

(ii) *The nature of the parameter w*

Suppose that the lowest energy level of a molecule A, when surrounded by z other A molecules, is $\epsilon_{A(0)}$; and that when this A molecule is surrounded by z B molecules, the lowest energy level is $\epsilon'_{A(0)}$ (see Fig. 7.4).

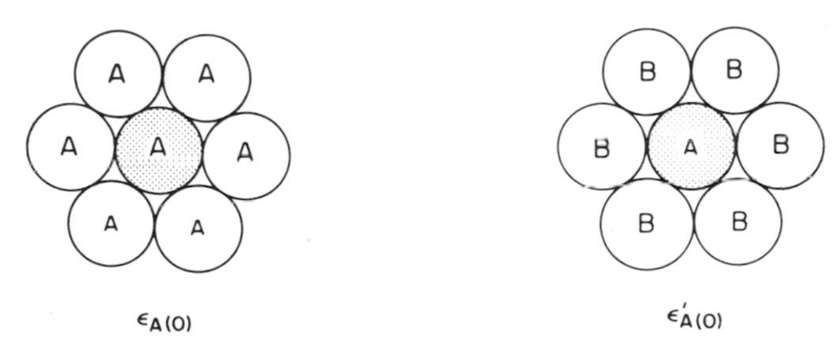

Fig. 7.4

According to our definition of w_{AA} and w_{AB}, the change in the zero-point energy of A, when it is transferred from one environment to the other, is then:

$$\epsilon'_{A(0)} - \epsilon_{A(0)} = z(w_{AB} - w_{AA})$$

In our account of the Bragg–Williams and Guggenheim theories, it was tacitly assumed that, for *all* levels i:

$$\epsilon'_{A(i)} - \epsilon_{A(i)} = z(w_{AB} - w_{AA}), \text{ a constant,}$$

i.e., that w_{AA} and w_{AB} are independent of the quantum state of A. It is *only* under this condition that the simple result 7.15 can be obtained. Actually, it is most unlikely that such will be the case. To illustrate this, let us suppose, for simplicity, that A is monatomic, and consider (as in the Einstein treatment) the lattice vibration energy. The ν-th vibrational energy level of A in the two environments has, respectively, the values:

$$\epsilon_{\text{vib A}(v)} = vh\omega$$

$$\epsilon'_{\text{vib A}(v)} = vh\omega'$$

so that:

$$\epsilon'_{\text{vib A}(v)} - \epsilon_{\text{vib A}(v)} = vh(\omega' - \omega)$$

a difference proportional to v, i.e., not constant, unless $(\omega' - \omega)$ happens to be zero. It is unlikely that $(\omega' - \omega)$ *will* be zero, because ω and ω' depend upon the force constants for vibration, which will, in general, have different values for A in the two environments.

The parameter w must therefore be regarded as some kind of average over many terms, each of which depends, in some way, upon the quantum states of the neighbouring atoms. Hence, its value will change as the occupancy of the quantum states is changed; that is, it will be dependent upon temperature.

The temperature dependence of w affects the entropy of mixing. To show this, we note firstly that:

$$\Delta S_{\text{mix}} = -\left(\frac{\partial \Delta G_{\text{mix}}}{\partial T}\right)_P$$

Equation 7.20 for ΔG_{mix} was obtained by assuming ΔS_{mix} to have the ideal value given by Equation 7.19; actually (as is shown in Section 14.5), Equation 7.20 can be derived without making this assumption, and therefore, one can without inconsistency, differentiate Equation 7.20 to *derive* the entropy of mixing. Thus:

$$\Delta S_{\text{mix}} = -\tilde{N}\frac{n_A n_B}{n_A + n_B} z \frac{\partial w}{\partial T} - R(n_A \ln x_A + n_B \ln x_B) \quad (7.30)$$

The second term in Equation 7.30 is the 'ideal' entropy of mixing. The difference between the observed ΔS_{mix} and this ideal value is referred to as the excess entropy of mixing $\Delta S^{\text{E}}_{\text{mix}}$. Thus:

$$\Delta S^{\text{E}}_{\text{mix}} = -N\frac{n_A n_B}{n_A + n_B} z \frac{\partial w}{\partial T} \quad (7.31)$$

Some values of $\Delta S^{\text{E}}_{\text{mix}}$ are given in Table 7.1.

To complete the treatment, one would aim to calculate w and $\partial w/\partial T$ theoretically, in terms of, say, intermolecular forces. Study of the general problems arising in such calculations constitutes an active field of current research, and numerous detailed investigations have been made. We shall mention just one type of approach, namely, that based upon the *cell theory* of Lennard-Jones and Devonshire. According to this, each molecule is free to move within a *cell*, i.e., a small volume enclosed by neighbouring molecules. The potential energy of the molecule at any position in the cell is determined by the resulting force exerted by the surrounding molecules. From the potential energy expression so obtained, the (classical) partition function for the motion of the molecule in the cell is obtained (see Chapter 11 for the general relationship between potential energy and partition function).

(iii) *Unsymmetrical mixtures*

All the equations for thermodynamic functions of mixing, obtained in Section 7.3, are symmetrical; that is to say, if n_A is interchanged with n_B, and x_A with x_B, the overall expression is unchanged. This means, for example, that the activity coefficient of B in a mixture for which $x_A = 0{\cdot}25$ and $x_B = 0{\cdot}75$ is the same as the activity coefficient of A in a mixture for which $x_A = 0{\cdot}75$ and $x_B = 0{\cdot}25$. Quite commonly, however, this is not found experimentally to be the case, i.e., the mixture is unsymmetrical.

It will be recalled that one of the assumptions, upon which the treatment of Section 7.3 is based, is that the molecules of the two components of the mixture are similar in size, and various arguments show that asymmetry occurs when this assumption is incorrect. G. Scatchard showed that a better description of the properties of unsymmetrical mixtures can be obtained by replacing the mole fractions x_A and x_B in Equations 7.19, 7.20, etc., by the *volume fractions*:

$$\phi_A = \frac{n_A \tilde{V}_A^\circ}{n_A \tilde{V}_A^\circ + n_B \tilde{V}_B^\circ} \quad \text{and} \quad \phi_B = \frac{n_B \tilde{V}_B^\circ}{n_B \tilde{V}_B^\circ + n_A \tilde{V}_A^\circ} \qquad (7.32)$$

where $\tilde{V}_A^\circ$ and $\tilde{V}_B^\circ$ are the molar volumes of A and B. ϕ_A and ϕ_B are only equal to the corresponding mole fractions when these molar volumes are identical. The entropy of mixing is, for example:

$$\Delta S_{mix} = -R(n_A \ln \phi_A + n_B \ln \phi_B)$$

Rigorous theoretical proof of these results has not been obtained for the general case. In the following section we shall, however, derive them for an extreme case of asymmetry, namely, polymer solutions.

7.5 Polymer solutions: the Flory–Huggins theory

The theory to be described was developed independently by

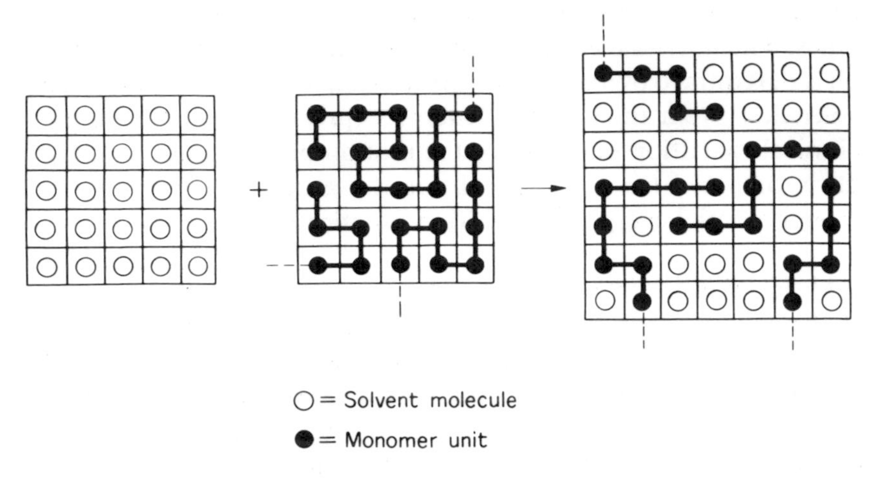

Fig. 7.5 Lattice model of a polymer solution

P.J. Flory and M.L. Huggins. It is based upon the lattice model discussed above; an important difference from the latter is, however, that it is assumed that each *monomer unit*, of which the polymer molecules are composed, occupies one lattice site, so that a *polymer molecule as a whole* occupies *many* (i.e., 10^3–10^4) lattice sites (see Fig. 7.5). It is, moreover, assumed that the polymer molecules are completely flexible, so that the polymer molecules occupy the available lattice sites in a purely random way.

Suppose that there N_2 polymer molecules. If each one of these consists of M monomer units, it will occupy M lattice sites. Therefore, the total number of sites in the lattice of the pure polymer is MN_2. If there are N_1 solvent molecules, the number of lattice sites in the solution is, correspondingly:

$$N_1 + MN_2 = N, \text{ say.}$$

The entropy of mixing is calculated, as in Section 7.1, by evaluating the respective numbers of complexions of the solvent, the solute, and the solution. As in Section 7.1 (compare with Equation 7.3), the number of complexions of the solution is:

$$\Omega_{\text{solution}} = \Omega_{D(1)} \times \Omega_{D(2)} \times \Omega_s$$

where $\Omega_{D(1)}$ and $\Omega_{D(2)}$ are, respectively, the numbers of complexions corresponding to the most probable distribution of solvent and polymer molecules among the energy levels, and Ω_s is the number of ways in which the polymer molecules can occupy the lattice sites of the solution. For the solvent:

$$\Omega_{\text{solvent}} = \Omega_{D(1)}$$

and for the pure polymer:

$$\Omega_{\text{polymer}} = \Omega_{D(2)} \times \Omega_{\text{p}}$$

The extra factor Ω_{p} in the last equation is the number of ways in which the polymer molecules can occupy the lattice sites of the pure polymer.

The entropy of mixing is then:

$$\begin{aligned} \Delta S_{\text{mix}} &= k \ln \Omega_{\text{solution}} - k \ln \Omega_{\text{solvent}} - k \ln \Omega_{\text{polymer}} \\ &= k \ln \Omega_{\text{s}} - k \ln \Omega_{\text{p}} \\ &= k \ln \frac{\Omega_{\text{s}}}{\Omega_{\text{p}}} \end{aligned} \tag{7.33}$$

The ratio $\Omega_{\text{s}}/\Omega_{\text{p}}$ is calculated by considering the numbers of ways in which the polymer molecules can be added one by one to the (initially empty) lattices of the solution and the pure polymer, respectively.

Suppose that, at some stage, r molecules have been added to both lattices. The number of occupied sites is, in both cases, rM.

The number of sites remaining unfilled in the *solution lattice* is now:

$$N - rM$$

and the fractional number, t_r, unfilled is:

$$t_r = \frac{N - rM}{N}. \tag{7.34}$$

Similarly, the fractional number, t_r', of sites remaining unfilled in the polymer lattice is:

$$t_r' = \frac{N_2M - rM}{N_2M} \tag{7.35}$$

Consider, now, the numbers of ways in which the next molecule can be added to the two lattices. Suppose that it is added unit by unit. The first monomer unit can be placed in any one of the vacant $(N - rM)$ sites in the solution, and $(N_2M - rM)$ sites in the polymer lattice. Thus, the numbers of places available for this unit are in the ratio:

$$\frac{\text{places in solution}}{\text{places in polymer}} = \frac{N - rM}{N_2M - rM} = \frac{Nt_r}{N_2Mt_r'}$$

The next monomer unit must go into one of the lattice sites which are adjacent to the first. Since only the fractions t_r or t_r' of these are not already occupied, the numbers of places available for the second unit are in the ratio:

$$\frac{\text{places in solution}}{\text{places in polymer}} = \frac{t_r}{t_r'}$$

The same is true for the third and subsequent monomer units, i.e., for each of these:

$$\frac{\text{places in solution}}{\text{places in polymer}} = \frac{t_r}{t_r'}$$

Finally, then, the numbers of places available to the $(r + 1)$-th molecule in the two lattices are in the ratio:

$$\frac{\text{places in solution}}{\text{places in polymer}} = \underset{\text{1st}}{\left(\frac{Nt_r}{N_2Mt_r'}\right)} \underset{\text{2nd}}{\left(\frac{t_r}{t_r'}\right)} \underset{\text{3rd}}{\left(\frac{t_r}{t_r'}\right)} \cdots \underset{M\text{-th}}{\left(\frac{t_r}{t_r'}\right)}$$

$$= \frac{N}{N_2M}\left(\frac{t_r}{t_r'}\right)^M$$

There is one term like this for each of the N_2 molecules, and Ω_s/Ω_p is the product of all such terms, i.e.:

$$\frac{\Omega_s}{\Omega_p} = \prod_{r=0}^{N_2-1}\left(\frac{N}{N_2M}\right)\left(\frac{t_r}{t_r'}\right)^M \tag{7.36}$$

It is a matter of elementary algebra (see below) to substitute Equation 7.36 into 7.33 to obtain the following result:

$$\Delta S_{\text{mix}} = k \ln \frac{\Omega_s}{\Omega_p}$$

$$= -k\left[N_1 \ln\left(\frac{N_1}{N_1 + MN_2}\right) + N_2 \ln\left(\frac{MN_2}{N_1 + MN_2}\right)\right] \tag{7.37}$$

Now:

$$\frac{N_1}{N_1 + MN_2} = \text{fraction of sites occupied by solvent molecules,}$$

$$\frac{MN_2}{N_1 + MN_2} = \text{fraction of sites occupied by polymer molecules.}$$

If, however, the volume of a substance is proportional to the number of lattice sites, these *site fractions* are the same as the volume fractions ϕ_1 and ϕ_2, defined in Equations 7.32. Furthermore, $N_1\boldsymbol{k} = n_1R$ and $N_2\boldsymbol{k} = n_2R$. Equation 7.37 thus simplifies to:

$$\Delta S_{\text{mix}} = -R(n_1 \ln\phi_1 + n_2 \ln\phi_2) \qquad (7.38)$$

The energy of mixing, ΔE_{mix}, of the polymer solution is derived by the same argument which led to Equation 7.18. Since, however, each of the N_2 molecules occupies M lattice sites, Equation 7.18a is modified to:

$$\Delta E_{\text{mix}} = \frac{N_1(MN_2)wz}{N} \qquad (7.39\text{a})$$

with $N = N_1 + MN_2$. Therefore, in terms of the site, or volume, fractions:

$$\Delta E_{\text{mix}} = N\phi_1\phi_2 wz \qquad (7.39\text{b})$$

and in terms of the numbers of moles present:

$$\Delta E_{\text{mix}} = \tilde{N}M\left(\frac{n_1 n_2}{n_1 + Mn_2}\right) wz \qquad (7.39\text{c})$$

The free energy of mixing is thus:

$$\Delta G_{\text{mix}} = RT(n_1 \ln\phi_1 + n_2 \ln\phi_2) + \tilde{N}M\left(\frac{n_1 n_2}{n_1 + Mn_2}\right) wz \qquad (7.40)$$

Hence the activity of the solvent (which is frequently the measurable quantity) is then (cf. page 128):

$$\ln a_1 = \frac{\mu_1 - \mu_1^\circ}{RT} = \left(\frac{\partial \Delta G_{\text{mix}}}{\partial n_1}\right)_{T,P,n_2}$$

$$= \ln(1 - \phi_2) + \left(1 - \frac{1}{M}\right)\phi_2 + \frac{\tilde{N}zw}{RT}\phi_2^2 \qquad (7.41)$$

The partial molar entropy of the solvent in the solution is defined in an analogous way to the chemical potential (p. 127):

$$\bar{S}_1 = \left(\frac{\partial S_{\text{m}}}{\partial n_1}\right)_{T,P,n_2}$$

and so

$$\bar{S}_1 = \left(\frac{\partial \Delta S_{\text{mix}}}{\partial n_1}\right)_{T,P,n_2} + \left(\frac{\partial S_1^\circ}{\partial n_1}\right)_{T,P}$$

$$= \left(\frac{\partial \Delta S_{\text{mix}}}{\partial n_1}\right)_{T,P,n_2} + \tilde{S}_1^\circ$$

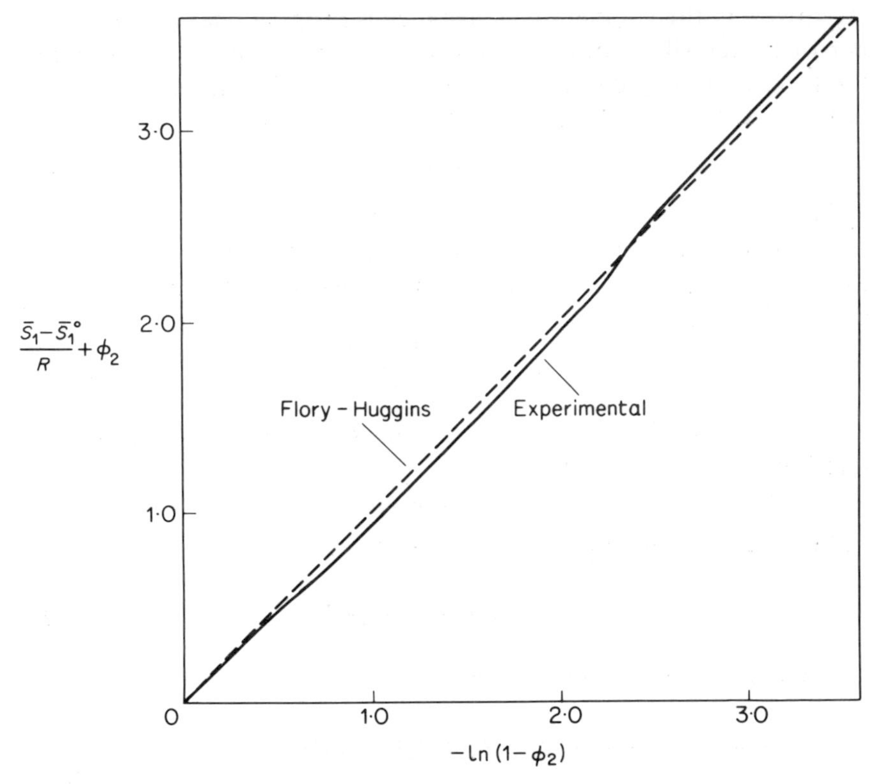

Fig. 7.6 Test of the Flory–Huggins theory for the benzene-rubber system. Experimental curve calculated from the results of G. Gee and L.R.G. Treloar, *Trans. Faraday Society*, 1942, *38*, 147, by assuming (as is almost the case) that the densities of rubber and benzene are the same.

where S_1° is the entropy *per mole* of pure solvent. Therefore,

$$\bar{S}_1 - \tilde{S}_1^\circ = -R\left(\ln(1-\phi_2) + \left(1 - \frac{1}{M}\right)\phi_2\right)$$

$$\approx -R(\ln(1-\phi_2) + \phi_2) \qquad (7.42)$$

if M is large.

Equations 7.38–7.42 are basic equations in the currently important field of polymer solution thermodynamics. Their validity has been extensively investigated by experiment, and it has been shown that they reproduce qualitatively, and sometimes quantitatively, the behaviour of actual solutions [see P.J. Flory, *Principles of Polymer Chemistry*, Cornell University Press (1953), Ch. 12]. Classical work in this field was the investigation of solutions of rubber in benzene by G. Gee and L.R.G. Treloar. Fig. 7.6 compares values of $(\bar{S}_1 - \tilde{S}_1^\circ)$ obtained from the experimental results of Gee and Treloar with values calculated from Equation 7.42.

Proof of Equation 7.37

Let $y_1 = N_1/M$ and $y_2 = N_2$. Then Equation 7.36 becomes:

$$\frac{\Omega_s}{\Omega_p} = \prod_{r=0}^{N_2-1} \frac{y_1+y_2}{y_2}\left(\frac{y_1+y_2-r}{y_1+y_2}\,\frac{y_2}{y_2-r}\right)^M$$

$$= \prod_{r=0}^{N_2-1}\left[\left(\frac{y_2}{y_1+y_2}\right)^{M-1}\left(\frac{y_1+y_2-r}{y_2-r}\right)^M\right]$$

$$= \left(\frac{y_2}{y_1+y_2}\right)^{N_2(M-1)}\left[\frac{\prod_{r=0}^{N_2-1}(y_1+y_2-r)}{\prod_{r=0}^{N_2-1}(y_2-r)}\right]^M$$

$$= \left(\frac{y_2}{y_1+y_2}\right)^{N_2(M-1)}\left[\frac{(y_1+y_2)!}{y_2!\,y_1!}\right]^M$$

Taking logarithms, and applying Stirling's theorem:

$$\ln\frac{\Omega_s}{\Omega_p} = (M-1)N_2\ln\left(\frac{y_2}{y_1+y_2}\right) - M\left[y_1\ln\left(\frac{y_1}{y_1+y_2}\right) + y_2\ln\left(\frac{y_2}{y_1+y_2}\right)\right]$$

$$= My_1\ln\left(\frac{y_1}{y_1+y_2}\right) - My_2\ln\left(\frac{y_2}{y_1+y_2}\right),$$

from which Equation 7.37 follows immediately.

Problems

7.1 Assuming that the mixtures are ideal, calculate the entropy of mixing at a constant pressure of 1 atmosphere of (a) 0·1 l nitrogen and 0·2 l oxygen, (b) 0·1 mol benzene and 0·2 mol toluene.

7.2 For the mixing of 0·5 mol carbon tetrachloride with 0·5 mol benzene at 298 K, the value of ΔE_{mix} is 81·5 J.
(a) Using Equation 7.18, calculate ΔE_{mix} for the mixing of 0·25 mol CCl_4 and 0·6 mol C_6H_6.
(b) Assuming that $z = 12$, calculate the parameter w, and compare its value with that of kT at 298 K. How is this result related to the randomness of the mixing?

7.3 For the mixing of 0·5 mol cyclohexane with 0·5 mol benzene at 293 K, the value of $\Delta S^{E}_{mix}/R$ is 0·21.
(a) From Equation 7.31, calculate $Nz\,\partial w/\partial T$ and, assuming that $z = 12$, $\partial w/\partial T$.
(b) Plot a graph showing the variation of ΔS^{E}_{mix} with composition for $n_A + n_B = 1$.
(c) Calculate the entropy of mixing of 0·25 mol cyclohexane and 0·6 mol benzene.

7.4 From the data in Table 7.1, plot a graph showing the variation of $\Delta G_{mix}/nRT$ (n is the total number of moles present) with composition, for carbon disulphide–acetone mixtures at 308 K.

7.5 Using the Gibbs–Helmholtz equation:

$$\Delta G = \Delta H + T\left(\frac{\partial \Delta G}{\partial T}\right)_P,$$

obtain an expression for the enthalpy of mixing of a regular solution, for the case in which $\partial w/\partial T$ is not zero. Use the data in Table 7.1 to calculate the enthalpy of mixing of 0·5 mol CCl_4 with 0·5 mol C_6H_6 at 298 K.

7.6 From thermodynamics and Equation 7.41, obtain an expression for the osmotic pressure of a polymer solution in terms of ϕ_2.

For a solution of 5 g natural rubber in 5 g benzene at 300 K estimate (a) the activity coefficient of benzene, (b) the osmotic pressure of the solution, and (c) $\bar{S}_1 - \tilde{S}_1^{\circ}$ for benzene. Assume that the densities of benzene and rubber are the same, and that $1/M$ and $\tilde{N}zw/RT$ are negligibly small.

CHAPTER EIGHT

Entropy

The introduction of the entropy function into classical thermodynamics is a consequence only of the second law, and the various thermodynamical equations in which S occurs are obtained without any reference to the physical meaning of entropy in terms of the molecular composition of matter. However, two equations have been derived by statistical mechanics.

Firstly (Equation 3.18):

$$\frac{S_2}{S_1} = k \ln \frac{\Omega_{D_0(2)}}{\Omega_{D_0(1)}} = k \ln \frac{\Omega_2}{\Omega_1}$$

or, equivalently:

$$S = k \ln \Omega_{D_0} = k \ln \Omega .$$

This was derived for systems of independent molecules, but, as we discuss in Chapter 14, it applies quite generally. It leads to the simple, qualitative interpretation of entropy in terms of the 'randomness' of the system (page 34); some applications were discussed in the preceding chapter.

Secondly (Equation 4.9):

$$S = Nk \ln\left(\frac{f\mathrm{e}}{N}\right) + \frac{E}{T}$$

which permits the calculation of the entropy of a gas from the molecular structure; it is this which we now consider.

8.1 Heat capacities of crystals at very low temperatures

Heat capacity measurement is an important part of the experimental determination of entropy, and it will be useful to begin the present discussion by considering the heat capacities of crystals at very low temperatures.

In Sections 6.3 and 6.4, the assumption was made that the only important contribution to the heat capacity of a monatomic crystal, at the temperatures under consideration, arises from the vibration of the atoms about their equilibrium positions, i.e., from the *lattice vibrations*. Both the Einstein and the Debye theories lead to the conclusion that, at temperatures below about 5 K, this vibrational heat capacity has negligible values. Consequently, *if it is true that only lattice vibrations contribute to C_V at such temperatures*, then the heat capacity of a crystal should be negligibly small at all temperatures below 5 K.

Actually, the heat capacities may have large values (of order R) at extremely low temperatures. Fig. 8.1 shows the heat capacities of chromium methylamine alum (a paramagnetic salt) plotted against the *logarithm* of temperature. At temperatures below about 1K, the heat capacity rises from the Debye value ($\sim 10^{-5} R$) to about R.

The explanation for the effect is roughly as follows. We have previously supposed the lowest energy level of an atom to be

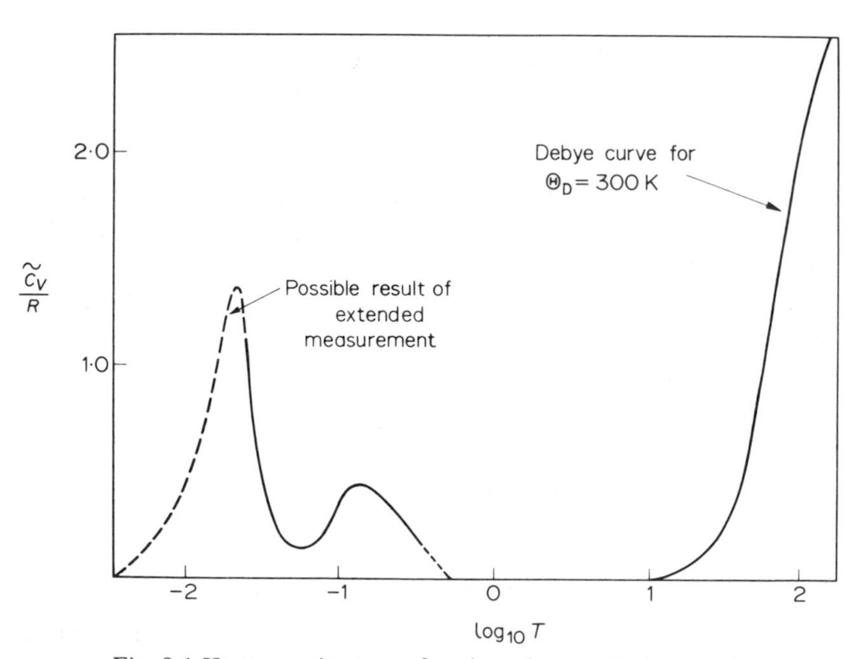

Fig. 8.1 Heat capacity curve for chromium methylamine alum

g_0-fold degenerate; in fact, magnetic and electrical forces in the crystal cause the g_0 quantum states to have slightly different energies. For example, suppose that $g_{el(0)} = 2$ and $g_{nuc(0)} = 1$. Then the relative energies of the two quantum states might be as follows:

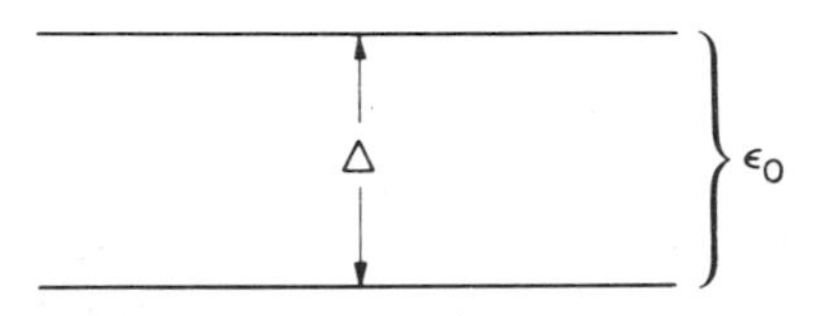

where Δ is the small splitting of the energy level. The only energies contributing significantly to the partition function at low temperatures are those with $\epsilon \approx \epsilon_0$. Thus:

$$f \approx 1 + e^{-\Delta/kT} + \ldots$$

at low temperatures. It is easily shown (problem 6.5) that the heat capacity is now:

$$\frac{\tilde{C}_V}{R} = \left(\frac{\Delta}{kT} \frac{1}{e^{\Delta/2kT} + e^{-\Delta/2kT}}\right)^2$$

and that this function has significant values only when kT is of the same order of magnitude as Δ. Thus, the rise in the C_V values at low temperatures occurs because, *at these temperatures, kT is of the same order of magnitude as the splitting of the ground-state energies.*

In a similar way, if $g_{nuc(0)} \neq 1$, the $g_{nuc(0)}$ spin states of the nucleus presumably have (because of magnetic and other interactions) very slightly different energies, and one would expect that *if* the experimentally determined curve in Fig. 8.1 could be extended down to, say, 10^{-5} K ($\log T = -5$), there would be a similar rise in the C_V values at these temperatures.

It is quite possible that there are other, as yet undiscovered, degeneracies of the lowest energy level. Such degeneracies would, of course, cancel in the calculation of the properties of the substance at ordinary temperatures, and need not be included in f. If, however, they are split by amounts of order δ, *however small*, then one would expect the value of C_V to rise as the temperature approached the value δ/k.

8.2 Calorimetric entropy

The equation:

$$\mathrm{d}S = \frac{\mathrm{d}Q}{T}$$

defining the entropy function, relates the *increase* in entropy of a system to the heat $\mathrm{d}Q$ absorbed reversibly by the system. Correspondingly, only *changes* in entropy, and not absolute values, can be measured experimentally (Lewis and Randall, Ch. 11).

If the temperature of a substance at constant pressure is raised from T_1 to T_2, and no phase change in the substance occurs between these temperatures, then the increase in entropy of the substance is:

$$\Delta S = \int_{T_1}^{T_2} \frac{C_P \mathrm{d}T}{T} \equiv \int_{\ln T_1}^{\ln T_2} C_P \mathrm{d}\ln T$$

and this can be evaluated as the area under the C_P/T *versus* T (or C_P *versus* $\ln T$) curve between $T = T_1$ and $T = T_2$.

If a phase change does occur in the substance at some temperature T_r intermediate between T_1 and T_2, there is an additional entropy increase, equal to $\Delta H_r/T_r$ where ΔH_r is the enthalpy of the phase change. The total increase in entropy of the substance, associated with the temperature change from T_1 to T_2 is then:

$$\Delta S = \int_{T_1}^{T_2} \frac{C_P \mathrm{d}T}{T} + \sum_r \frac{\Delta H_r}{T_r}$$

where the sum is over all phase changes which can occur between T_1 and T_2. If $T_1 = 0\,\mathrm{K}$, then ΔS is the difference $S_{T_2\mathrm{K}} - S_{0\mathrm{K}}$, i.e., it is the entropy at T_2 relative to the entropy at 0K. Since it is experimentally difficult to make thermal measurements below about 12K, let us suppose that T_l is the lowest temperature at which C_P has been determined for the substance. Then $S_{T_2\mathrm{K}} - S_{0\mathrm{K}}$ can be expressed as:

$$S_{T_2\mathrm{K}} - S_{0\mathrm{K}} = \int_0^{T_l} \frac{C_P \mathrm{d}T}{T} + \left\{ \int_{T_l}^{T_2} \frac{C_P \mathrm{d}T}{T} + \sum_r \frac{\Delta H_r}{T_r} \right\}$$

The quantity in brackets is measured experimentally. *If, for $T < T_l$, there were no contributions to C_P other than that from the lattice vibrations*, the value of the first term on the right-hand side could be derived by extrapolating the C_P *versus* $\ln T$ curve to 0K.

This is most conveniently done by using the Debye equation (Section 6.4) for C_V. The difference between C_P and C_V is negligible, and, from Equation 6.24, for low temperatures, $C_P = cT^3$, where c is a constant. Hence:

$$\int_0^{T_l} \frac{C_P \mathrm{d}T}{T} = \int_0^{T_l} \frac{cT^3 \mathrm{d}T}{T} = \frac{1}{3}\, cT_l^3 = \frac{(C_P)_l}{3}$$

where $(C_P)_l$ is the value of C_P at T_l.

The resulting estimate of $S_{T_2\,\mathrm{K}} - S_{0\,\mathrm{K}}$ is called the *calorimetric entropy* of the substance, because it is obtained experimentally from calorimetric measurements. One thus *defines* the calorimetric entropy of a substance as:

$$S_{\mathrm{cal}} = \underbrace{\left[\int_0^{T_l} \frac{C_P \mathrm{d}T}{T}\right]}_{\text{Estimated}} + \int_{T_l}^{T} \frac{C_P \mathrm{d}T}{T} + \sum_r \frac{\Delta H_r}{r} \tag{8.1}$$

S_{cal} is, then, the difference between the entropies at T_2 K and 0 K *as estimated by extrapolating the* C_P *versus* T *curve from* T_l *to* 0 K. Clearly, such an extrapolation neglects any rise in C_P which, as we saw in Section 8.1, might occur at extremely low temperatures.

8.3 Spectroscopic entropy

Suppose that at the lowest temperature T_l of calorimetric measurements, g_0^c atomic or molecular quantum states of a substance can be regarded (cf. Section 8.1) as having the energy ϵ_0.

Now consider two states of the substance:

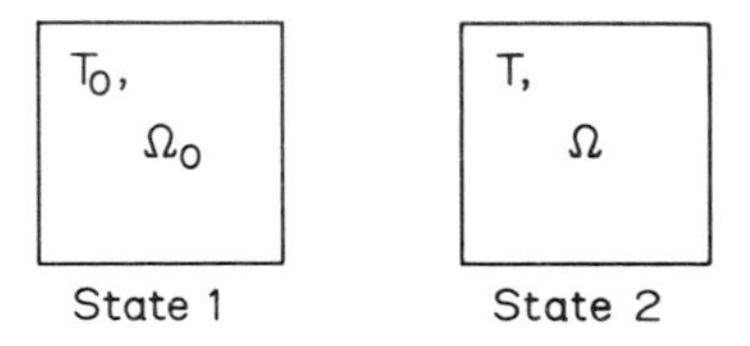

State 2 is the gaseous substance at TK. State 1 is the crystalline substance at a temperature T_0 *sufficiently close to* 0 K *so that only a negligible fraction of the molecules are in levels other than* ϵ_0, *but not low enough for the splitting of the* g_0^c *components of* ϵ_0 *to be important*. Let the numbers of complexions in state 1 be Ω_0 and in state 2, Ω.

According to our definition of calorimetric entropy, $S_{\mathrm{cal}} = S_{\mathrm{state\ 2}} - S_{\mathrm{state\ 1}}$:

$$k \ln \Omega - k \ln \Omega_0 = k \ln \frac{\Omega}{\Omega_0}$$

Our present aim is to calculate the ratio Ω/Ω_0, and therefore to obtain a theoretical value of S_{cal}.

Let the *theoretical estimates* of Ω and Ω_0 be, respectively, Ω' and Ω'_0. We shall obtain these estimates by making three assumptions concerning the substance.

Assumption 1: $\ln \Omega'$ *is evaluated by the methods of Chapters 4 and 5.*

Assumption 2: Ω'_0 *is the number of ways in which N molecules can be allocated to the* g_0^c *quantum states*, i.e.:

$$\Omega'_0 = (g_0^c)^N \tag{8.2}$$

(we shall find that, in a few cases, this underestimates the value of Ω_0).

The energy of a crystal at low temperatures can usually be expressed as the sum of nuclear, electronic, and vibrational energies, and we suppose that:

$$g_0^c = g_{nuc(0)}^c g_{el(0)}^c g_{vib(0)}^c \tag{8.3}$$

where the superscript c indicates that the terms apply to the crystal, not the gas (hydrogen is exceptional, in that there can also be molecular *rotation* at very low temperatures; see Section 8.4). The nuclear term is, of course, independent of whether the molecule is in the solid or the gas, i.e.:

$$g_{nuc(0)}^c = g_{nuc(0)} \tag{8.4a}$$

For all known cases, theory and experiment are consistent with the supposition that:

$$g_{vib(0)}^c = 1 \tag{8.4b}$$

and, usually:

$$g_{el(0)}^c = 1 \tag{8.4c}$$

Note that $g_{el(0)}^c$ need not equal the $g_{el(0)}$ for the gaseous molecule.

The reasons for Equation 8.4c are various.

(i) The lowest electronic energy level of most stable molecules and ions is non-degenerate. An exception is the O_2 molecule, for which $g_{el(0)} = 3$ (see Section 5.3). The value of S_{cal} for oxygen can only be interpreted, however, if it be assumed that, for molecules in the crystal, $g_{el(0)} = 1$, which presumably implies the coupling of electron spins between neighbouring molecules. A similar effect occurs in the NO crystal (cf. page 156).

(ii) In crystals of elements with covalent bonding, $g_{el(0)} = 1$, because, principally, all the electron spins are paired.

(iii) Crystals of metallic elements may be regarded as consisting of positive ions, e.g., Na^+, Mg^{++}, and free electrons. The free electrons do not contribute significantly to the entropy at low temperatures (see Section 13.4), and, if the metal ions have closed-shell ground states (^1S), $g^c_{el(0)} = 1$.

In view of the Equations 8.4, we make

Assumption 3: *In general*:

$$g^c_0 = g_{nuc(0)}$$

We shall, however, encounter a few cases for which this assumption is invalid.

The theoretical entropy, obtained on the basis of the three above assumptions, will be called the *spectroscopic entropy* S_{spec} of the substance, because its evaluation by the methods of Chapter 5 involves the use of spectroscopic data. Thus:

$$S_{spec} = k \ln \frac{\Omega'}{\Omega'_0} = k \ln \frac{\Omega'}{(g_{nuc(0)})^N} .$$

Now:

$$k \ln \Omega' = Nk \ln \left(f \frac{e}{N} \right) + \frac{E}{T} ,$$

for a gaseous substance, and therefore:

$$S_{spec} = Nk \ln \left(\frac{f \frac{e}{N}}{g_{nuc(0)}} \right) + \frac{E}{T} \tag{8.5}$$

The partition function f contains $g_{nuc(0)} f_0$ as a factor (see Section 5.1), and so:

$$S_{spec} = Nk \ln \left(f' \frac{e}{N} \right) + Nk \ln f_0 + \frac{E}{T} \tag{8.6}$$

where:

$$f' = \frac{f}{f_0 g_{nuc(0)}}$$

Since:

$$Nk \ln f_0 = Nk \left(-\frac{\epsilon_0}{kT} \right) = -\frac{N\epsilon_0}{T} = -\frac{E_0}{T}$$

where E_0 is the zero-point energy, Equation 8.6 simplifies to:

$$S_{\text{spec}} = Nk \ln\left(f' \frac{\mathrm{e}}{N}\right) + \frac{E - E_0}{T} \tag{8.7}$$

Equation 8.7 can be written in the alternative form:

$$\tilde{S}_{\text{spec}} = \tilde{S}_{\text{el}} + \tilde{S}_{\text{vib}} + \tilde{S}_{\text{rot}} + \tilde{S}_{\text{i.r.}} + \tilde{S}_{\text{trans}} \tag{8.8}$$

where:

$$\begin{aligned}
\tilde{S}_{\text{el}} &= R \ln f_{\text{el}} + \tilde{E}_{\text{el}}/T \\
\tilde{S}_{\text{vib}} &= R \ln f_{\text{vib}} + \tilde{E}_{\text{vib}}/T \\
\tilde{S}_{\text{rot}} &= R \ln f_{\text{rot}} + \tilde{E}_{\text{rot}}/T \qquad (8.9)\\
\tilde{S}_{\text{i.r.}} &= R \ln f_{\text{i.r.}} + \tilde{E}_{\text{i.r.}}/T \\
\tilde{S}_{\text{trans}} &= R \ln\left(f_{\text{trans}} \frac{\mathrm{e}}{\tilde{N}}\right) + \tilde{E}_{\text{trans}}/T
\end{aligned}$$

the values of $\tilde{E}_{\text{trans}}$, $\tilde{E}_{\text{rot}}$, etc. being found from the corresponding partition functions, as discussed in Section 5.7. Substitution for f_{trans}, f_{rot}, etc. in the Equations 8.9 yields explicit formulae for the various contributions to S_{spec}. We here give only the calculation of S_{trans} (all the formulae are listed in Chapter 10).

Since (cf. page 57):

$$f_{\text{trans}} = \frac{(2\pi mkT)^{3/2} V}{h^3}$$

the molar translational entropy is:

$$\tilde{S}_{\text{trans}} = R \ln\left[\frac{(2\pi mkT)^{3/2}}{h^3} \tilde{V} \frac{\mathrm{e}}{\tilde{N}}\right] + \frac{\tilde{E}_{\text{trans}}}{T}$$

Therefore, putting $\tilde{E}_{\text{trans}} = (3/2)RT$ and $\tilde{V} = \tilde{N}kT/P$:

$$\begin{aligned}
\frac{\tilde{S}_{\text{trans}}}{R} &= \ln\left[\frac{(2\pi m)^{3/2}(kT)^{5/2}\mathrm{e}}{h^3 P}\right] + \frac{3}{2} \\
&= \ln\left[\left(\frac{2\pi\tilde{M}}{\tilde{N}}\right)^{3/2} \frac{(kTe)^{5/2}}{h^2 P}\right] \qquad (8.10\text{a})
\end{aligned}$$

where $\tilde{M}$ is the mass per mole (note that $\frac{3}{2} = \frac{3}{2}\ln e = \ln e^{3/2}$). The units of P must, of course, be chosen so as to make the quantity inside the square brackets dimensionless. With this requirement understood, the logarithm can be expanded:

$$\frac{\tilde{S}_{\text{trans}}}{R} = \ln\left[\left(\frac{2\pi}{\tilde{N}}\right)^{3/2} \frac{(ke)^{5/2}}{h^2}\right] + \frac{3}{2}\ln\tilde{M} + \frac{5}{2}\ln T - \ln P \tag{8.10b}$$

A more convenient form of this equation is

$$\frac{\tilde{S}_{trans}}{R} = \frac{3}{2}\ln M + \frac{5}{2}\ln T - \ln\frac{P}{P^\circ} - 1{\cdot}165 \qquad (8.10c)$$

in which M is the molecular weight and P° the value of 1 atmosphere pressure expressed in the same units as P (P° = 1 atmosphere = 760 mm Hg = 101·3 kN m^{-2}, etc.). The three equations (8.10) are equivalent forms of a result known as the *Sackur–Tetrode equation* for translational entropy.

A possible contribution to the entropy, which we have not yet discussed, is the *entropy of isotope mixing*. For example, natural chlorine contains the molecules $^{35}Cl_2$, $^{37}Cl_2$, and $^{37}Cl^{35}Cl$, and it might be thought that there will be a corresponding entropy of mixing contribution to $\tilde{S}_{cal}$. As we have already emphasized, however, $\tilde{S}_{cal}$ is an entropy *difference*, not an absolute value. There is an entropy of mixing of the solid, given by Equation 7.4; but there is also an entropy of mixing of the gas at T, given by Equation 7.10, which is identical with 7.4. Thus, when the entropy *difference* $\tilde{S}_{cal}$ is evaluated, the mixing term always cancels. No mixing terms are included in the theoretical entropy $\tilde{S}_{spec}$.

8.4 Comparison of $\tilde{S}_{cal}$ and $\tilde{S}_{spec}$

In Table 8.1 are shown the results of applying Equations 8.1 and 8.9 to the determination of $\tilde{S}_{cal}$ and $\tilde{S}_{spec}$ for carbonyl sulphide.* In this case, the values of $\tilde{S}_{cal}$ and $\tilde{S}_{spec}$ agree within the limits of experimental error. A similar correspondence between $\tilde{S}_{cal}$ and $\tilde{S}_{spec}$ has been found for numerous other substances; typical results are shown in Table 8.2.

In a few cases, $\tilde{S}_{spec}$ is larger than $\tilde{S}_{cal}$ by an amount which is outside the limits of experimental error. Thus:

$$\tilde{S}_{spec} = \tilde{S}_{cal} + s \qquad (8.11)$$

where s is a positive number. Examples of this type are given in Table 8.3.

Two types of behaviour leading to non-zero values of s can be recognized:

* The entropy data discussed in this chapter are taken from K.K. Kelley and E.G. King, *U.S. Bureau of Mines Bulletin 592* (1961).

Table 8.1 Standard* Molar Entropy $\tilde{S}^\circ$ of Carbonyl Sulphide at 298·15 K

	ΔS ($J\,K^{-1}\,mol^{-1}$)
Calorimetric	
0–15 K Debye extrapolation of C_P	2·30
15–134·31 K graphical integration $C_P\,d\ln T$	62·59
Melting at 134·31 K	35·19
134·31–222·87 K graphical integration of $C_P\,d\ln T$	36·23
Vaporization at 222·87 K	83·05
†Correction to ideal gas state	·54
222·87–298·15 K integration of $C_P\,d\ln T$ for gas	11·34
$\tilde{S}^\circ_{cal} =$	231·24
Spectroscopic‡	
$\tilde{S}^\circ_{trans}$	159·83
$\tilde{S}_{rot}(I = 138 \times 10^{-47}\,kg\,m^2)$	65·94
$\tilde{S}_{vib}\ (\omega_1 = \omega_2 = 521{\cdot}5,\ \omega_3 = 859{\cdot}2,\ \omega_4 = 2050\,cm^{-1})$	5·77
$\tilde{S}^\circ_{spec} =$	231·54

* Experimental data were obtained for pressure = 1 atmosphere; correspondingly, $\tilde{S}_{trans}$ is calculated with $P = 1$ atmosphere.

† The correction for non-ideality can be made as follows:

$$\left(\frac{\partial \tilde{S}}{\partial P}\right)_T = -\left(\frac{\partial \tilde{V}}{\partial T}\right)_P = -\phi(\tilde{V}), \text{ say.}$$

The function $\phi(\tilde{V})$ is known from experimental $P-V-T$ data for the gas. For ideal gases, and all gases at low pressures, $P\tilde{V} = RT$, so that $\phi(\tilde{V}) = R/P$. Therefore, at any pressure P:

$$\tilde{S}_{ideal} - \tilde{S}_{actual} = \int_{P*}^{P} \left[\phi(\tilde{V}) - \frac{R}{P}\right] dP,$$

where $P*$ is some pressure low enough for the gas to behave ideally.

‡ Explicit formulae for these calculations are given in Chapter 10.

(i) *Frozen-in randomness of orientation in the crystal*

Carbon monoxide molecules are so nearly symmetrical that, when a carbon monoxide crystal ‘grows’ in the presence of the liquid, each additional molecule can condense with one of two equally probable orientations (represented by CO and OC). Consequently there is, in the crystal, a randomness between the two kinds of orientation, represented diagrammatically as follows:

Table 8.2 Comparison of $\tilde{S}_{spec}$ and $\tilde{S}_{cal}$ for ideal* gases at 1 atmosphere pressure and 298·15 K

Molecule	$\tilde{S}_{spec}$ ($J\,K^{-1}\,mol^{-1}$)	$\tilde{S}_{cal}$ ($J\,K^{-1}\,mol^{-1}$)	T (K)
O_2	205·1	205·4	11·8
N_2	191·5	192·2	10
Cl_2	223·0	223·0	15
HCl	186·8	186·1	16
A	154·7	154·6	10
CO_2	213·7	213·8	15

* See footnotes to Table 8.1.

Table 8.3 Comparison of $\tilde{S}_{spec}$ and $\tilde{S}_{cal}$ for ideal gases at 1 atmosphere pressure and 298·15 K (units of entropy are $J\,K^{-1}\,mol^{-1}$)

Molecule	$\tilde{S}_{spec}$	$\tilde{S}_{cal}$	T (K)	s observed	s calculated
H_2	130·6	124·2	14	6·4	6·85
D_2	144·9	141·8	12	3·1	3·05
CO	197·9	193·4	11·7	4·5	5·76
N_2O	219·9	215·2	14	4·7	5·76
NO	210·6	208·0	14·4	2·6	2·88
H_2O(g)	188·7	185·4	10	3·3	3·37

CO CO OC OC

CO OC CO CO

CO OC OC CO

OC CO CO OC

When the solid is cooled towards 0 K this randomness remains 'frozen in'.

Since each of the N lattice sites in the crystal can be occupied in either of two ways, the number of complexions of the *crystal* is increased, through this effect, by a factor 2^N.

Consider how this will affect the theoretical calculation of S_{cal} for the gas. As on page 150:

$$S_{cal} = k \ln \frac{\Omega}{\Omega_0}$$

where Ω and Ω_0 are actual numbers of complexions of the gas and the crystal, respectively. The theoretical estimates of these numbers are Ω' and Ω'_0. Now, the effect of randomness in the crystal was not allowed for in the calculation of Ω'_0, i.e., Ω'_0 *will be too low by the factor* 2^N. If this is the only error in the theoretical estimation of Ω and Ω_0, therefore:

$$\frac{\Omega'}{2^N \Omega'_0} = \frac{\Omega}{\Omega_0}$$

i.e.,

$$k \ln \frac{\Omega'}{\Omega'_0} - Nk \ln 2 = k \ln \frac{\Omega}{\Omega_0}$$

i.e.,

$$S_{\text{spec}} - Nk \ln 2 = S_{\text{cal}}$$

so that:

$$s = R \ln 2$$
$$= 5{\cdot}76 \text{ J K}^{-1} \text{ mol}^{-1}$$

which is of the same order as the observed difference:

$$\tilde{S}_{\text{spec}} - \tilde{S}_{\text{cal}} = 4{\cdot}60 \text{ J K}^{-1} \text{ mol}^{-1}$$

The above argument should apply to any *nearly* symmetrical *linear* molecule; nitrous oxide is another such example. The carbonyl sulphide molecule, although linear, is not sufficiently symmetrical to behave in this way, and no significant discrepancy is found between the $\tilde{S}_{\text{spec}}$ and $\tilde{S}_{\text{cal}}$ values in Table 8.1.

In the case of CH_3D, there are four nearly equivalent ways in which each molecule can be oriented. Hence:

$$\tilde{S}_{\text{spec}} - \tilde{S}_{\text{cal}} = k \ln 4^{\tilde{N}} = R \ln 4$$
$$= 11{\cdot}52 \text{ J K}^{-1} \text{ mol}^{-1}$$

The value of s found experimentally for CH_3D is $11{\cdot}6 \text{ J K}^{-1} \text{ mol}^{-1}$.

The nitric oxide crystal is composed of dimeric N_2O_2 molecules. As in the case of carbon monoxide, the similarity of the nitrogen and oxygen atoms permits randomness between the two kinds of orientation:

```
N—O              O—N
|  |     and     |  |
O—N              N—O
```

Since there are $\tilde{N}/2$ dimeric molecules per mole of NO:

$$\tilde{S}_{\text{spec}} - \tilde{S}_{\text{cal}} = k \ln 2^{\tilde{N}/2}$$
$$= \tfrac{R}{2} \ln 2 = 2{\cdot}88 \text{ J K}^{-1} \text{ mol}^{-1}$$

The ice crystal has a tetrahedral structure, arising from hydrogen bonding between adjacent molecules, and each hydrogen atom belonging to a given molecule can occupy one of four alternative positions. Pauling showed (Fowler and Guggenheim, p. 214) that the randomness with which these positions can be occupied leads to the value:

$$\tilde{S}_{\text{spec}} - \tilde{S}_{\text{cal}} = R \ln \tfrac{3}{2} = 3{\cdot}37\,\text{J K}^{-1}\,\text{mol}^{-1}$$

(ii) *Wrong value of* g_0^c *assumed for temperature* T_1

It was shown in Section 6.2 that ordinary hydrogen must be treated as a metastable mixture of two distinct species, i.e., as 3 : 1 mixture of orthohydrogen and parahydrogen. There is a corresponding entropy of mixing; however, as this contributes equally at all temperatures, *it does not affect the observed value of* S_{cal} (cf. the similar case of isotope mixing, discussed on page 153). No entropy of mixing is therefore included in the *theoretical* estimate of S_{cal}. The spectroscopic entropy of the mixture is thus just the sum of two terms:

$$\tilde{S}_{\text{spec}} = \tfrac{3}{4}\tilde{S}_{\text{spec}}^{\text{ortho}} + \tfrac{1}{4}\tilde{S}_{\text{spec}}^{\text{para}} \tag{8.12}$$

where $\tilde{S}_{\text{spec}}^{\text{ortho}}$ and $\tilde{S}_{\text{spec}}^{\text{para}}$ are, respectively, the molar spectroscopic entropies of pure ortho- and pure para-hydrogen. Each of these can be calculated by Equation 8.8, and this leads to the value of $\tilde{S}_{\text{spec}}$ given in the second column of Table 8.3. There is a discrepancy between the value so obtained and the calorimetric entropy, which we must now consider.

Hydrogen and its isotopes are exceptional in that their molecules, because of their small moments of inertia and nearly spherical external field of force, continue to rotate, even in the crystal at low temperatures. Consequently, instead of Equation 8.3, we have:

$$\begin{aligned} g_0^c &= g_{\text{nuc}(0)}^c\, g_{\text{el}(0)}^c\, g_{\text{vib}(0)}^c\, g_{\text{rot}(0)}^c \\ &= g_{\text{nuc}(0)}\, g_{\text{rot}(0)}^c \end{aligned}$$

For an isolated orthohydrogen molecule:

$$g_{\text{rot}(0)}^{\text{ortho}} = 2J + 1 = 3,$$

since the lowest value of the rotational quantum number J is 1, for *ortho* states. If the effects of intermolecular forces could be neglected, this would also be the degeneracy $g_{\text{rot}(0)}^{c\,(\text{ortho})}$ of the lowest rotational level of *ortho* molecules in the *crystal*. Actually, because of interactions between neighbouring molecules, the three rotational wave functions associated with the $J = 1$ level have, for the crystal, slightly different energies; but at 10 K, the splitting of the three components of the lowest level is negligible compared with kT, and to evaluate the partition function at this temperature one takes $g_{\mathrm{rot}(0)}^{\mathrm{c(ortho)}} = 3$. In this case:

$$g_0^{\mathrm{c(ortho)}} = 3g_{\mathrm{nuc}(0)}$$

instead of $g_{\mathrm{nuc}(0)}$, as was previously assumed. Therefore:

$$\Omega_0' = (g_{\mathrm{nuc}(0)})^N$$

underestimates Ω_0 by the factor 3^N, for N molecules of orthohydrogen.

The molar spectroscopic entropy of orthohydrogen, calculated from:

$$\tilde{S}_{\mathrm{spec}} = k \ln \frac{\Omega'}{(g_{\mathrm{nuc}(0)})^{\tilde{N}}}$$

is thus larger than the corresponding $\tilde{S}_{\mathrm{cal}}$ by an amount:

$$k \ln 3^{\tilde{N}} = R \ln 3$$

Hence the molar spectroscopic entropy of ordinary hydrogen calculated from Equation 8.12 is larger than $\tilde{S}_{\mathrm{cal}}$ by the amount:*

$$s = \tfrac{3}{4}R \ln 3 = 6{\cdot}85\ \mathrm{J\,K^{-1}\,mol^{-1}}.$$

Quite recently, heat capacity measurements on hydrogen have been extended down to about 1 K, and the value of C_V has been found to rise from the Debye value to approximately $7R$, at around 1·6 K (cf. Fig. 8.1). The area under this peak in the heat capacity *versus* $\ln T$ curve corresponds roughly to an entropy $\frac{3}{4}R \ln 3$, and it arises because the energy separation of the three components of the $J = 1$ rotational level is of the order of kT at 1·6 K. Below 1 K, all the molecules can be regarded as occupying the lowest of the three sub-levels, and it is then correct to take:

$$g_{\mathrm{rot}(0)}^{\mathrm{c(ortho)}} = 1 \quad \text{and} \quad g_0^{\mathrm{c}} = g_{\mathrm{nuc}(0)}.$$

To conclude: there are two equivalent ways of viewing the discrepancy between the values of $\tilde{S}_{\mathrm{spec}}$ and $\tilde{S}_{\mathrm{cal}}$ for hydrogen. These are:

* Note that there is no corresponding effect for parahydrogen molecules, because the lowest ($J = 0$) level is non-degenerate.

(i) $\tilde{S}_{\text{cal}}$ is evaluated by extrapolation from $T_1 = 10\,\text{K}$. For *ortho*-hydrogen at this temperature, $g_0^c = 3g_{\text{nuc}(0)}$, not $g_{\text{nuc}(0)}$, as assumed in the calculation of $\tilde{S}_{\text{spec}}$.

(ii) $\tilde{S}_{\text{spec}}$ is calculated on the assumption that $g_0^c = g_{\text{nuc}(0)}$; but this value of g_0^c can only be assumed if $T_1 \approx 1\,\text{K}$, not $10\,\text{K}$, is used in the extrapolation to obtain $\tilde{S}_{\text{cal}}$.

An essentially similar discussion is applicable to chromium methylamine alum, mentioned in Section 8.1, and to other paramagnetic salts such as $FeCl_2$, $CoCl_2$, $NiCl_2$, and $MnCl_2$. In these cases it is, of course, the electronic degeneracy $g_{\text{el}(0)}^c$, not the rotational degeneracy, which has to be considered. The general principle is, however, the same, namely, that g_0^c will equal $g_{\text{nuc}(0)}$ only at very low temperatures inaccessible to normal calorimetric measurements.

8.5 The third law of thermodynamics

(i) *The Nernst Heat Theorem*

Consider the reaction:

$$\text{A} + \text{B} \longrightarrow \text{AB}. \tag{8.13}$$

The associated entropy change is temperature dependent. Let its value at some temperature T be ΔS, and its value at the temperature T_0 (defined on page 149) be ΔS_0.

To investigate the relationship between ΔS and ΔS_0, consider two paths by means of which A and B, initially both at T_0 can be converted into AB at temperature T:

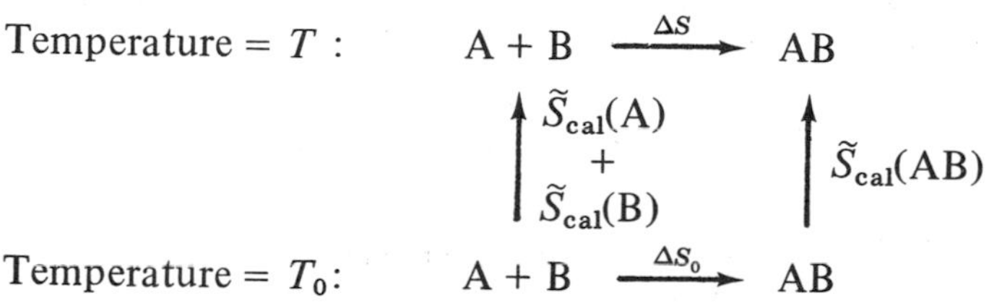

$$\begin{array}{lccc}
\text{Temperature} = T: & \text{A} + \text{B} & \xrightarrow{\Delta S} & \text{AB} \\
 & \uparrow \; \tilde{S}_{\text{cal}}(\text{A}) + \tilde{S}_{\text{cal}}(\text{B}) & & \uparrow \; \tilde{S}_{\text{cal}}(\text{AB}) \\
\text{Temperature} = T_0: & \text{A} + \text{B} & \xrightarrow{\Delta S_0} & \text{AB}
\end{array}$$

The first process consists of heating A and B from T_0 and T, and allowing them to react at T; in the second, A and B react at T_0, and the product AB is heated to T. The overall entropy change is the same for both paths, and therefore:

$$\tilde{S}_{\text{cal}}(\text{A}) + \tilde{S}_{\text{cal}}(\text{B}) + \Delta S = \Delta S_0 + \tilde{S}_{\text{cal}}(\text{AB})$$

where $\tilde{S}_{\text{cal}}(\text{X})$ is the molar calorimetric entropy of X. Thus:

$$\Delta S = \tilde{S}_{\text{cal}}(\text{AB}) - \tilde{S}_{\text{cal}}(\text{A}) - \tilde{S}_{\text{cal}}(\text{B}) + \Delta S_0 \tag{8.14}$$

There are two ways in which ΔS_0 can be obtained experimentally:

(a) by measuring ΔS, $\tilde{S}_{\text{cal}}(\text{AB})$, $\tilde{S}_{\text{cal}}(\text{A})$, and $\tilde{S}_{\text{cal}}(\text{B})$, and substituting these values into Equation 8.14;

(b) by extrapolating values of ΔS, observed at a series of temperatures, down to T_0.

Note that neither method strictly gives ΔS_0; rather, both give the value of ΔS *extrapolated to* T_0, i.e., one makes the identification:

$$\Delta S_0 = \underset{T \to T_0}{\text{Lt}} \Delta S$$

It is found experimentally that for many chemical reactions:

$$\Delta S_0 = 0 \tag{8.15}$$

The value of $\underset{T \to T_0}{\text{Lt}} \Delta S$ has also been found to be zero for other types of process. Examples are:

(a) *The transition between solid and liquid helium*, which, by a suitable change of pressure, can be carried out at temperatures far below 1K. The transition temperature T is related to the pressure P by:

$$\frac{\mathrm{d}P}{\mathrm{d}T} = \frac{\Delta S}{\Delta V}$$

It is found experimentally that, when $T \longrightarrow T_0$, $\mathrm{d}P/\mathrm{d}T \longrightarrow 0$ and therefore $\Delta S \longrightarrow 0$.

(b) *Coefficients of expansion*. The effect of a small increase in pressure on the entropy of a stable phase at a constant temperature is:

$$\delta S = \left(\frac{\partial S}{\partial P}\right)_T \delta P$$

Now:

$$\left(\frac{\partial S}{\partial P}\right)_T = \left(\frac{\partial V}{\partial T}\right)_P = \alpha, \text{ the coefficient of expansion.}$$

It has been found experimentally in a number of cases (e.g., for Cu, Ag, Al) that, as $T \longrightarrow T_0$, $\alpha \longrightarrow 0$, which therefore implies that $\delta S \longrightarrow 0$.

The generalization that, *for any process*:

$$\Delta S_0 = 0$$

constitutes the *Nernst heat theorem*. Actually, as we see below, this simple statement is too broad a generalization. Modern formulations of the third law of thermodynamics consist, however, of the statement of the Nernst heat theorem together with specification of the restrictions on its applicability.

(ii) *The Entropy of Chemical Reactions*

The Nernst heat theorem, if it were true, would be of great importance in chemistry, because Equation 8.14 would then simplify to:

$$\Delta S = \tilde{S}_{cal}(AB) - \tilde{S}_{cal}(A) - \tilde{S}_{cal}(B) \qquad (8.16)$$

i.e., the entropy change of the chemical reaction *could be deduced entirely from the calorimetrically determined* quantities $\tilde{S}_{cal}$. Similarly, for the general chemical reaction:

$$aA + bB + \ldots \longrightarrow lL + mM + \ldots \qquad (8.17)$$

one would obtain:

$$\Delta S = [l\tilde{S}_{cal}(L) + m\tilde{S}_{cal}(M) + \ldots] - [a\tilde{S}_{cal}(A) + b\tilde{S}_{cal}(B) + \ldots] \qquad (8.18)$$

Unfortunately, although Equations 8.16 and 8.18 are true for *many* chemical reactions, they do not hold for *all*. Statistical mechanics now provides the answer to the question: *under what circumstances is the Nernst heat theorem applicable?*

Consider the reaction 8.13. It is possible to envisage a process by which the (initially separated) reactants are transformed reversibly into products. Thus, state I (A + B) and state II (AB) are really two states of the same thermodynamic system. Corresponding to each state is a value of Ω, and *the change in entropy on reaction* is:

$$\Delta S = k \ln \frac{\Omega_{II}}{\Omega_{I}}$$

Since Ω_{II} is the number of complexions of the product AB at the temperature and pressure under consideration, we shall denote this number by $\Omega(AB)$. Correspondingly, Ω_I, the number of complexions of the reactants, is:

$$\Omega_I = \Omega(A) \times \Omega(B)$$

since each of the $\Omega(A)$ complexions of A can be taken with each of the complexions of B. (A and B are, in the initial state, unmixed, so there is no mixing factor of the type 7.2.)

Hence:

$$\Delta S = k \ln \frac{\Omega(AB)}{\Omega(A)\Omega(B)}$$

The limiting value of this entropy change at $T \to T_0$ is:

$$\Delta S_0 = k \ln \frac{\Omega_0(AB)}{\Omega_0(A)\Omega_0(B)} \,. \tag{8.19}$$

To investigate whether or not ΔS_0 is zero, we seek a theoretical value by means of statistical mechanics, i.e., we obtain the quantity:

$$k \ln \frac{\Omega_0'(AB)}{\Omega_0'(A)\Omega_0'(B)}$$

According to Section 8.3:

$$\frac{\Omega_0'(AB)}{\Omega_0'(A)\Omega_0'(B)} = \frac{[g_0^c(AB)]^{\tilde{N}}}{[g_0^c(A)]^{\tilde{N}}[g_0^c(B)]^{\tilde{N}}}$$

in which $g_0^c(AB)$ is the value of g_0^c for AB, and so on. The right-hand expression can be simplified in two ways. Firstly, in accordance with assumption 3 (p. 151), g_0^c is replaced by $g_{nuc(0)}$. Secondly, in the molecule AB, the nucleus A can be in any one of $g_{nuc(0)}(A)$ states, and nucleus B in any one of $g_{nuc(0)}(B)$ states, so that $g_{nuc(0)}(AB) = g_{nuc(0)}(A) \times g_{nuc(0)}(B)$. Consequently,

$$\frac{\Omega_0'(AB)}{\Omega_0'(A)\Omega_0'(B)} = \frac{[g_{nuc(0)}(A)\, g_{nuc(0)}(B)]^{\tilde{N}}}{[g_{nuc(0)}(A)]^{\tilde{N}}\, [g_{nuc(0)}(B)]^{\tilde{N}}} = 1 \tag{8.20}$$

so that:

$$k \ln \frac{\Omega_0'(AB)}{\Omega_0'(A)\Omega_0'(B)} = 0 \,,$$

from which Nernst's theorem follows directly. The extension of the argument to the general reaction 8.18 is obvious.

It was, however, shown in Section 8.4 that there are a few cases for which the above method of calculating Ω_0 needs to be modified. There are, correspondingly, cases for which Nernst's theorem does not hold. For example, ΔS_0 is *not* zero for the reaction:

$$2CO + O_2 \longrightarrow 2CO_2 \tag{8.21}$$

It is, however, easy to express ΔS_0 in terms of the parameters s defined by Equation 8.11. Thus:

$$s = \tilde{S}_{\text{spec}} - \tilde{S}_{\text{cal}}$$
$$= k \ln \frac{\Omega'}{\Omega'_0} - k \ln \frac{\Omega}{\Omega_0}$$
$$= k \ln \frac{\Omega' \Omega_0}{\Omega \Omega'_0}$$

In all the cases discussed in Section 8.4, this simplifies to:

$$s = k \ln \frac{\Omega_0}{\Omega'_0}$$

i.e., the discrepancy between $\tilde{S}_{\text{spec}}$ and $\tilde{S}_{\text{cal}}$ arises from the incorrect estimation of Ω_0 rather than of Ω. Hence, denoting the value of s for AB by $s(\text{AB})$, and so on, we find:

$$s(\text{AB}) - s(\text{A}) - s(\text{B}) = k \ln \frac{\Omega_0(\text{AB})}{\Omega'_0(\text{AB})} - k \ln \frac{\Omega_0(\text{A})}{\Omega'_0(\text{A})} - k \ln \frac{\Omega_0(\text{B})}{\Omega'_0(\text{B})}$$
$$= k \ln \left(\frac{\Omega_0(\text{AB})}{\Omega_0(\text{A})\Omega_0(\text{B})} \frac{\Omega'_0(\text{A})\Omega'_0(\text{B})}{\Omega'_0(\text{AB})} \right)$$

Now:

$$\frac{\Omega'_0(\text{A})\Omega'_0(\text{B})}{\Omega'_0(\text{AB})} = 1, \text{ by Equation 8.20}$$

and

$$k \ln \frac{\Omega_0(\text{AB})}{\Omega_0(\text{A})\Omega_0(\text{B})} = \Delta S_0, \text{ by Equation 8.19}$$

Hence:

$$\Delta S_0 = s(\text{AB}) - s(\text{A}) - s(\text{B})\,. \tag{8.22}$$

For example, ΔS_0 for reaction 8.21 is $-2R \ln 2$.

(iii) *Statement of the Third Law of Thermodynamics*

Equation 8.22 can be generalized. For the reaction 8.17:

$$\Delta S_0 = \underset{T \to T_0}{\text{Lt}} \Delta S = [ls(\text{L}) + ms(\text{M}) + \ldots] - [as(\text{A}) + bs(\text{B}) + \ldots] \tag{8.23}$$

or, more concisely:

$$\underset{T \to T_0}{\text{Lt}} \Delta S = \sum_f s_f - \sum_i s_i \tag{8.24}$$

where the s_f and the s_i are the s-parameters of, respectively, the final state (products) and the initial state (reactants).

Equation 8.24 is, for practical purposes, a statement of the third law of thermodynamics. To complete the enunciation of this law,

we add that s for any substance is zero, *provided that*:
(a) extrapolation of calorimetric measurements is made from a suitably low temperature, and
(b) there is no frozen-in orientational randomness in the solid upon which the calorimetric measurements are made.

The following are examples of reactions with non-zero values of ΔS_0.

(a) $$H_2 + Cl_2 \longrightarrow 2HCl$$

if ΔS_0 is determined by extrapolation from measurements made above 10K.

(b) $$2CO + O_2 \longrightarrow 2CO_2$$

(c) $$\text{glycerol (glass)} \longrightarrow \text{glycerol (crystal)}$$

because there is frozen-in randomness in the glass.

(d) $AgBr\text{(cryst)} + AgCl\text{(cryst)} \longrightarrow AgBr, AgCl\text{(mixed crystal)}$

because there is frozen-in randomness in the mixture (the entropy of mixing, given by Equation 7.4).

The following are examples of reactions for which ΔS_0 is zero.

(a) $$2Ag + Hg_2Cl_2 \longrightarrow 2AgCl + 2Hg$$

because s is zero for each of the substances involved.

(b) $$H_2 + Cl_2 \longrightarrow 2HCl$$

if ΔS_0 is determined by extrapolation from measurements down to 1K.

(c) CO (crystal, pressure P) $\longrightarrow$ CO (crystal, pressure $P + \Delta P$)

because the s-terms cancel; i.e., the orientational randomness in the CO crystal is unaffected by the pressure change.

(iv) *Conventional Entropies*

It follows from Equation 8.22 that 8.14 may be written as:

$$\Delta S = [\tilde{S}_{\mathbf{cal}}(AB) + s(AB)] - [\tilde{S}_{\mathbf{cal}}(A) + s(A)] - [\tilde{S}_{\mathbf{cal}}(B) + s(B)]$$

i.e., quite generally, Equation 8.16 should be replaced by:

$$\Delta S = \tilde{S}_{\mathbf{spec}}(AB) - \tilde{S}_{\mathbf{spec}}(A) - \tilde{S}_{\mathbf{spec}}(B) \qquad (8.25)$$

and, similarly, 8.18 should be replaced by:

$$\Delta S = [l\tilde{S}_{\mathbf{spec}}(L) + m\tilde{S}_{\mathbf{spec}}(M) + \ldots] - [a\tilde{S}_{\mathbf{spec}}(A) + b\tilde{S}_{\mathbf{spec}}(B) + \ldots] \qquad (8.26)$$

$\widetilde{S}_{spec}$ values (or values of $\widetilde{S}_{cal}$ corrected by adding on the appropriate term s) have been extensively tabulated, and Equation 8.26 then permits the value of ΔS to be calculated for any arbitrary reaction at any temperature. The quantities so tabluated are often referred to as *absolute entropies*. $\widetilde{S}_{spec}$ and $\widetilde{S}_{cal}$ are, however, entropy differences; a better name is therefore *conventional entropy*, and henceforth the symbol S will refer to the conventional entropy of a substance. To summarize:

$$\widetilde{S} \equiv \widetilde{S}_{spec} = \widetilde{S}_{cal} + s = k \ln \frac{\Omega'}{[g_{nuc(0)}]^{\widetilde{N}}}$$

and is calculated directly from Equations 8.7–8.9.

Further Comments on the Formulation of the Third Law.

Nernst enunciated a principle, known as the *principle of the unattainability of the absolute zero*, which Fowler and Guggenheim (p. 224) adopted as a statement of the third law. This is a much more elegant formulation than the one which we have given; the latter seems, however, to be more suited to the aims of the present book.

An apparently simpler statement of the third law, due essentially to Planck, is:

$$\operatorname*{Lt}_{T \to 0\,K} S = 0 \tag{8.27}$$

for all perfect crystals, a perfect crystal being one in which all the constituent atoms, molecules, or ions are arranged in a perfectly ordered way. This is obviously consistent with our formulation of the third law. We have, however, two objections to the use of Equation 8.27. Firstly, the theoretical basis of this equation is rather less obvious than that of Equation 8.24. Secondly, there is a danger that the essentially conventional nature of such entropy zeros might be overlooked, and an unnecessary emphasis placed upon the idea of 'absolute entropy'. The latter idea, as Fowler (p. 231) says, has caused much confusion and been of very little assistance in the development of the subject. It is certainly not true, although it has often been asserted, that agreement between $\widetilde{S}_{cal}$ and $\widetilde{S}_{spec}$ for a single substance is in itself evidence for zero entropy at 0 K.

Problems

8.1 The normal melting point of carbon monoxide is 68·09 K (ΔH = 835·6 J mol^{-1}), and the boiling point 81·61 K (ΔH = 6042 J mol^{-1}). The phase transition

$$\text{crystal II} \longrightarrow \text{crystal I}$$

occurs at 61·55 K (ΔH = 633·1 J mol^{-1}) at 1 atmosphere pressure, II being the low-temperature form. The following heat capacity contributions to the calorimetric entropy of CO(g) have been reported (note that they are here given in units of R):

	$\tilde{S}/R$
0–11·7 K (extrapolation)	0·23
11·7–61·55 K (crystals II)	0·47
61·55–68·09 K (crystals I)	0·62
68·09–81·61 K (liquid)	1·30
81·61–298 K (gas)	4·48

Calculate the molar calorimetric entropy of CO(g) at 1 atmosphere pressure and 298 K, correcting for the non-ideality of the gas by adding 0·88 J K^{-1} to the entropy derived from the above data.

8.2 The moment of inertia of CO is $14{\cdot}50 \times 10^{-47}$ kg m^2. Neglecting the contribution (actually $< 0{\cdot}01$ J K^{-1}) from vibration, calculate $\tilde{S}_{\text{spec}}$ for CO(g) at 298 K and compare with the value of $\tilde{S}_{\text{cal}}$ found in the preceding problem. (Note: the value of $\tilde{S}_{\text{spec}}$ given in Table 8.1 was found by a somewhat more accurate method.)

8.3 Carry through the arguments in Section 8.4(ii) to interpret the value of s given in Table 8.3 for D_2.

8.4 Show that the entropy is independent of f_0, and hence of ϵ_0', the zero-point energy.

8.5 Show, from the Sackur–Tetrode equation, that, if the vapour is monatomic, the vapour pressure $\bar{P}$ of a crystalline element is given by:

$$\ln \bar{P} = -\frac{\tilde{S}_s}{R} - \frac{\Delta H}{RT} + \left[\frac{5}{2}\ln T + \ln\left(\frac{(2\pi m)^{3/2} k^{5/2} g_{\text{el}(0)}}{h^3}\right) + \frac{5}{2}\right]$$

where $\tilde{S}_s$ is the calorimetric entropy of the solid at temperature T, and ΔH is the enthalpy of evaporation.

8.6 (Note the result of the previous problem, and Equation 8.10)

At 84 K, the entropy of solid argon is 38·3 J K^{-1} mol^{-1}, and the heat of sublimation 7940 J mol^{-1}. Calculate the vapour pressure of solid argon at 84 K.

8.7 At 25 K, the entropy of solid neon is 14·7 J K^{-1} mol^{-1}, and the vapour pressure of the crystal is 0·426 atmosphere. Calculate the enthalpy of sublimation of neon at 25 K.

8.8 Explain why it is that the vapour pressure of thallium is partly determined by the fact that the lowest level of the Tl atom is $^2P_{\frac{1}{2}}$.

CHAPTER NINE

Equilibrium Constants

9.1 Thermodynamical preliminaries

Suppose that the gas-phase reaction:

$$aA + bB + \dots \longrightarrow lL + mM + \dots$$

has proceeded to equilibrium. Let the mole fractions of A, B, ..., L, M, ... in the equilibrium mixture be x_A, x_B, ..., x_L, x_M, ... The corresponding *partial pressures* P_A, P_B, ..., P_L, P_M, ... in the mixture are respectively defined as

$$P_A = x_A P, \quad P_B = x_B P, \quad \text{etc.,}$$

where P is the total pressure of the mixture.

Taking some value P° as a standard, or reference pressure (e.g. $P^\circ = 1$ atmosphere, or $101{\cdot}325\ \mathrm{kN\,m^{-2}}$, etc.), let:

$$P'_A = \frac{P_A}{P^\circ}, \quad P'_B = \frac{P_B}{P^\circ}, \quad \text{etc.}$$

If all the species are perfect gases, the value P'_A, P'_B, ..., P'_L, P'_M, ... for an equilibrium composition are related by:

$$\frac{(P'_L)^l (P'_M)^m \dots}{(P'_A)^a (P'_B)^b \dots} = K_p. \tag{9.1}$$

K_p is a quantity dependent only upon temperature; it is called the *equilibrium constant in terms of partial pressures*. If n_A is the number of moles present in the mixture:

$$\frac{P_A V}{n_A T} = \frac{P^\circ \tilde{V}^\circ}{1 \times T}$$

where $\tilde{V}^\circ$ is the volume of one mole of gas at pressure P° and temperature T. Hence:

$$P'_A = \frac{P_A}{P^\circ} = C_A \, \tilde{\tilde{V}}^\circ$$

where c_A is the concentration (n_A/V), in *moles per litre*, of A in the mixture, and $\tilde{\tilde{V}}^\circ$ is the molar volume at P° and T, *expressed in litres.* Therefore,

$$K_p = \frac{(c_L \, \tilde{\tilde{V}}^\circ)^l \, (c_M \, \tilde{\tilde{V}}^\circ)^m \, ...}{(c_A \, \tilde{\tilde{V}}^\circ)^a (c_B \, \tilde{\tilde{V}}^\circ)^b \, ...}$$

$$= \frac{(c_L)^l \, (c_M)^m \, ...}{(c_A)^a (c_B)^b \, ...} \, (\tilde{\tilde{V}}^\circ)^{\Delta\nu} \tag{9.2}$$

where: $$\Delta\nu = (l + m + ...) - (a + b + ...) \tag{9.3}$$

i.e., $\Delta\nu$ is the change in the number of molecules represented by the chemical equation. The concentration ratio on the right-hand side of Equation 9.2 is called the *equilibrium constant in terms of concentration*, K_c:

$$K_c = \frac{(c_L)^l (c_M)^m \, ...}{(c_A)^a (c_B)^b \, ...} \tag{9.4}$$

Thus defined, K_c has the units* $(\text{mol}\,\text{l}^{-1})^{\Delta\nu}$. Equation 8.2 can now be written:

$$K_p = K_c (\tilde{\tilde{V}}^\circ)^{\Delta\nu} \tag{9.5}$$

The standard molar Gibbs free energy $\tilde{G}_i$ of any component of a gaseous mixture is the free energy per mole of the pure substance i at pressure P°. The standard free energy change of a chemical reaction is defined to be:

$$\Delta G^\circ = (l\tilde{G}_L^\circ + m\tilde{G}_M^\circ + ...) - (a\tilde{G}_A^\circ + b\tilde{G}_B^\circ + ...) \tag{9.6}$$

This quantity is related to the equilibrium constant by (Lewis and Randall, p. 173):

* More properly, K_c should, like K_p, be defined as a dimensionless ratio involving quantities such as c/c° where c° is a unit of concentration. Since the $\text{mol}\,\text{l}^{-1}$ is allowed in conjunction with the SI system, and is the unit of concentration used throughout this book, we have adopted definition 9.4 for simplicity.

$$\Delta G^\circ = -RT \ln K_p \tag{9.7}$$

If values of $\tilde{G}_i^\circ$ are tabulated for elements and compounds at various temperatures, the values of the equilibrium constants of all conceivable reactions at these temperatures can be deduced from Equations 9.6 and 9.7. In practice, because $\tilde{G}_i^\circ$ often varies rapidly with temperature (making interpolation between tabulated values difficult), it is more convenient to tabulate the *free energy function*.*

$$\frac{\tilde{G}^\circ - \tilde{E}_0^\circ}{T} \tag{9.8}$$

$\tilde{E}_0^\circ$ being the standard molar internal energy at 0K, i.e., the value of $\tilde{E}$ for the pure substance in the same phase (gas, liquid or solid), in the hypothetical state of 0K and pressure P°.

The molar internal energy of a perfect gas is independent of volume, and hence of pressure: i.e., $\tilde{E} \equiv \tilde{E}^\circ$ at all pressures. As we discussed in Chapter 8, at 0 K all the molecules are in their lowest energy level. The molar internal energy of a perfect gas in the hypothetical state of 0 K and P° is therefore $\tilde{E}_0$, where $\tilde{E}_0$ is the molar zero-point energy, defined in Equation 5.40. According to Equation 9.6 and 9.8 the standard free energy change is given by:

$$\begin{aligned}\frac{\Delta G^\circ}{T} = & \left[l\left(\frac{\tilde{G}^\circ - \tilde{E}_0}{T}\right)_L + m\left(\frac{\tilde{G}^\circ - \tilde{E}_0}{T}\right)_M + \ldots\right] \\ & - \left[a\left(\frac{\tilde{G}^\circ - \tilde{E}^\circ}{T}\right)_A + b\left(\frac{\tilde{G}^\circ - \tilde{E}_0}{T}\right)_B + \ldots\right] \\ & + \frac{\Delta E_0}{T}\end{aligned} \tag{9.9}$$

* Lewis and Randall, p. 166 *et seq.* These authors define the free energy function in terms of $\tilde{H}_0^\circ$ rather than $\tilde{E}_0^\circ$, but in practice the difference between $\tilde{H}_0^\circ$ and $\tilde{E}_0^\circ$ can be neglected; $\tilde{H}_0^\circ = \tilde{E}_0^\circ$ exactly for perfect gases, and approximately for solids and liquids. The alternative function $(\tilde{G}^\circ - \tilde{H}_{298K}^\circ/T$ is often tabulated. This is related to 9.8 by:

$$\begin{aligned}\frac{\tilde{G}^\circ - \tilde{H}_{298K}^\circ}{T} &= \frac{\tilde{G}^\circ - \tilde{E}_0^\circ}{T} - \frac{\tilde{H}_{298K}^\circ - \tilde{E}_0^\circ}{T} \\ &= \frac{\tilde{G}^\circ - \tilde{E}_0^\circ}{T} - \frac{1}{T}\int_0^{298} C_p \, dT\end{aligned}$$

If this function is used, $\Delta E_0^\circ/T$ in Equation 9.9 is replaced by $\Delta H_{298K}^\circ/T$.

where ΔE_0 is the change in zero-point energy, defined by:

$$\Delta E_0 = (l\tilde{E}_{0L} + m\tilde{E}_{0M} + \dots) - (a\tilde{E}_{0A} + b\tilde{E}_{0B} + \dots) \quad (9.10)$$

9.2 The calculation of Equilibrium constants: method 1

There are two equivalent ways of formulating the statistical-mechanical calculation of equilibrium constants. The first method is to compute the free energy function 9.8 for each of the reactants and products, and then to calculate $\Delta G°$ from Equation 9.9. The equilibrium constant is then obtained from Equation 9.7. The second method of calculating K_p is to express it, through $\Delta G°$, directly in terms of the partition functions of reactants and products.

The second method has the advantage that it enables one to write down K_p or K_c explicitly in terms of molecular parameters. It becomes unwieldy for any but the simplest reactions, however, and the first method is to be preferred for numerical work.

Now:

$$\frac{\tilde{G}° - \tilde{E}_0°}{T} = \frac{\tilde{H}° - T\tilde{S}° - \tilde{E}_0°}{T} \quad (9.11)$$

and so, for a perfect gas:

$$\frac{\tilde{G}° - \tilde{E}_0°}{T} = \frac{\tilde{E} - \tilde{E}_0°}{T} + R - \tilde{S}° \quad (9.12)$$

As explained above, we do not distinguish between $\tilde{E}_0°$ and $\tilde{E}_0$. Therefore, according to Equation 5.39:

$$\tilde{E} - \tilde{E}_0° = \tilde{E}_{trans} + \tilde{E}_{vib} + \tilde{E}_{el} + \tilde{E}_{rot}$$

(plus a term for internal rotation, if this is present) the energies on the right-hand side being evaluated in terms of molecular energies *relative to* the lowest level.

Applying Equation 8.15 to the standard molar entropy:

$$\tilde{S}° = \tilde{S}°_{trans} + \tilde{S}_{vib} + \tilde{S}_{el} + \tilde{S}_{rot}$$

plus a term for internal rotation, if this is present. (The superscript ° is applied only to $\tilde{S}_{trans}$, because the other terms on the right-hand side are independent of pressure.)

Substituting for $\tilde{E} - \tilde{E}_0$ and $\tilde{S}°$ in Equation 9.12:

$$\frac{\tilde{G}^\circ - \tilde{E}_0^\circ}{T} = \left(\frac{\tilde{G}^\circ - \tilde{E}_0}{T}\right)_{\text{trans}} + \left(\frac{\tilde{G}^\circ - \tilde{E}_0}{T}\right)_{\text{vib}} + \left(\frac{\tilde{G}^\circ - \tilde{E}_0}{T}\right)_{\text{el}} + \left(\frac{\tilde{G}^\circ - \tilde{E}_0}{T}\right)_{\text{rot}} \tag{9.13}$$

where:

$$\left(\frac{\tilde{G}^\circ - \tilde{E}_0}{T}\right)_{\text{rot}} = \frac{\tilde{E}_{\text{rot}}}{T} - \tilde{S}_{\text{rot}}$$

with similar expressions for vibrational and electronic terms, and:

$$\left(\frac{\tilde{G}^\circ - \tilde{E}_0}{T}\right)_{\text{trans}} = \frac{\tilde{E}_{\text{trans}}}{T} - \tilde{S}^\circ_{\text{trans}} + R$$

(it being conventional to include R of Equation 9.12 in the translational term).

Formulae are given in Chapter 10 for the numerical evaluation of the free energy function. From the values of this function for reactants and products, ΔG° given by Equation 9.6 and hence K_p, can be calculated. The following examples show how the free energy function is used.

Example 9.2(a)

Calculate the equilibrium constant for the reaction:

$$S_2\,(g) \longrightarrow 2S(g)$$

at 2000 K. The dissociation energy of S_2, found spectroscopically, is 429,700 J mol^{-1}, and statistical mechanics gives the following values at 2000 K:

$$S(g): \quad \frac{G^\circ - E_0}{T} = -\,191{\cdot}4 \text{ J K}^{-1}\text{ mol}^{-1}$$

$$S_2(g): \quad \frac{G^\circ - E_0}{T} = -\,265{\cdot}5 \text{ J K}^{-1}\text{ mol}^{-1}$$

Substituting these values into Equation 9.9:

$$\frac{\Delta G^\circ}{T} = 2(-\,191{\cdot}4) - (-\,265{\cdot}5) + \frac{429{,}700}{2000} = 97{\cdot}6 \text{ J K}^{-1}\text{ mol}^{-1}$$

Therefore:

$$\ln K_p = \frac{\Delta G^\circ}{RT} = -\frac{97{\cdot}6}{8{\cdot}314} = -\,11{\cdot}73$$

and

$$K_p = 8{\cdot}08 \times 10^{-6}.$$

□ □ □

Example 9.2 (b)
Evaluate ΔE_0° for the reaction:

$$F_2(g) \longrightarrow 2F(g)$$

At 1000 K, dissociation constant measurements gave $\Delta G^\circ = 38{,}400$ J mol^{-1}. Statistical mechanics gives the following values:

$$F_2(g): \quad \frac{G^\circ - E_0^\circ}{T} = -211{\cdot}0 \text{ J K}^{-1} \text{ mol}^{-1}$$

$$F(g): \quad \frac{G^\circ - E_0^\circ}{T} = -163{\cdot}5 \text{ J K}^{-1} \text{ mol}^{-1}$$

From Equation 9.9:

$$\frac{38{,}400}{1000} = 2(-163{\cdot}5) - (-211{\cdot}10) + \frac{\Delta E_0^\circ}{1000}$$

Hence:

$$\Delta E_0^\circ = 154{,}000 \text{ J mol}^{-1}$$

□ □ □

9.3 The calculation of equilibrium constants: method 2

The equations which we shall now obtain actually follow directly from Equation 9.12; but to emphasize the principles of the method, we shall not obtain them in this way, preferring to return instead to basic equations.

For a perfect gas:

$$\tilde{G} = \tilde{A} + RT$$

$$= -RT \ln\left(f \frac{e}{\tilde{N}}\right) + RT$$

by Equation 4.10. Since $\ln e = 1$, it follows that:

$$\tilde{G} = -RT \ln\left(\frac{f}{\tilde{N}}\right)$$

Hence the *standard* molar free energy is:

$$\tilde{G}^\circ = -RT \ln\left(\frac{f^\circ}{\tilde{N}}\right) \qquad (9.14)$$

f° being the partition function of the gas at standard pressure:

$$f^\circ = f_0 \, f^\circ_{\text{trans}} \, f_{\text{el}} \, f_{\text{vib}} \, f_{\text{nuc}} \, f_{\text{rot}}$$

in which $f_0 = e^{-\epsilon_0/kT}$ (the zero-point energy term), and:

$$f^\circ_{\text{trans}} = \frac{(2\pi mkT)^{3/2} \, \tilde{V}^\circ}{h^3}$$

where $\tilde{V}^\circ$ is the molar volume at P°. Let:

$$f' = f^\circ_{\text{trans}} \, f_{\text{nuc}} \, f_{\text{vib}} \, f_{\text{el}} \, f_{\text{rot}} \qquad (9.15)$$

so that:

$$f^\circ = f_0 f' = f' e^{-\epsilon_0/kT}$$

From equations 9.14 and 9.6:

$$\Delta G^\circ = -RT \left[\left(l \ln\frac{f'_L}{\tilde{N}} + m \ln\frac{f'_M}{\tilde{N}} + \ldots\right) - \left(a \ln\frac{f'_A}{\tilde{N}} + b \ln\frac{f'_B}{\tilde{N}} + \ldots\right) - \frac{1}{kT}(l\epsilon_{0L} + m\epsilon_{0M} + \ldots - a\epsilon_{0A} - b\epsilon_{0B} - \ldots)\right]$$

$$= -RT\left[(l \ln f'_L + m \ln f'_M + \ldots) - (a \ln f'_A + b \ln f'_B + \ldots) + \ln\left(\frac{1}{\tilde{N}}\right)^{\Delta\nu} - \frac{\Delta\epsilon_0}{kT}\right]$$

where $\Delta\nu$ is defined by Equation 9.3, and:

$$\Delta\epsilon_0 = (l\epsilon_{0L} + m\epsilon_{0M} + \ldots) - (a\epsilon_{0A} + b\epsilon_{0B} + \ldots)$$

Thus:

$$\Delta G^\circ = -RT\left(\ln\left[\frac{(f'_L)^l \, (f'_M)^m \ldots}{(f'_A)^a \, (f'_B)^b \ldots}\left(\frac{1}{\tilde{N}}\right)^{\Delta\nu}\right] - \frac{\Delta\epsilon_0}{kT}\right)$$

Hence:

$$K_p = \frac{(f'_L)^l \, (f'_M)^m \ldots}{(f'_A)^a \, (f'_B)^b \ldots}\left(\frac{1}{\tilde{N}}\right)^{\Delta\nu} e^{-\Delta\epsilon_0/kT} \qquad (9.16\text{ a})$$

Alternatively, since:

$$\tilde{E}_{0i} = \tilde{N}\epsilon_{0i}$$

K_p can be expressed in terms of the change in molar zero-point energy, ΔE_0:

$$K_p = \frac{(f'_L)^l (f'_M)^m \ \ldots}{(f'_A)^a (f'_B)^b \ \ldots} \left(\frac{1}{\tilde{N}}\right)^{\Delta\nu} e^{-\Delta E_0/RT} \qquad (9.16\text{ b})$$

From Equation 9.5,

$$K_c = \frac{(f'_L)^l (f'_M)^m \ \ldots}{(f'_A)^a (f'_B)^b \ \ldots} \left(\frac{1}{\tilde{\tilde{V}}^\circ \tilde{N}}\right)^{\Delta\nu} e^{-\Delta E_0/RT} \qquad (9.17)$$

9.4 Equilibrium constants of simple systems

Following are three examples of the application of method 2 to the calculation of equilibrium constants.

(i) *Ionization of Metal Atoms.*
At high temperatures, a significant number of atoms in the vapour of a metal may be ionized according to:

$$M \longrightarrow M^+ + e$$

The equilibrium constant for this process is given by Equation 9.16:

$$K_p = \frac{f'_{M^+} f'_e}{f'_M} \left(\frac{1}{\tilde{N}}\right) e^{-\Delta\epsilon_0/kT} \qquad (9.18)$$

where $\Delta\epsilon_0$ is the least energy required to remove one electron from the metal atom in its ground state, i.e., the ionization potential.

The energy of the free electron is purely translational kinetic energy, and the partition function for the free electron is, accordingly:

$$f'_e = 2\frac{(2\pi m_e kT)^{3/2}\ \tilde{V}^\circ}{h^3}$$

where m_e is the mass of an electron. The extra factor 2 occurs because the electron may have either of two possible spins.
For the metal atom and ion:

$$f'_M = g_{\text{nuc}\,M}\ f_{\text{el}\,M}\ f^\circ_{\text{trans}\,M}$$

$$f'_{M^+} = g_{\text{nuc}\,M^+}\ f_{\text{el}\,M^+}\ f^\circ_{\text{trans}\,M^+}$$

Now, $g_{\text{nuc}\,M}$ will cancel with $g_{\text{nuc}\,M^+}$ and $f^\circ_{\text{trans}\,M}$ will cancel with $f^\circ_{\text{trans}\,M^+}$. Hence, Equation 9.18 simplifies to:

$$K_p = 2\frac{(2\pi m_e kT)^{3/2}}{h^3}\frac{f'_{elM^+}}{f_{elM}}\frac{\tilde{V}^\circ}{\tilde{N}} e^{-\Delta\epsilon_0/kT}$$

For the alkali-metal and hydrogen atoms:

$$f_{elM} = g_{el(o)M} = 2$$

and

$$f_{elM^+} = g_{el(o)M^+} = 1$$

Furthermore (see Table 10.7):

$$\frac{\Delta\epsilon_0}{kT} = 1{\cdot}161 \times 10^4 \frac{\Delta\epsilon_0}{T}$$

when $\Delta\epsilon_0$ is the ionization potential in electron volts. Finally, putting $M = 1/1837$ in Equation 4.2 for $f^\circ_{trans}/\tilde{N}$, we obtain:

$$K_p = 0{\cdot}02559\left(\frac{1}{1837}\right)^{3/2} T^{5/2} \exp\left(\frac{-1{\cdot}161 \times 10^4 \Delta\epsilon_0}{T}\right) \quad (9.19)$$

The degree of ionization α is defined by:

$$\frac{c_{M^+}}{c_M} = \frac{\alpha}{1-\alpha}$$

From Equation 9.1:

$$K_p = \frac{\alpha^2}{1-\alpha}\left(\frac{P}{P^\circ}\right)$$

where P is the pressure of the equilibrium mixture. If α is small, $1 - \alpha \approx 1$, and then:

$$\alpha = \left(\frac{P^\circ}{P}\right)^{1/2} K_p^{1/2}$$

From Equation 9.19, therefore:

$$\alpha = 5{\cdot}694 \times 10^{-4}\left(\frac{P^\circ}{P}\right)^{1/2} T^{5/4} \exp\left(\frac{-0{\cdot}581 \times 10^4 \Delta\epsilon_0}{T}\right) \quad (9.20)$$

Equation 9.20 has been tested experimentally for the process $Cs \rightarrow Cs^+ + e$.

(ii) *Dissociation of Diatomic Molecules*

The equilibrium constant for the process:

$$A_2 \rightarrow 2A$$

is, according to Equation 9.16:

$$K_p = \frac{(f'_A)^2}{f'_{A_2}} \left(\frac{1}{\tilde{N}}\right) e^{-\Delta E_0/RT}$$

The nuclear factors in f'_A and f'_{A_2} cancel, and:

$$\frac{(f'_A)^2}{f_{A_2}} = \left(\frac{(f_{el\,A})^2}{f_{el\,A_2}}\right) \left(\frac{1}{f_{rot\,A_2}}\right) \left(\frac{1}{f_{vib\,A_2}}\right) \left(\frac{(f^\circ_{trans\,A})^2}{f^\circ_{trans\,A_2}}\right)$$

If the mass of A is m, then the mass of A_2 is $2m$, and hence the ratio of the translational factors is:

$$\frac{(f^\circ_{trans\,A})^2}{f^\circ_{trans\,A_2}} = \frac{[(2\pi mkT)^{3/2}\,\tilde{V}^\circ/h^3]^2}{(4\pi mkT)^{3/2}\,\tilde{V}^\circ/h^3} \simeq \frac{(\pi mkT)^{3/2}\,\tilde{V}^\circ}{h^3}$$

The moment of inertia of A_2 is, according to Equation 5.17:

$$I = \frac{mr^2}{2}$$

and the symmetry factor is 2. Hence:

$$\frac{1}{f_{rot\,A_2}} = \frac{2h^2}{8\pi^2 IkT} = \frac{h^2}{2\pi^2 mr^2 kT}$$

the vibrational factor is:

$$\frac{1}{f_{vib\,A_2}} = 1 - e^{-h\omega/kT}$$

Assuming that $f_{el} = g_{el(0)}$, we obtain finally:

$$K_p = \frac{1}{2}\left(\frac{g^2_{el(0)\,A}}{g_{el(0)\,A_2}}\right)\left(\frac{mkT}{\pi}\right)^{1/2} \frac{1}{hr^2}(1 - e^{-h\omega/kT})\left(\frac{\tilde{V}^\circ}{\tilde{N}}\right) e^{-\Delta E_0/RT} \quad (9.21)$$

Consider, for example, the dissociation of hydrogen. Since Θ_{vib} is about 6000 K for hydrogen, $f_{vib} \approx 1$, if $T \leqslant 3000$ K. The dissociation energy, obtained spectroscopically, is $\Delta E_0 =$ 432,200 J mol^{-1}, and $r = 0{\cdot}74 \times 10^{-10}$ m, $g_{el(0)H} = 2$, $g_{el(0)H_2} = 1$. Equation 9.21 then simplifies to:

$$\log_{10} K_p = 0{\cdot}21 + \frac{3}{3}\log_{10} T - \frac{2{\cdot}257 \times 10^4}{T}$$

Some results are shown in Table 9.1. The experimental values of the degree of dissociation, shown in this table, were obtained by Langmuir from his work on conduction of heat from tungsten filaments by gases.

Table 9.1. Dissociation of H_2 Molecules

T (K)	K_p	a(calculated)	a(observed)
1000	$1{\cdot}4 \times 10^{-18}$	$1{\cdot}2 \times 10^{-9}$	
2000	$3{\cdot}9 \times 10^{-7}$	$6{\cdot}2 \times 10^{-4}$	$1{\cdot}7 \times 10^{-3}$
2500	$1{\cdot}9 \times 10^{-4}$	$1{\cdot}4 \times 10^{-2}$	$1{\cdot}6 \times 10^{-2}$
3000	$7{\cdot}9 \times 10^{-3}$	$8{\cdot}5 \times 10^{-2}$	$7{\cdot}2 \times 10^{-2}$

a, Degree of dissociation for $P^\circ/P = 1$.

(iii) *Isotopic Exchange Equilibria.*

Consider the equilibrium:

$$A_2 + B_2 \rightleftharpoons 2AB$$

in which A and B are isotopes.

Since A and B have the same nuclear charge, the molecules A_2, B_2, and AB have, to a good degree of approximation,* the same electronic energies and charge distributions, and hence the same interatomic distances and vibrational force constants.

The equilibrium constant is:

$$K_p = \frac{(f'_{AB})^2}{f'_{A_2}\, f'_{B_2}}\, e^{-\Delta\epsilon_0/kT} \tag{9.22}$$

Since the electronic energies of reactants and products are the same, the only contributions to $\Delta\epsilon_0$ arise from the zero-point vibrational terms ($h\omega/2$ per molecule):

$$\Delta\epsilon_0 = \frac{h}{2}(2\omega_{AB} - \omega_{A_2} - \omega_{B_2})$$

Since the force constants of A_2, B_2, and AB are the same, it follows from Equations 5.5 and 5.6 that:

$$\omega_{AB} = \omega_{A_2}\left(\frac{\mu_{A_2}}{\mu_{AB}}\right)^{1/2} = \omega_{A_2}\left(\frac{m_A + m_B}{2m_B}\right)^{1/2}$$

where m_A and m_B are the masses of A and B. Similarly:

$$\omega_{B_2} = \omega_{A_2}\left(\frac{m_A}{m_B}\right)^{1/2}$$

* i.e., exactly the same values, within the Born–Oppenheimer approximation.

Therefore:

$$\Delta\epsilon_0 = \frac{\Theta_{\text{vib A}_2}}{2}\left[\sqrt{2}\left(\frac{m_A + m_B}{m_B}\right)^{1/2} - \left(\frac{m_A}{m_B}\right)^{1/2} - 1\right] \quad (9.23)$$

where

$$\Theta_{\text{vib A}_2} = h\omega_{A_2}/k.$$

The f' terms appearing in Equation 9.22 can be evaluated in the usual way. For simplicity, let us consider the behaviour at temperatures for which:

$$\Theta_{\text{rot}} < T \ll \Theta_{\text{vib}}$$

so that the classical rotational partition function of Equation 5.22 can be used, and $f'_{\text{vib}} = 1$. The nuclear and electronic factors cancel, and:

$$\frac{(f'_{AB})^2}{f'_{A_2}\, f'_{B_2}} = \left(\frac{(f_{\text{rot AB}})^2}{f_{\text{rot A}_2}\, f_{\text{rot B}_2}}\right)\left(\frac{(f^{\circ}_{\text{trans AB}})^2}{f^{\circ}_{\text{trans A}_2}\, f^{\circ}_{\text{trans B}_2}}\right)$$

Since:

$$f_{\text{rot}} = \frac{8\pi^2 IkT}{\sigma h^2}$$

with $\sigma_{A_2} = \sigma_{B_2} = 2$ and $\sigma_{AB} = 1$, the rotational factor becomes:

$$\frac{4I^2_{AB}}{I_{A_2}\, I_{B_2}} = \frac{16 m_A m_B}{(m_A + m_B)^2} \quad (9.24)$$

after the substitution $I = \mu r^2$. Since:

$$f^{\circ}_{\text{trans}} = \frac{(2\pi mkT)^{3/2}\, \tilde{V}^{\circ}}{h^3}$$

the translational factor is:

$$\frac{(m_A + m_B)^3}{(2m_A)^{3/2}\,(2m_B)^{3/2}} = \frac{1}{8}\frac{(m_A + m_B)^3}{(m_A\, m_B)^{3/2}} \quad (9.25)$$

Substituting Equations 9.23–9.25 in to Equation 9.22:

$$K_p = \frac{2(m_A + m_B)}{(m_A\, m_B)^{1/2}}\exp\left\{-\frac{\Theta_{\text{vib A}_2}}{2T}\left[\sqrt{2}\left(\frac{m_A + m_B}{m_B}\right) - \left(\frac{m_A}{m_B}\right)^{1/2} - 1\right]\right\} \quad (9.26)$$

Consider, for example, the deuterium–hydrogen system. In this case, $m_A = 2m$, $m_B = m$, where m is the mass of the hydrogen atom. Hence:

$$K_p = 3\sqrt{2}\exp\left[-\frac{\Theta_{\text{vib D}_2}}{2T}(\sqrt{6}-\sqrt{2}-1)\right]$$

Using the value $\Theta_{\text{vib D}_2} = 4307$ K, we obtain:

$$K_p = 4{\cdot}24e^{-75/T} \qquad (9.27)$$

Some results are shown in Table 9.2. In all cases, the calculated and experimental values agree to within the likely experimental error.

Table 9.2 K_p for $H_2 + D_2 \rightleftharpoons 2HD$

T (K)	273	298	383	543	670	741
K_p(observed)	3·24	3·28	3·50	3·85	3·8	3·70
K_p(calculated)	3·22	3·30	3·49	3·69	3·79	3·83

The experimental values are given by Fowler and Guggenheim, p. 169. The calculated values are derived from Equation 9.27.

For atoms somewhat heavier than the hydrogen isotopes, $m_A/m_B \approx 1$, and K_p takes the limiting value 4. This value can be understood by reference to Equation 9.22. When m_A and m_B are nearly equal, f'_{AB} differs from f'_{A_2} and f'_{B_2} only in the symmetry number. Hence, Equation 9.24 simplifies to:

$$K_p = \frac{\sigma_{A_2}\,\sigma_{B_2}}{\sigma_{AB}^2} = 4$$

Problems

9.1 The following values have been obtained from spectroscopic data (cf. problem 10.12):

	$(\tilde{G}^\circ - \tilde{E}_0^\circ)/T$, J K^{-1} mol^{-1}
CO_2	222·1
H_2	134·0
CO	201·3
H_2O	193·0

for $T = 900$ K. Calculate the equilibrium constant K_p of the water gas reaction:

$$H_2 + CO_2 \rightleftharpoons H_2O + CO$$

at 900 K given that $\Delta E_0^\circ = 40{,}330$ J. (The experimental value of K_p is 0·46 at 900 K.)

9.2 Show that the translational contribution to the free energy function is given by:

$$\left(\frac{\tilde{G}^\circ - \tilde{E}_0^\circ}{T}\right)_{\text{trans}} = R(-\tfrac{3}{2}\ln M - \tfrac{5}{2}\ln T + 3{\cdot}657)$$

Evaluate this term for I_2 and I with $T = 1275$ K.

9.3 Using the data and/or results of problems 5.6, 5.9 and 9.2, and taking $f_{el} = 1$ for I_2, 4 for I, calculate the free energy functions of I_2 and I at 1275 K. Hence calculate the dissociation energy ΔE_0° of I_2, given that the value of the equilibrium constant K_p of the reaction:

$$I_{2(g)} \rightleftharpoons 2I$$

is 0·168 at 1275 K.

9.4 For the first excited ($^2P_{\frac{1}{2}}$) level of the iodine atom, $\epsilon_{el(1)} = 1{\cdot}42 \times 10^{-19}$ J. Repeat the preceding problem, including the contribution from the $^2P_{\frac{1}{2}}$ level to the free energy function of I.

9.5 It is found from the spectrum of $Cs_{(g)}$ that the ionization potential $\Delta\epsilon_0$ of Cs is 3·893 eV. Calculate (a) the equilibrium constant for the ionization of Cs at 2000 K, and (b) the degree of ionization when $P/P^\circ = 10^{-4}$.

9.6 Express the ratio of the dissociation constants of H_2 and D_2 at a given temperature in terms only of the vibrational frequency of H_2 and universal constants.

9.7 The equilibrium internuclear distance in Na_2 is $3{\cdot}077 \times 10^{-10}$ m, and the vibrational frequency is $4{\cdot}734 \times 10^{12}$ s^{-1}. The dissociation energy ΔE_0° is 73,200 J mol^{-1}. The lowest level of the Na atom is $^2S_{\frac{1}{2}}$. Calculate the equilibrium constant of the reaction:

$$Na_{2(g)} \rightleftharpoons 2Na \qquad \text{at 1200 K.}$$

9.8 Suppose that A and B are two molecules differing only in isotopic composition. Then it can be shown that (cf. problem 11.9):

$$\frac{f_A^\circ}{f_B^\circ} = \frac{\sigma_B}{\sigma_A}\prod_\alpha\left(\frac{m_{\alpha A}}{m_{\alpha B}}\right)^{3/2}\prod_i \frac{u_{iA}}{u_{iB}}\frac{e^{-u_{iA}/2}}{e^{-u_{iB}/2}}\frac{1-e^{-u_{iB}}}{1-e^{-u_{iA}}}$$

where σ_A, σ_B are the symmetry numbers of A and B, $u_{iA} = h\omega_{iA}/kT$, the first product is over all the nuclear masses, and the second over the vibrational modes.

(a) Show that Equation 9.26 is consistent with this result.

(b) Calculate the equilibrium constant of the reaction:

$$H_2{}^{18}O(g) + N^{16}O(g) \rightleftharpoons H_2{}^{16}O(g) + N^{18}O(g)$$

at 300 K from the following normal frequencies:

	ω_i (s^{-1})		ω_i (s^{-1})
$H_2{}^{16}O$:	$11{\cdot}4759 \times 10^{13}$	$H_2{}^{18}O$:	$11{\cdot}4465 \times 10^{13}$
	$4{\cdot}9617 \times 10^{13}$		$4{\cdot}9434 \times 10^{13}$
	$11{\cdot}8068 \times 10^{13}$		$11{\cdot}7582 \times 10^{13}$
$N^{16}O$:	$5{\cdot}7185 \times 10^{13}$		

[Notes: (i) Assume that vibrations are unexcited.

(ii) $$\frac{e^{-u_{iA}/2}}{e^{-u_{iB}/2}} = e^{-(u_{iA} - u_{iB})/2} \approx 1 - (u_{iA} - u_{iB})/2$$

(iii) u_{iA} and u_{iB} for NO are related to the corresponding reduced masses.]

CHAPTER TEN

Applications of Spectroscopic Data: The Computation of Thermodynamic Functions

The theme underlying most of our previous discussion has been this: if the spacings between the energies associated with the various quantum states of a molecule are known, then the molecular partition function can be evaluated, and, from the latter, the thermodynamic functions can be calculated. Knowledge of molecular energy levels comes chiefly from spectroscopy, and it is timely at this point to pause in our development of statistical thermodynamics, to consider precisely how the spectroscopic data are used.

There are two methods of employing spectroscopic data for this purpose. The simpler way is to deduce from the spectra the values of *molecular constants,* of which the most important are the moments of inertia and vibrational frequencies (the others are small correction terms such as anharmonicity constants). It was shown in Chapter 5 how the molecular partition function is related to these. In the present chapter, we summarize the statistical mechanical results in the form of convenient tables which give the thermodynamic properties directly in terms of molecular and numerical constants.

The second way (which we discuss in Section 10.7) of using spectroscopic data is to calculate the partition function as the sum of $g_i e^{-\epsilon_i/kT}$ terms, in which the ϵ_i values are evaluated directly from the frequencies of the observed spectral lines. This, of course, is the only method which can be used for the electronic partition function.

Such calculations can be described as *applications of spectroscopy to statistical mechanics.* Before dealing with them, however, we shall digress a little to consider, in Section 10.1, an *application of statistical mechanics to spectroscopy;* strictly speaking, this item does not constitute a part of statistical thermodynamics, but is included because of its importance and because it forms a convenient introduction to later sections.

10.1 Intensities of rotational lines in molecular spectra (Herzberg I, p. 121)

Excitation of a molecule from energy level ϵ_1 to level ϵ_2 can be induced by absorption of light of frequency ν where

$$h\nu = \epsilon_2 - \epsilon_1 , \qquad (10.1)$$

or by Raman scattering, in which case:

$$h(\nu'' - \nu') = \epsilon_2 - \epsilon_1 , \qquad (10.2)$$

$(\nu'' - \nu')$ being the difference between the frequencies of the incident and scattered radiation (the Raman shift).

Let us consider the excitation of a diatomic molecule, according to Equation 10.1, in which the vibrational and rotational energies change, but the electronic state remains the same. Such excitations are associated with the absorption of radiation in the *infrared* region, and are restricted to molecules which has a permanent dipole moment, in this case to *heteronuclear* diatomic molecules. We consider, initially, the simplified case in which the vibration is strictly simple harmonic motion; the (important) complications arising from anharmonicity are discussed in the closing paragraphs of this section.

Let the vibrational and rotational quantum numbers be, respectively, v and J in the initial level ϵ_1, and v' and J' in the final level ϵ_2. According to Equation 5.8, the increase in *vibrational* energy is:

$$\epsilon_{\text{vib}(2)} - \epsilon_{\text{vib}(1)} = (v' - v)h\omega$$

For simple harmonic motion, the *vibrational selection rule* restricts the permitted transitions to those for which:

$$v' - v = 1$$

In such case:

$$\epsilon_{\text{vib}(2)} - \epsilon_{\text{vib}(1)} = h\omega \qquad (10.3)$$

The change in *rotational* energy is, according to Equation 5.17:

$$\epsilon_{\text{rot}(2)} - \epsilon_{\text{rot}(1)} = J'(J' + 1)A - J(J + 1)A$$

where:

$$A = \frac{h^2}{8\pi^2 I}$$

is the same for both vibrational levels in *simple harmonic* motion. The selection rule for transitions within the lowest electronic level is (except for nitric oxide; Herzberg I, p. 121):

$$J' - J = \pm 1 \qquad (10.4)$$

Thus:

$$\epsilon_{\text{rot}(2)} - \epsilon_{\text{rot}(1)} = (2J + 2)A \quad \text{if} \quad J' - J = 1$$

or

$$\epsilon_{\text{rot}(2)} - \epsilon_{\text{rot}(1)} = -2JA \qquad \text{if} \quad J' - J = -1.$$

By adding these values to the change in vibrational energy given by Equation 10.3, we see that the energy corresponding to a line in the infrared absorption spectrum associated with the transition from the $v = 0$ level to the $v = 1$ level is either:

$$h\omega - 2JA \qquad (\text{if } v' - v = 1 \text{ and } J' - J = -1) \qquad (10.5a)$$

or

$$h\omega + (2J + 2)A \qquad (\text{if } v' - v = 1 \text{ and } J' - J = 1) \qquad (10.5b)$$

The frequencies $[= (\epsilon_2 - \epsilon_1)/h]$ corresponding to Equation 10.5a are given by:

$$\nu = \omega - 2JB \ (J = 1,2,3... \textit{ but not } 0) \qquad (10.6a)$$

where:

$$B = \frac{A}{h} = \frac{h}{8\pi^2 I} \qquad (10;7)$$

These frequenceis are smaller than ω, and collectively are said to constitute the *P-branch* of the vibration–rotation spectrum.

The frequencies corresponding to Equation 10.5 are:

$$\nu = \omega + (2J + 2)\, B\ (J = 0,1,2,3, \dots) \qquad (10.6b)$$

They are greater than ω and are said to constitute the *R-branch* in the vibration–rotation spectrum.

The relationship between the P and R branches is shown in Fig. 10.1.

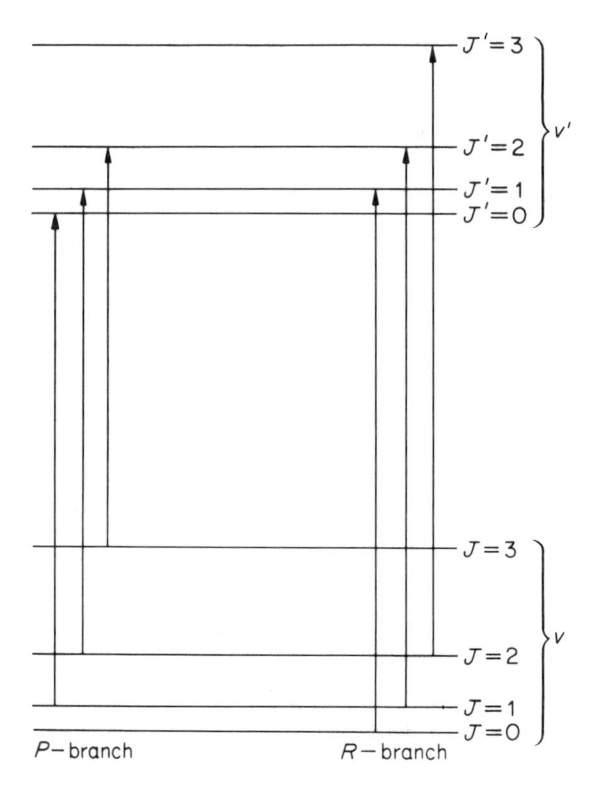

Fig. 10.1 Vibration–rotation transitions

The first member of the P-branch (i.e., that for which $J = 1$) has the frequence $\omega - 2B$, the second ($J = 2$) the frequency $\omega - 4B$, and so on. Similarly, the first members of the R-branch have frequencies $(\omega + 2B)$, $(\omega + 4B)$, $(\omega + 6B)$, and so on. Each branch thus consists of a sequence of *equally spaced* lines, the frequency separation between successive lines being:

$$2B = \frac{h}{4\pi^2 I}$$

These lines, although equally spaced, are not equally intense, and the interpretation of the relative intensities, which we now consider, is an important application of statistical theory.

As in Section 5.1, the value of the J-th term in the rotational partition function:

$$f_{rot} = g_{rot(0)}\,e^{-\epsilon_{rot(0)}/kT} + g_{rot(1)}\,e^{-\epsilon_{rot(1)}/kT} + \ldots$$
$$+ g_{rot(J)}\,e^{-\epsilon_{rot(J)}/kT} + \ldots$$

is proportional to the equilibrium number, $\bar{n}_J$, of molecules with the rotational energy $\epsilon_{rot(J)}$. Thus:

$$\bar{n}_J \propto (2J + 1)e^{-J(J+1)\,\Theta_{rot}/T} \qquad (10.8)$$

Fig. 10.2 shows the variation of $\bar{n}_J$ with J for $H^{37}Cl$ at 150 K. For the $H^{37}Cl$ molecule, Θ_{rot} is 15·00 K, and the proportionality 10.8 becomes:

$$\bar{n}_J \propto (2J + 1)e^{-J(J+1)/10}$$

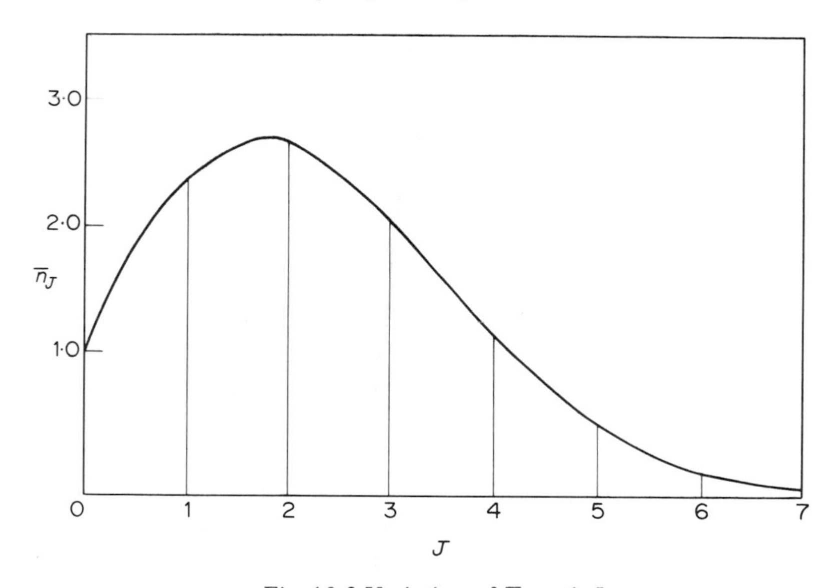

Fig. 10.2 Variation of $\bar{n}_J$ with J

It is seen, from the figure, that the lowest ($J = 0$) level does *not* have the largest value of $\bar{n}_J$; in fact, it is easy to show, by differentiation, that the maximum of the function 10.8 occurs when:

$$J = \sqrt{\frac{T}{2\Theta_{rot}}} - \frac{1}{2} \qquad (10.9)$$

This has an important effect upon the intensity distribution in the vibration–rotation spectrum. The intensity of a spectral line associated with the excitation of molecules from the J-th level is determined mainly by the number of molecules in this level, i.e., it is proportional to $\bar{n}_J$. Hence (see Fig. 10.3) the intensities of the lines in the P- and R-branches *fall on curves of the same type* as that in Fig.* 10.2.

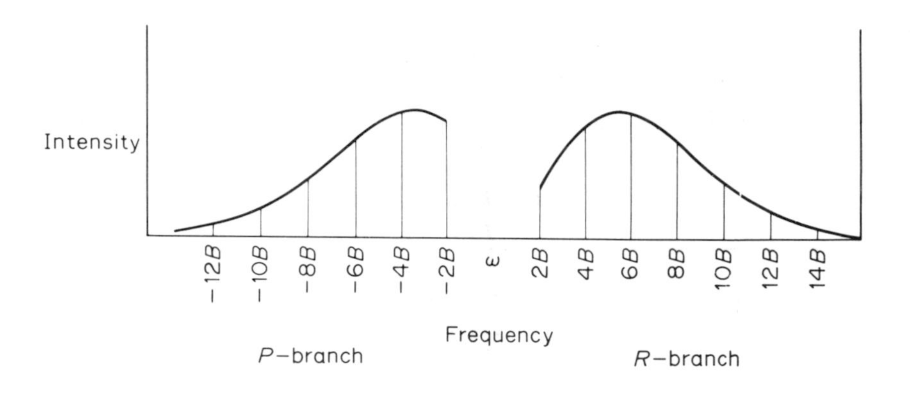

Fig. 10.3 Intensities in a vibration–rotation spectrum

The frequency separation between the maxima is found (see problem 10.3) from Equation 10.9 to be:

$$\sqrt{\frac{8T}{\Theta_{\mathrm{rot}}}}$$

This expression has a practical application: even if individual rotational lines have not been resolved, the approximate value of Θ_{rot}, and hence of I, can be calculated from the separation of the maximum intensities of the P- and R-branches.

Considerations such as we have outlined above are not restricted to the infrared spectra of heteronuclear diatomic molecules. The same type of effect is manifested in the rotational lines of the electronic, Raman, and pure rotation spectra of all

*Actually, the intensity also depends slightly upon ΔJ, and a more exact relationship is

$$I \propto (2J + 1 + \Delta J)e^{-J(J+1)\Theta_{\mathrm{rot}}/T}.$$

This means that a line in the P-branch ($\Delta J = -1$) is rather weaker than the line in the R-branch ($\Delta J = 1$) associated with the same J.

molecules, in so far as these spectra are observable. In considering the general case, however, the following should be noted.

(i) At ordinary temperatures, the first term in the vibrational partition function is often by far the most important; in such cases, only transitions from the lowest vibrational level need be considered (the same is true *a fortiori* of electronic transitions). For vibrations with low values of ω (see Table 5.2), the population of the second and higher vibrational levels may be important; for example, the ratio of the number of molecules in the $v = 1$ level to the number in the $v = 0$ level is 0·36 for I_2 at 300 K. In such circumstances, transitions *from* the higher vibrational levels contribute to the observed spectrum.

(ii) *Homonuclear* diatomic molecules do not exhibit a vibration–rotation spectrum. Changes in vibration–rotation energy can, however, be observed from electronic or Raman spectra. The intensities of such spectra are dependent upon the effect discussed above. There is also, however, a very important superimposed *alternation* of intensities of successive lines, arising from the occurrence of *ortho* and *para* states; we discuss this in Section 13.7.

(iii) Molecular vibrations are actually *anharmonic* (see page 70). This profoundly affects the molecular spectrum in two ways. Firstly, the vibrational (but *not* the rotational) selection rule is relaxed, so that $v' - v$ can have any integer value; the intensities of the vibrational bands diminish rapidly, however, with increasing values of $v' - v$.

Secondly, the parameter A appearing in Equation 10.4 has different values for different vibrational levels; in fact, to a good degree of approximation (see page 75)

$$A = A_e + \alpha(v + \tfrac{1}{2})$$

where α is a parameter for the molecule and A_e is calculated from the moment of inertia corresponding to the equilibrium internuclear distance. Equation 10.6a for the frequencies of the P-branch is now modified to:

$$\nu = \omega - 2JB + J(J-1)\alpha' v'$$

where $\alpha' = \alpha/h$ and $B = A_e/h + \alpha'/2$. The separation between successive lines in the P-branch is thus:

$$2B - 2J\alpha'v',$$

where J is the smaller of the two J-values involved. Similarly, Equation 10.6b for the R-branch is modified to:

$$\nu = \omega + 2(J+1)B + (J+1)(J+2)\alpha'v'$$

and the separation between successive lines is:

$$2B + 2(J+1)\alpha'v',$$

where J is the larger of the two J-values involved.

10.2 Energy units

Molecular energy levels are determined from the observed spectral frequencies by means of Equations 10.1 or 10.2. Now:

$$\nu = \frac{c}{\lambda}$$

where c is the speed of light:

$$c = 2{\cdot}998 \times 10^{8}\ \mathrm{m\,s^{-1}},$$

and λ the wavelength. The dimension of ν is thus $(\text{time})^{-1}$, and the unit of frequency is the hertz (1 Hz = 1 cycle per second or, in brief, 1 s^{-1}), Since Planck's constant is:

$$h = 6{\cdot}626 \times 10^{-34}\ \mathrm{J\,s}$$

Equation 10.1 becomes:

$$\epsilon_2 - \epsilon_1 = \frac{hc}{\lambda} = \frac{6{\cdot}626 \times 10^{-34} \times 2{\cdot}998 \times 10^{8}}{\lambda}$$

$$= \frac{1{\cdot}986 \times 10^{-25}}{\lambda}\ \mathrm{J}$$

when λ is expressed in *metres.*

The position of a line in the spectrum is often expressed in terms of $1/\lambda$, which is called the *wave number* of the line, and for this purpose $1/\lambda$ is expressed in *reciprocal centimetres* (cm^{-1}). If $\tilde{\nu}$ is the wave number in cm^{-1}, then:

$$\lambda = \frac{1}{100\tilde{\nu}}\ \mathrm{m}$$

and so:

$$\epsilon_2 - \epsilon_1 = 1{\cdot}986 \times 10^{-23}\,\tilde{\nu}\ \mathrm{J}$$

where $\tilde{\nu}$ is expressed in cm^{-1}.

Other units of energy besides the joule are in common use. The main ones are:

the electron volt (eV) = $1{\cdot}6021 \times 10^{-19}$ J

the thermochemical kilocalorie (k cal) = $4{\cdot}184 \times 10^{3}$ J

the erg = 10^{-7} J

For example:

$$\epsilon_2 - \epsilon_1 = \frac{1{\cdot}986 \times 10^{-23}}{1{\cdot}602 \times 10^{-19}} \tilde{\nu} \text{ eV}$$

$$= 1{\cdot}240 \times 10^{-4} \tilde{\nu} \text{ eV},$$

when $\tilde{\nu}$ is in cm^{-1}.

It is also common usage to express *molecular* energies in terms of the corresponding *energies per mole*:

$$\text{energy per mole} = \tilde{N} \times \text{molecular energy},$$

where $\tilde{N}$ is the Avogadro number.
For example, an energy of ϵ eV per molecule is equivalent to:

$$(6{\cdot}023 \times 10^{23} \times 1{\cdot}602 \times 10^{-19})\epsilon \text{ J mol}^{-1}$$

$$= 96{,}490\epsilon \text{ J mol}^{-1}$$

$$= 23{\cdot}06\epsilon \text{ k cal mol}^{-1}.$$

The relationship between the various units is shown in Table 10.1.

Table 10.1 Conversion factors.

		J molecule^{-1}	eV	cm^{-1}	J mol^{-1}	kcal mol^{-1}
1 J molecule^{-1}	≡	1	$6{\cdot}242 \times 10^{18}$	$5{\cdot}035 \times 10^{22}$	$6{\cdot}023 \times 10^{23}$	$1{\cdot}440 \times 10^{20}$
1 eV	≡	$1{\cdot}602 \times 10^{-19}$	1	8066	96490	23·06
1 cm^{-1}	≡	$1{\cdot}986 \times 10^{-23}$	$1{\cdot}240 \times 10^{-4}$	1	11·96	$2{\cdot}859 \times 10^{-3}$
1 J mol^{-1}	≡	$1{\cdot}660 \times 10^{-24}$	$1{\cdot}036 \times 10^{-5}$	$8{\cdot}359 \times 10^{-2}$	1	$2{\cdot}390 \times 10^{-4}$
1 kcal mol^{-1}	≡	$6{\cdot}947 \times 10^{-21}$	$4{\cdot}336 \times 10^{-2}$	349·7	$4{\cdot}184 \times 10^{3}$	1

We conclude this section by giving the orders of magnitude of the quantities involved. Table 10.2 shows the approximate wavelengths and energies associated with various regions of the electromagnetic spectrum.

Table 10.2 Approximate wavelengths and energies corresponding to various spectral regions

Region	m	cm^{-1}	J	eV
Microwave	10^{-3}	10	2×10^{-22}	$1{\cdot}2 \times 10^{-3}$
Far infrared	2×10^{-5}	5×10^{2}	10^{-20}	6×10^{-2}
Near infrared	7×10^{-7}	$1{\cdot}4 \times 10^{4}$	$2{\cdot}8 \times 10^{-19}$	$1{\cdot}7$
Ultraviolet and visible	10^{-7}	10^{5}	2×10^{-18}	12

Typical orders of magnitude of the vibrational and rotational constants for diatomic molecules are:

$$\omega = 10^{13}\,\mathrm{s}^{-1} \approx 10^{2}\,\mathrm{cm}^{-1} \approx 0{\cdot}1\ \mathrm{eV} \approx 10^{-20}\,\mathrm{J};$$

$$B = \frac{h}{8\pi^2 I} \approx 10^{10}\,\mathrm{s}^{-1} \approx 0{\cdot}1\ \mathrm{cm}^{-1} \approx 10^{-4}\ \mathrm{eV} \approx 10^{-23}\,\mathrm{J}$$

Hence, vibration–rotation spectra occur in the infrared region, the spacing between successive lines in the *P*- and *R*-branches being about 10^{-3} of the separation between successive vibrational bands. Pure rotation spectra ($v = 0$) are observed at much longer wavelengths, i.e., in the microwave region. Electronic excitation energies are usually greater than about 5 eV, and so electronic transitions give rise to spectra in the visible and ultraviolet regions.

10.3 Tables of thermodynamic functions in terms of molecular constants

Having completed our opening remarks, we proceed to the main object of the present chapter, namely, discussion of the calculation of thermodynamic properties from spectroscopic constants. We begin by giving, in Tables 10.3–10.6, the formulae for the calculation of heat capacities, entropies, and free energy functions. The tables do not include contributions from excited electronic levels; these, if they are important, can be evaluated by the method given in Section 10.7.

Each of the tables is divided into two parts. The first part gives the contributions to the thermodynamic property arising from translation, rotation, and the lowest electronic level. The second part expresses the contributions from vibration and restricted internal rotation; these are obtained from quantities tabulated separately in Sections 10.4 and 10.5.

The symbols in the tables have the following meanings.

$g_{el(0)}$: the degeneracy of the lowest electronic level.

M: the molecular weight (i.e., the mass per mole $= \tilde{M} = M$ g $= 10^{-3} M$ kg). This should, strictly speaking, be the molecular weight for an individual nuclide composition (e.g., for $^{35}Cl_2$ or $^{37}Cl_2$). In practice (see problem 10.5), the 'chemical' molecular weight of a substance containing isotopes can be used.

I: moment of inertia (linear or diatomic molecules).

I_{red}: reduced moment of inertia for an internal rotation (page 84).

A,B,C: the principal moments of inertia of non-linear molecules. The product (ABC) is easily evaluated as D, the determinant in Equation 5.26.

σ: the symmetry number for rotation (page 81).

σ_{int}: the symmetry number for internal rotation (page 84).

$\gamma = 47$ if the moments of inertia are evaluated in SI or M.K.S. units;

$\gamma = 40$ if they are evaluated in c.g.s. units.

The meaning and evaluation of the quantities $c_{vib}(u_i)$, $c_{i.r.}(f_{free}, V_m/RT)$, etc., is discussed in Sections 10.4 and 10.5.

Finally, note that the tables give the dimensionless quantities $\tilde{C}_V/R$, $\tilde{S}/R$, $[(\tilde{G}^\circ - \tilde{E}_0^\circ)/T]/R$. The corresponding thermodynamic function is obtained by multiplying these by R:

Units	R	$\log_{10}R$
cal K^{-1} mol^{-1}	1·987	0·2982
J K^{-1} mol^{-1}	8·314	0·9198

Table 10.3 Molar heat capacity, $\tilde{C}_V/R$

Contribution	Term in $\tilde{C}_V/R$
Translation	3/2
Rotation	
diatomic and linear	1
non-linear	3/2
Tables	
10.8 vibration	$\Sigma\, c_{\text{vib}}(u_l)$
10.11 internal rotation	$\Sigma\, c_{\text{i.r.}}(f_{\text{free}}, V_m/RT)$

Table 10.4 Standard Molar Conventional Entropy, $\tilde{S}^\circ/R$

Contribution	Term in $\tilde{S}^\circ/R$
Electronic	$2{\cdot}303 \log_{10} g_{\text{el}(0)}$
Translation	$2{\cdot}303(\frac{3}{2}\log_{10} M + \frac{5}{2}\log_{10} T - 0{\cdot}5058)$
Rotation	
diatomic and linear	$2{\cdot}303(\log_{10}[10^{\gamma} I] + \log_{10} T - \log_{10}\sigma - 1{\cdot}1706)$
non-linear	$2{\cdot}303(\frac{1}{2}\log_{10}[10^{3\gamma} ABC] + \frac{3}{2}\log_{10} T - \log_{10}\sigma - 1{\cdot}5072)$
free internal	$2{\cdot}303\Sigma(\frac{1}{2}\log_{10}[10^{\gamma} I_{\text{red}}] + \frac{1}{2}\log_{10} T - \log_{10}\sigma_{\text{int}} - 0{\cdot}3366)$
Tables	
10.9 vibration	$\Sigma\, s_{\text{vib}}(u_l)$
10.12 restriction on internal rotation	$-\Sigma\, S_{i.r.}(f_{\text{free}}, V_m/RT)$

Table 10.5 Entropy at non-standard pressures

$\tilde{S}/R = \tilde{S}^\circ/R - 2{\cdot}303 \log_{10} P$	P in atmospheres
$\tilde{S}/R = \tilde{S}^\circ/R - 2{\cdot}303(\log_{10} P - 2{\cdot}0056)$	P in kN m^{-2}
$\tilde{S}/R = \tilde{S}^\circ/R - 2{\cdot}303(\log_{10} T - \log_{10}\tilde{V} - 1{\cdot}0859)$	$\tilde{V}$ in litres

10.4 Vibrational contributions to thermodynamic functions

Tables 10.3, 10.4, and 10.6 contain, respectively, the sums:

$$\sum c_{\text{vib}}(u_l), \quad \sum s_{\text{vib}}(u_l), \quad \text{and} \quad \sum g_{\text{vib}}(u_l)$$

The sums are over terms arising from the individual normal modes of vibration. Hence, as already explained, if there are $\mathfrak{N}$ atoms in the molecule and if there is no internal rotations, there will be $3\mathfrak{N} - 5$ (for linear molecules) or $3\mathfrak{N} - 6$ (non-linear molecules)

Table 10.6 Free energy function, $[(\tilde{G}^\circ - \tilde{E}_0^\circ)/T]/R$

Contribution	Term in $[(\tilde{G}^\circ - \tilde{E}_0^\circ)/T]/R$
Electronic	$-2{\cdot}303 \log_{10} g_{el(o)}$
Translation	$-2{\cdot}303(\frac{3}{2}\log_{10} M + \frac{5}{2}\log_{10} T - 1{\cdot}592)$
Rotation	
diatomic and linear	$-2{\cdot}303(\log_{10}[10^{\gamma}I] + \log_{10} T - \log_{10}\sigma - 1{\cdot}6049)$
non-linear	$-2{\cdot}303(\frac{1}{2}\log_{10}[10^{3\gamma}ABC] + \frac{3}{2}\log_{10} T - \log_{10}\sigma - 2{\cdot}1582)$
free internal	$-2{\cdot}303\Sigma(\frac{1}{2}\log_{10}[10^{\gamma}I_{red}] + \frac{1}{2}\log_{10} T - \log_{10}\sigma_{int} - 0{\cdot}5533)$
Tables	
10.10 vibration	$\Sigma g_{vib}(u_l)$
10.13 restriction on internal rotation	$+\ \Sigma g_{i.r.}(f_{free}, V_m/RT)$

terms in the sum. These numbers are reduced by 1 for each internal rotation present.

The factor in the partition function associated with one normal mode is (cf. page 70):

$$f_{vib\,(l)} = \frac{1}{1 - e^{-u_l}}$$

where:

$$u_l = \frac{h\omega_l}{kT}$$

in which ω_l is the vibrational frequency of the normal mode.

The contributions $c_{vib}(u_l)$, $s_{vib}(u_l)$, and $g_{vib}(u_l)$ to, respectively, C_V, $\tilde{S}_0^\circ$, and $(\tilde{G}^\circ - \tilde{E}_0^\circ)/T$, arising from the factor $f_{vib\,(l)}$ have been considered previously (see Equation 6.11 and problem 5.5). The values are given in Tables 10.8–10.10. To use these tables, one first calculates u_l for the given frequency; since the latter may be expressed in different ways, the conversion factors are given in Table 10.7.

Table 10.7 Conversion factors for the calculation of u_i

Given	u_i
$\omega_i\,(s^{-1})$	$4{\cdot}800 \times 10^{-11}\ \omega_i/T$
$\tilde{\nu}_i\,(cm^{-1})$	$1{\cdot}439\ \tilde{\nu}_i/T$
$\epsilon = h\omega_i$ (eV)	$1{\cdot}161 \times 10^{4}\ \epsilon/T$
$\epsilon = h\omega_i$ (J)	$7{\cdot}245 \times 10^{22}\ \epsilon/T$

Table 10.8. Contribution to $\widetilde{C}_V/R$ from a harmonic oscillator: c_{vib} a function of $u = h\omega/kT$.

u	0.0	0.01	0.02	0.03	0.04	0.05	0.06	0.07	0.08	0.09	$\Delta c_v/\Delta u$
0.3	0.9925	0.9920	0.9915	0.9910	0.9904	0.9899	0.9893	0.9887	0.9881	0.9874	-0.06
0.4	0.9868	0.9861	0.9854	0.9847	0.9840	0.9833	0.9826	0.9818	0.9810	0.9802	-0.07
0.5	0.9794	0.9786	0.9778	0.9769	0.9761	0.9752	0.9743	0.9734	0.9724	0.9715	-0.09
0.6	0.9705	0.9696	0.9686	0.9676	0.9666	0.9655	0.9645	0.9634	0.9623	0.9613	-0.10
0.7	0.9601	0.9590	0.9579	0.9568	0.9556	0.9544	0.9532	0.9520	0.9508	0.9496	-0.12
0.8	0.9483	0.9471	0.9458	0.9445	0.9432	0.9419	0.9406	0.9392	0.9379	0.9365	-0.13
0.9	0.9351	0.9338	0.9324	0.9309	0.9295	0.9281	0.9266	0.9251	0.9237	0.9222	-0.14
1.0	0.9207	0.9192	0.9176	0.9161	0.9145	0.9130	0.9114	0.9098	0.9082	0.9066	-0.16
1.1	0.9050	0.9034	0.9017	0.9001	0.8984	0.8967	0.8950	0.8933	0.8916	0.8899	-0.17
1.2	0.8882	0.8864	0.8847	0.8829	0.8811	0.8794	0.8776	0.8758	0.8740	0.8721	-0.18
1.3	0.8703	0.8685	0.8666	0.8648	0.8629	0.8610	0.8591	0.8572	0.8553	0.8534	-0.19
1.4	0.8515	0.8496	0.8476	0.8457	0.8437	0.8418	0.8398	0.8378	0.8358	0.8339	-0.20
1.5	0.8318	0.8298	0.8278	0.8258	0.8238	0.8217	0.8197	0.8176	0.8156	0.8135	-0.20
1.6	0.8114	0.8093	0.8073	0.8052	0.8031	0.8010	0.7989	0.7967	0.7946	0.7925	-0.21
1.7	0.7903	0.7882	0.7861	0.7839	0.7817	0.7796	0.7774	0.7752	0.7731	0.7709	-0.22
1.8	0.7687	0.7665	0.7643	0.7621	0.7599	0.7577	0.7555	0.7532	0.7510	0.7488	-0.22
1.9	0.7466	0.7443	0.7421	0.7399	0.7376	0.7354	0.7331	0.7308	0.7286	0.7263	-0.22
2.0	0.7241	0.7218	0.7195	0.7172	0.7150	0.7127	0.7104	0.7081	0.7058	0.7036	-0.23
2.1	0.7013	0.6990	0.6967	0.6944	0.6921	0.6898	0.6875	0.6852	0.6829	0.6806	-0.23
2.2	0.6783	0.6760	0.6737	0.6713	0.6690	0.6667	0.6644	0.6621	0.6598	0.6575	-0.23
2.3	0.6552	0.6528	0.6505	0.6482	0.6459	0.6436	0.6413	0.6389	0.6366	0.6343	-0.23
2.4	0.6320	0.6297	0.6274	0.6251	0.6227	0.6204	0.6181	0.6158	0.6135	0.6112	-0.23
2.5	0.6089	0.6066	0.6043	0.6020	0.5997	0.5974	0.5951	0.5928	0.5905	0.5882	-0.23
2.6	0.5859	0.5836	0.5813	0.5790	0.5767	0.5745	0.5722	0.5699	0.5676	0.5653	-0.23
2.7	0.5631	0.5608	0.5585	0.5563	0.5540	0.5517	0.5495	0.5472	0.5450	0.5427	-0.23
2.8	0.5405	0.5382	0.5360	0.5338	0.5315	0.5293	0.5271	0.5249	0.5226	0.5204	-0.22
2.9	0.5182	0.5160	0.5138	0.5116	0.5094	0.5072	0.5050	0.5028	0.5006	0.4984	-0.22
	0.0	0.1	0.2	0.3	0.4	0.5	0.6	0.7	0.8	0.9	
3.0	0.4963	0.4747	0.4536	0.4330	0.4129	0.3933	0.3743	0.3558	0.3380	0.3207	-0.20
4.0	0.3041	0.2881	0.2726	0.2578	0.2436	0.2300	0.2170	0.2046	0.1928	0.1815	-0.14
5.0	0.1707	0.1605	0.1508	0.1416	0.1329	0.1246	0.1168	0.1094	0.1025	0.0959	-0.08
6.0	0.0897	0.0838	0.0783	0.0732	0.0683	0.0637	0.0594	0.0554	0.0516	0.0481	-0.05
7.0	0.0448	0.0417	0.0388	0.0360	0.0335	0.0311	0.0289	0.0269	0.0249	0.0232	-0.02

Table 10.9. Contribution to $\widetilde{S}/R$ from a harmonic oscillator: s_{vib} as a function of $u = \hbar\omega/kT$.

u	0.0	0.01	0.02	0.03	0.04	0.05	0.06	0.07	0.08	0.09	$\Delta s/\Delta u$
0.3	2.2077	2.1752	2.1437	2.1132	2.0836	2.0549	2.0270	1.9999	1.9736	1.9479	-2.44
0.4	1.9229	1.8986	1.8748	1.8516	1.8290	1.8069	1.7853	1.7642	1.7435	1.7233	-1.79
0.5	1.7035	1.6841	1.6651	1.6465	1.6282	1.6103	1.5928	1.5755	1.5586	1.5420	-1.38
0.6	1.5257	1.5097	1.4939	1.4784	1.4632	1.4482	1.4335	1.4190	1.4047	1.3907	-1.11
0.7	1.3768	1.3632	1.3498	1.3366	1.3236	1.3108	1.2982	1.2857	1.2734	1.2613	-0.91
0.8	1.2494	1.2376	1.2260	1.2145	1.2032	1.1921	1.1811	1.1702	1.1595	1.1489	-0.75
0.9	1.1384	1.1281	1.1179	1.1078	1.0979	1.0881	1.0784	1.0688	1.0593	1.0499	-0.64
1.0	1.0407	1.0315	1.0225	1.0135	1.0047	0.9959	0.9873	0.9787	0.9703	0.9619	-0.54
1.1	0.9536	0.9454	0.9374	0.9293	0.9214	0.9136	0.9058	0.8982	0.8906	0.8830	-0.47
1.2	0.8756	0.8682	0.8609	0.8537	0.8466	0.8395	0.8325	0.8256	0.8187	0.8119	-0.40
1.3	0.8052	0.7985	0.7919	0.7854	0.7789	0.7725	0.7662	0.7599	0.7537	0.7475	-0.35
1.4	0.7414	0.7353	0.7293	0.7234	0.7175	0.7117	0.7059	0.7002	0.6945	0.6889	-0.31
1.5	0.6833	0.6778	0.6723	0.6669	0.6615	0.6562	0.6509	0.6457	0.6405	0.6354	-0.27
1.6	0.6303	0.6252	0.6202	0.6153	0.6103	0.6055	0.6006	0.5958	0.5911	0.5864	-0.24
1.7	0.5817	0.5771	0.5725	0.5679	0.5634	0.5590	0.5545	0.5501	0.5458	0.5414	-0.21
1.8	0.5371	0.5329	0.5287	0.5245	0.5203	0.5162	0.5122	0.5081	0.5041	0.5001	-0.19
1.9	0.4962	0.4923	0.4884	0.4845	0.4807	0.4769	0.4732	0.4694	0.4657	0.4621	-0.17
2.0	0.4584	0.4548	0.4513	0.4477	0.4442	0.4407	0.4372	0.4338	0.4304	0.4270	-0.15
2.1	0.4237	0.4203	0.4170	0.4138	0.4105	0.4073	0.4041	0.4009	0.3978	0.3947	-0.13
2.2	0.3916	0.3885	0.3855	0.3824	0.3794	0.3765	0.3735	0.3706	0.3677	0.3648	-0.12
2.3	0.3619	0.3591	0.3563	0.3535	0.3507	0.3480	0.3452	0.3425	0.3399	0.3372	-0.11
2.4	0.3345	0.3319	0.3293	0.3267	0.3242	0.3216	0.3191	0.3166	0.3141	0.3117	-0.10
2.5	0.3092	0.3068	0.3044	0.3020	0.2996	0.2973	0.2949	0.2926	0.2903	0.2880	-0.09
2.6	0.2858	0.2835	0.2813	0.2791	0.2769	0.2747	0.2726	0.2704	0.2683	0.2662	-0.08
2.7	0.2641	0.2620	0.2600	0.2579	0.2559	0.2539	0.2519	0.2499	0.2479	0.2460	-0.07
2.8	0.2440	0.2421	0.2402	0.2383	0.2364	0.2346	0.2327	0.2309	0.2291	0.2272	-0.06
2.9	0.2255	0.2237	0.2219	0.2202	0.2184	0.2167	0.2150	0.2133	0.2116	0.2099	-0.06
	0.0	0.1	0.2	0.3	0.4	0.5	0.6	0.7	0.8	0.9	
3.0	0.2083	0.1923	0.1776	0.1640	0.1513	0.1396	0.1288	0.1188	0.1096	0.1010	-0.03
4.0	0.0931	0.0858	0.0790	0.0728	0.0670	0.0617	0.0568	0.0523	0.0481	0.0442	-0.01
5.0	0.0407	0.0374	0.0344	0.0316	0.0290	0.0267	0.0245	0.0225	0.0206	0.0190	-0.00
6.0	0.0174	0.0160	0.0146	0.0134	0.0123	0.0113	0.0104	0.0095	0.0087	0.0080	-0.00
7.0	0.0073	0.0067	0.0061	0.0056	0.0051	0.0047	0.0043	0.0039	0.0036	0.0033	-0.00

Table 10.10. Contribution to $-\left(\frac{\tilde{G}^\circ - \tilde{E}_0^\circ}{T}\right)\Big/R$ from a harmonic oscillator: $-g_{\text{vib}}$ as a function of $u = h\omega/kT$.

u	0.0	0.01	0.02	0.03	0.04	0.05	0.06	0.07	0.08	0.09	$\Delta/\Delta u$
0.3	1.3502	1.3222	1.2952	1.2691	1.2440	1.2197	1.1963	1.1736	1.1516	1.1303	-2.89
0.4	1.1096	1.0896	1.0702	1.0513	1.0329	1.0151	0.9977	0.9808	0.9644	0.9484	-2.22
0.5	0.9328	0.9175	0.9027	0.8882	0.8741	0.8603	0.8468	0.8336	0.8207	0.8082	-1.79
0.6	0.7959	0.7838	0.7721	0.7606	0.7493	0.7382	0.7274	0.7168	0.7065	0.6963	-1.50
0.7	0.6863	0.6766	0.6670	0.6576	0.6484	0.6394	0.6305	0.6218	0.6132	0.6049	-1.28
0.8	0.5966	0.5885	0.5806	0.5728	0.5651	0.5576	0.5502	0.5429	0.5358	0.5287	-1.12
0.9	0.5218	0.5150	0.5084	0.5018	0.4953	0.4890	0.4827	0.4766	0.4705	0.4645	-0.98
1.0	0.4587	0.4529	0.4472	0.4416	0.4361	0.4307	0.4253	0.4201	0.4149	0.4098	-0.88
1.1	0.4048	0.3998	0.3949	0.3901	0.3854	0.3807	0.3761	0.3716	0.3671	0.3627	-0.78
1.2	0.3584	0.3541	0.3499	0.3457	0.3416	0.3376	0.3336	0.3297	0.3258	0.3220	-0.71
1.3	0.3182	0.3145	0.3108	0.3072	0.3036	0.3001	0.2966	0.2932	0.2898	0.2864	-0.64
1.4	0.2832	0.2799	0.2767	0.2735	0.2704	0.2673	0.2643	0.2613	0.2583	0.2554	-0.58
1.5	0.2525	0.2496	0.2468	0.2440	0.2413	0.2386	0.2359	0.2333	0.2306	0.2281	-0.53
1.6	0.2255	0.2230	0.2205	0.2181	0.2156	0.2133	0.2109	0.2086	0.2063	0.2040	-0.49
1.7	0.2017	0.1995	0.1973	0.1951	0.1930	0.1909	0.1888	0.1867	0.1847	0.1827	-0.45
1.8	0.1807	0.1787	0.1768	0.1748	0.1729	0.1711	0.1692	0.1674	0.1656	0.1638	-0.41
1.9	0.1620	0.1603	0.1585	0.1568	0.1551	0.1535	0.1518	0.1502	0.1486	0.1470	-0.38
2.0	0.1454	0.1439	0.1423	0.1408	0.1393	0.1378	0.1363	0.1349	0.1335	0.1320	-0.35
2.1	0.1306	0.1292	0.1279	0.1265	0.1252	0.1238	0.1225	0.1212	0.1200	0.1187	-0.32
2.2	0.1174	0.1162	0.1150	0.1138	0.1126	0.1114	0.1102	0.1090	0.1079	0.1068	-0.30
2.3	0.1056	0.1045	0.1034	0.1024	0.1013	0.1002	0.0992	0.0981	0.0971	0.0961	-0.28
2.4	0.0951	0.0941	0.0931	0.0922	0.0912	0.0902	0.0893	0.0884	0.0875	0.0865	-0.25
2.5	0.0857	0.0848	0.0839	0.0830	0.0822	0.0813	0.0805	0.0796	0.0788	0.0780	-0.24
2.6	0.0772	0.0764	0.0756	0.0748	0.0740	0.0733	0.0725	0.0718	0.0710	0.0703	-0.22
2.7	0.0696	0.0689	0.0681	0.0674	0.0667	0.0661	0.0654	0.0647	0.0640	0.0634	-0.20
2.8	0.0627	0.0621	0.0615	0.0608	0.0602	0.0596	0.0590	0.0584	0.0578	0.0572	-0.19
2.9	0.0566	0.0560	0.0554	0.0549	0.0543	0.0538	0.0532	0.0527	0.0521	0.0516	-0.17
	0.0	0.1	0.2	0.3	0.4	0.5	0.6	0.7	0.8	0.9	
3.0	0.0511	0.0461	0.0416	0.0376	0.0339	0.0307	0.0277	0.0250	0.0226	0.0204	-0.12
4.0	0.0185	0.0167	0.0151	0.0137	0.0124	0.0112	0.0101	0.0091	0.0083	0.0075	-0.05
5.0	0.0068	0.0061	0.0055	0.0050	0.0045	0.0041	0.0037	0.0034	0.0030	0.0027	-0.02
6.0	0.0025	0.0022	0.0020	0.0018	0.0017	0.0015	0.0014	0.0012	0.0011	0.0010	-0.01
7.0	0.0009	0.0008	0.0007	0.0007	0.0006	0.0006	0.0005	0.0005	0.0004	0.0004	-0.00

10.5 Internal rotation

The partition function for a single free, or unrestricted, internal rotation is, according to Equation 5.35:

$$f_{\text{free}} = \frac{1}{\sigma_{\text{int}}}\left(\frac{8\pi^3 I_{\text{red}} kT}{h^2}\right)^{1/2}$$
$$= \frac{0{\cdot}2794}{\sigma_{\text{int}}}(10^{\gamma} I_{\text{red}}\, T)^{1/2} \qquad (10.10)$$

where γ is as defined in Section 10.3.

It was explained in Section 5.6 that, in many cases, the internal rotation cannot be regarded as free; the partition function then depends upon *both* I_{red} and $\mathbf{V}_{\text{m}}/T$, where $\mathbf{V}_{\text{m}}$ is the potential energy barrier to internal rotation. Tables* 10.11–10.13 give the contributions to $\tilde{C}_V$, $\tilde{S}$, and $(\tilde{G}^\circ - \tilde{E}_0^\circ)/T$, which then arise from one internal rotation. In the use of the tables, the following points should be noted.

(i) If there is more than one internal rotation in the molecule, the sum over the contributions of individual internal rotations is taken, subject to the limitations mentioned in Section 5.6.

(ii) For ease of tabulation, the contributions are given in terms of $\mathbf{V}_{\text{m}}/kT$ and $1/f_{\text{free}}$, so that one must first calculate f_{free} by means of Equation 10.10.

(iii) $\mathbf{V}_{\text{m}}$ is often expressed in units of kcal mol^{-1} or J mol^{-1}; in such cases, the ratio $\mathbf{V}_{\text{m}}/kT$ is replaced by $\mathbf{V}_{\text{m}}/RT$.

(iv) $c_{\text{i.r.}}\ (f_{\text{free}},\ \mathbf{V}_{\text{m}}/RT)$ is the *total* contribution to $\tilde{C}_V$ arising from one internal rotation. On the other hand (for easier interpolation), $S_{\text{i.r.}}\ (f_{\text{free}},\ \mathbf{V}_{\text{m}}/RT)$ and $g_{\text{i.r.}}\ (f_{\text{free}},\ \mathbf{V}_{\text{m}}/RT)$ *are only the correction terms arising when* $\mathbf{V}_{\text{m}} \neq 0$. For this reason, Tables 10.4 and 10.6 contain expressions for the contributions from free internal rotations, and these are included for *all* internal rotations.

* These are based upon tables given by K.S. Pitzer and co-workers: Pitzer and Gwinn, *J. Chem. Phys.*, **10**, 428 (1942). Li and Pitzer, *J. Phys. Chem.*, **60**, 466 (1956). Lewis and Randall, pp.441–446.

Table 10.11. Contribution to C_V/R from internal rotation: $c_{i.r.}$.

$\frac{V_m}{kT}$	$1/f_{free}$									
	0.0	0.05	0.10	0.15	0.20	0.25	0.30	0.35	0·40	0.45
0.0	0.500	0.500	0.500	0.500	0.500	0.500	0.500	0.500	0.500	0.500
0.2	0.505	0.505	0.505	0.504	0.504	0.503	0.503	0.502	0.502	0.502
0.4	0.520	0.520	0.519	0.518	0.517	0.516	0.515	0.514	0.513	0.512
0.6	0.544	0.543	0.543	0.541	0.540	0.537	0.536	0.533	0.531	0.529
0.8	0.575	0.575	0.574	0.573	0.570	0.568	0.564	0.561	0.557	0.553
1.0	0.614	0.613	0.612	0.610	0.607	0.603	0.599	0.594	0.588	0.582
1.5	0.730	0.729	0.727	0.722	0.716	0.708	0.700	0.689	0.678	0.666
2.0	0.844	0.853	0.849	0.842	0.833	0.821	0.808	0.792	0.775	0.757
2.5	0.967	0.965	0.960	0.950	0.939	0.926	0.906	0.884	0.864	0.840
3.0	1.056	1.054	1.048	1.038	1.023	1.004	0.982	0.956	0.929	0.903
3.5	1.118	1.116	1.109	1.097	1.080	1.060	1.034	1.004	0.973	0.940
4.0	1.157	1.154	1.145	1.132	1.114	1.091	1.062	1.031	0.996	0.960
4.5	1.175	1.172	1.163	1.147	1.126	1.102	1.071	1.038	1.001	0.962
5.0	1.180	1.176	1.166	1.150	1.128	1.100	1.067	1.035	0.992	0.951
6.0	1.165	1.161	1.149	1.130	1.103	1.072	1.036	0.996	0.953	0.907
7.0	1.140	1.135	1.121	1.099	1.070	1.034	0.993	0.948	0.899	0.849
8.0	1.115	1.110	1.094	1.069	1.036	0.996	0.950	0.900	0.847	0.793
9.0	1.095	1.089	1.072	1.044	1.006	0.961	0.910	0.855	0.799	0.742
10.0	1.080	1.073	1.054	1.023	0.982	0.933	0.878	0.820	0.758	0.695
12.0	1.059	1.051	1.028	0.992	0.944	0.887	0.823	0.756	0.687	0.620
14.0	1.047	1.038	1.011	0.968	0.913	0.848	0.778	0.704	0.631	0.560
16.0	1.039	1.029	0.998	0.950	0.888	0.816	0.739	0.660	0.582	0.508
18.0	1.034	1.022	0.987	0.932	0.864	0.786	0.703	0.620	0.538	0.462
20.0	1.030	1.016	0.978	0.919	0.844	0.760	0.671	0.583	0.499	0.421

$\frac{V_m}{kT}$	$1/f_{free}$									
	0.50	0.55	0.60	0.65	0.70	0.75	0.80	0.85	0.90	0.95
0.0	0.500	0.500	0.500	0.500	0.500	0.500	0.500	0.500	0.500	0.500
0.2	0.503	0.503	0.503	0.503	0.503	0.503	0.503	0.503	0.503	0.503
0.4	0.512	0.512	0.511	0.510	0.509	0.508	0.507	0.507	0.506	0.505
0.6	0.528	0.526	0.524	0.521	0.519	0.516	0.514	0.512	0.510	0.509
0.8	0.549	0.545	0.541	0.537	0.532	0.528	0.523	0.519	0.516	0.513
1.0	0.576	0.569	0.563	0.556	0.549	0.542	0.536	0.529	0.523	0.519
1.5	0.654	0.641	0.627	0.613	0.600	0.586	0.574	0.561	0.548	0.538
2.0	0.737	0.717	0.695	0.675	0.654	0.633	0.613	0.594	0.577	0.560
2.5	0.815	0.786	0.757	0.729	0.701	0.675	0.649	0.623	0.599	0.577
3.0	0.872	0.837	0.804	0.771	0.738	0.705	0.673	0.642	0.612	0.586
3.5	0.907	0.869	0.832	0.795	0.758	0.721	0.685	0.651	0.617	0.586
4.0	0.923	0.883	0.842	0.802	0.761	0.722	0.684	0.647	0.611	0.578
4.5	0.922	0.880	0.837	0.794	0.753	0.711	0.671	0.634	0.596	0.561
5.0	0.910	0.864	0.821	0.776	0.733	0.691	0.650	0.611	0.574	0.537
6.0	0.589	0.812	0.765	0.719	0.675	0.632	0.590	0.552	0.514	0.480
7.0	0.799	0.748	0.699	0.652	0.607	0.564	0.523	0.484	0.448	0.416
8.0	0.739	0.687	0.635	0.586	0.540	0.497	0.457	0.420	0.385	0.354
9.0	0.685	0.629	0.576	0.527	0.481	0.437	0.397	0.361	0.328	0.298
10.0	0.635	0.579	0.526	0.475	0.428	0.385	0.346	0.311	0.280	0.251
12.0	0.557	0.498	0.441	0.389	0.343	0.302	0.266	0.233	0.205	0.180
14.0	0.492	0.445	0.374	0.324	0.279	0.241	0.207	0.177	0.152	0.132
16.0	0.439	0.377	0.322	0.273	0.230	0.195	0.163	0.137	0.115	0.098
18.0	0.392	0.331	0.276	0.229	0.190	0.157	0.130	0.108	0.088	0.072
20.0	0.353	0.292	0.240	0.196	0.159	0.129	0.105	0.085	0.068	0.055

Table 10.12. Correction term $S_{i.r.}$ to be subtracted from entropy of internal rotation.

$\frac{V_m}{kT}$	$1/f_{free}$									
	0.0	0.05	0.10	0.15	0.20	0.25	0.30	0.35	0.40	0.45
0.0	0.000	0.000	0.000	0.000	0.000	0.000	0.000	0.000	0.000	0.000
0.2	0.002	0.003	0.002	0.002	0.002	0.003	0.002	0.002	0.002	0.001
0.4	0.010	0.010	0.009	0.009	0.009	0.009	0.008	0.008	0.007	0.006
0.6	0.022	0.022	0.022	0.022	0.020	0.020	0.020	0.019	0.017	0.017
0.8	0.039	0.039	0.039	0.038	0.036	0.035	0.034	0.034	0.033	0.031
1.0	0.060	0.059	0.059	0.058	0.056	0.056	0.054	0.053	0.051	0.048
1.5	0.127	0.127	0.126	0.125	0.122	0.119	0.116	0.114	0.108	0.103
2.0	0.210	0.210	0.209	0.206	0.202	0.198	0.192	0.187	0.179	0.171
2.5	0.302	0.301	0.299	0.294	0.290	0.286	0.277	0.268	0.257	0.246
3.0	0.395	0.394	0.391	0.386	0.381	0.373	0.362	0.352	0.340	0.326
3.5	0.486	0.485	0.482	0.475	0.467	0.458	0.446	0.434	0.421	0.402
4.0	0.571	0.570	0.567	0.559	0.550	0.539	0.525	0.509	0.493	0.473
4.5	0.650	0.649	0.644	0.637	0.626	0.614	0.597	0.581	0.562	0.538
5.0	0.722	0.720	0.715	0.706	0.694	0.681	0.663	0.645	0.622	0.598
6.0	0.844	0.842	0.836	0.827	0.813	0.796	0.776	0.752	0.727	0.699
7.0	0.945	0.943	0.936	0.924	0.909	0.888	0.866	0.840	0.810	0.779
8.0	1.029	1.027	1.018	1.005	0.987	0.965	0.940	0.910	0.877	0.843
9.0	1.100	1.097	1.088	1.074	1.054	1.029	1.001	0.968	0.932	0.894
10.0	1.162	1.159	1.149	1.133	1.111	1.085	1.052	1.016	0.978	0.937
12.0	1.266	1.262	1.250	1.231	1.205	1.173	1.138	1.095	1.050	1.003
14.0	1.351	1.347	1.333	1.312	1.282	1.245	1.204	1.156	1.104	1.051
16.0	1.423	1.419	1.403	1.379	1.346	1.304	1.256	1.204	1.148	1.090
18.0	1.487	1.481	1.464	1.437	1.399	1.354	1.301	1.243	1.183	1.121
20.0	1.543	1.537	1.518	1.487	1.445	1.396	1.338	1.277	1.212	1.146

$\frac{V_m}{kT}$	$1/f_{free}$									
	0.50	0.55	0.60	0.65	0.70	0.75	0.80	0.85	0.90	0.95
0.0	0.000	0.000	0.000	0.000	0.000	0.000	0.000	0.000	0.000	0.000
0.2	0.001	0.001	0.003	0.001	0.001	0.001	0.002	0.002	0.004	0.002
0.4	0.006	0.006	0.006	0.006	0.006	0.005	0.005	0.005	0.005	0.005
0.6	0.016	0.015	0.014	0.013	0.013	0.012	0.011	0.011	0.010	0.010
0.8	0.028	0.029	0.026	0.025	0.024	0.021	0.019	0.016	0.016	0.014
1.0	0.046	0.044	0.041	0.038	0.037	0.033	0.030	0.027	0.025	0.022
1.5	0.100	0.093	0.088	0.083	0.076	0.069	0.064	0.058	0.055	0.048
2.0	0.163	0.155	0.146	0.137	0.127	0.117	0.107	0.098	0.090	0.080
2.5	0.234	0.223	0.211	0.198	0.185	0.172	0.157	0.144	0.131	0.119
3.0	0.310	0.305	0.278	0.262	0.244	0.227	0.209	0.192	0.175	0.159
3.5	0.383	0.364	0.344	0.324	0.303	0.282	0.261	0.239	0.218	0.198
4.0	0.451	0.430	0.407	0.384	0.360	0.334	0.310	0.286	0.262	0.238
4.5	0.515	0.489	0.464	0.439	0.412	0.383	0.356	0.329	0.300	0.274
5.0	0.573	0.546	0.517	0.487	0.457	0.427	0.397	0.367	0.336	0.307
6.0	0.670	0.637	0.604	0.571	0.536	0.502	0.467	0.433	0.398	0.364
7.0	0.745	0.708	0.673	0.635	0.597	0.560	0.521	0.483	0.445	0.408
8.0	0.806	0.766	0.726	0.686	0.644	0.604	0.562	0.521	0.481	0.441
9.0	0.854	0.811	0.768	0.726	0.681	0.637	0.593	0.550	0.507	0.465
10.0	0.893	0.848	0.803	0.756	0.709	0.663	0.617	0.572	0.528	0.484
12.0	0.954	0.903	0.852	0.802	0.749	0.699	0.650	0.602	0.555	0.509
14.0	0.998	0.942	0.887	0.832	0.778	0.724	0.673	0.620	0.571	0.523
16.0	1.031	0.972	0.912	0.853	0.797	0.741	0.685	0.633	0.581	0.532
18.0	1.057	0.995	0.931	0.870	0.810	0.751	0.695	0.641	0.588	0.538
20.0	1.078	1.012	0.946	0.882	0.820	0.760	0.702	0.646	0.593	0.542

Table 10.13. Correction term $g_{i.r.}$ to be added to free energy function for internal rotation.

$\frac{V_m}{kT}$	$1/f_{free}$									
	0.0	0.05	0.10	0.15	0.20	0.25	0.30	0.35	0.40	0.45
0.0	0.000	0.000	0.000	0.000	0.000	0.000	0.000	0.000	0.000	0.000
0.2	0.097	0.077	0.059	0.043	0.031	0.023	0.017	0.013	0.009	0.006
0.4	0.190	0.164	0.138	0.113	0.089	0.066	0.049	0.036	0.028	0.022
0.6	0.278	0.246	0.213	0.182	0.150	0.119	0.093	0.072	0.057	0.045
0.8	0.360	0.322	0.285	0.248	0.211	0.176	0.144	0.116	0.093	0.073
1.0	0.438	0.395	0.352	0.310	0.270	0.231	0.196	0.162	0.132	0.105
1.5	0.614	0.561	0.508	0.457	0.407	0.360	0.313	0.271	0.227	0.190
2.0	0.764	0.702	0.642	0.583	0.526	0.471	0.417	0.365	0.316	0.269
2.5	0.892	0.823	0.755	0.690	0.627	0.566	0.505	0.447	0.391	0.340
3.0	1.001	0.925	0.852	0.781	0.712	0.645	0.580	0.517	0.456	0.399
3.5	1.095	1.013	0.934	0.857	0.783	0.712	0.642	0.575	0.513	0.449
4.0	1.176	1.088	1.004	0.922	0.843	0.770	0.694	0.624	0.558	0.490
4.5	1.247	1.154	1.065	0.979	0.896	0.816	0.739	0.665	0.596	0.525
5.0	1.309	1.212	1.118	1.028	0.940	0.856	0.777	0.700	0.626	0.555
6.0	1.414	1.308	1.206	1.108	1.014	0.924	0.838	0.755	0.676	0.601
7.0	1.501	1.386	1.277	1.171	1.071	0.974	0.884	0.797	0.714	0.635
8.0	1.575	1.452	1.335	1.224	1.117	1.016	0.920	0.828	0.742	0.660
9.0	1.639	1.509	1.385	1.268	1.156	1.051	0.950	0.855	0.765	0.680
10.0	1.695	1.558	1.429	1.305	1.189	1.079	0.975	0.876	0.783	0.696
12.0	1.791	1.642	1.501	1.368	1.242	1.124	1.013	0.909	0.811	0.720
14.0	1.872	1.711	1.559	1.417	1.284	1.159	1.042	0.933	0.831	0.736
16.0	1.942	1.770	1.609	1.458	1.317	1.187	1.065	0.952	0.846	0.748
18.0	2.002	1.821	1.650	1.492	1.346	1.210	1.083	0.966	0.857	0.757
20.0	2.057	1.865	1.687	1.522	1.369	1.228	1.098	0.977	0.867	0.764

$\frac{V_m}{kT}$	$1/f_{free}$									
	0.50	0.55	0.60	0.65	0.70	0.75	0.80	0.85	0.90	0.95
0.0	0.000	0.000	0.000	0.000	0.000	0.000	0.000	0.000	0.000	0.000
0.2	0.005	0.003	0.003	0.002	0.001	0.001	0.001	0.001	0.001	0.000
0.4	0.018	0.012	0.009	0.007	0.005	0.003	0.003	0.003	0.002	0.001
0.6	0.034	0.026	0.021	0.015	0.011	0.008	0.006	0.004	0.003	0.002
0.8	0.056	0.045	0.034	0.026	0.020	0.014	0.010	0.007	0.005	0.003
1.0	0.082	0.066	0.052	0.040	0.031	0.023	0.016	0.011	0.007	0.004
1.5	0.155	0.126	0.101	0.079	0.061	0.046	0.032	0.023	0.016	0.009
2.0	0.227	0.187	0.152	0.121	0.095	0.072	0.053	0.037	0.025	0.014
2.5	0.290	0.243	0.201	0.162	0.128	0.099	0.073	0.052	0.034	0.020
3.0	0.343	0.292	0.244	0.200	0.160	0.125	0.093	0.066	0.044	0.025
3.5	0.390	0.334	0.281	0.232	0.188	0.148	0.111	0.079	0.053	0.030
4.0	0.428	0.369	0.313	0.260	0.212	0.167	0.128	0.092	0.061	0.034
4.5	0.460	0.398	0.340	0.285	0.232	0.185	0.142	0.104	0.068	0.038
5.0	0.488	0.423	0.361	0.303	0.249	0.199	0.153	0.111	0.074	0.041
6.0	0.530	0.460	0.395	0.334	0.276	0.222	0.171	0.125	0.083	0.045
7.0	0.559	0.488	0.420	0.356	0.295	0.237	0.184	0.134	0.089	0.047
8.0	0.582	0.508	0.438	0.371	0.308	0.248	0.193	0.141	0.093	0.049
9.0	0.600	0.523	0.450	0.383	0.317	0.256	0.199	0.146	0.096	0.049
10.0	0.613	0.535	0.461	0.391	0.325	0.263	0.205	0.149	0.098	0.050
12.0	0.633	0.552	0.475	0.404	0.336	0.271	0.211	0.153	0.100	0.051
14.0	0.647	0.563	0.485	0.412	0.342	0.276	0.215	0.156	0.102	0.051
16.0	0.657	0.571	0.491	0.417	0.346	0.280	0.217	0.158	0.103	0.051
18.0	0.664	0.577	0.496	0.420	0.349	0.282	0.219	0.159	0.104	0.051
20.0	0.669	0.581	0.499	0.422	0.351	0.283	0.220	0.161	0.104	0.051

10.6 Examples of the use of the tables

Example 10.6(a)

Calculate the standard molar entropy of argon at 300 K.

Argon is a monatomic gas, and $g_{el(0)} = 1$, so the only contribution to $\tilde{S}^\circ$ is $\tilde{S}^\circ_{trans}$. Since $M = 39{\cdot}95$:

$$\begin{aligned}\tilde{S}^\circ &= 2{\cdot}303R(\tfrac{3}{2} \times 1{\cdot}6015 + \tfrac{5}{2} \times 2{\cdot}4771 - 0{\cdot}5058)\\ &= 18{\cdot}63R\\ &= 154{\cdot}9\ \mathrm{J\,K^{-1}\,mol^{-1}}\end{aligned}$$

□ □ □

Example 10.6(b)

Calculate the standard molar entropy of carbon dioxide at 300 K, from the following data. The observed vibrational frequencies are 667·3, 1337, and 2349·3 $\mathrm{cm^{-1}}$, the 667·3 $\mathrm{cm^{-1}}$ vibration being doubly degenerate. The moment of inertia is $71{\cdot}1 \times 10^{-47}\ \mathrm{kg\,m^2}$ and the molecular weight 44.

The lowest electronic level of CO_2 is non-degenerate. The translational and rotational contribution to $\tilde{S}^\circ$ is (with $\sigma = 2$):

$$\begin{aligned}\tilde{S}^\circ_{trans} + \tilde{S}_{rot} &= 2{\cdot}303R(\tfrac{3}{2}\log_{10} 44 + \tfrac{5}{2}\log_{10} 300 - 0{\cdot}5058\\ &\quad + \log_{10} 71{\cdot}1 + \log_{10} 300 - \log_{10} 2 - 1{\cdot}1706)\\ &= 25{\cdot}36R.\end{aligned}$$

CO_2 is a linear molecule, hence there are:

$$9 - 5 = 4$$

normal modes of vibration. The fact that only three different values of ω_l are observed arises from the symmetry of the molecule, and it can be deduced that two of the normal modes have the same frequency, 667·3 $\mathrm{cm^{-1}}$. From Table 10.7, the values of u_i are:

$$u_1 = u_2 = \frac{1{\cdot}439 \times 667{\cdot}3}{300} = 3{\cdot}200$$

$$u_3 = \frac{1{\cdot}439 \times 1337}{300} = 6{\cdot}413$$

$$u_4 = \frac{1{\cdot}439 \times 2349{\cdot}3}{300} = 11{\cdot}27.$$

The corresponding entropy contributions are found from Table 10.4:

$$\tilde{S}_{vib} = R(2 \times 0{\cdot}1776 + 0{\cdot}0122 + 0{\cdot}0000)$$
$$= 0{\cdot}367R\,.$$

Finally, therefore:

$$\tilde{S}^{\circ} = 25{\cdot}73R$$
$$= 213{\cdot}9\,\mathrm{J\,K^{-1}\,mol^{-1}}$$

□ □ □

Example 10.6(c)

A reported value of the calorimetric entropy of ethane is 229·5 ± 0·8 J K^{-1}mol^{-1} at 298 K. The spectroscopic entropy, calculated on the assumption that internal rotation is free, is 235·8 J K^{-1} mol^{-1}. Assuming that the difference between the two values is due to restriction of the internal rotation, calculate (i) the barrier to internal rotation, and (ii) the contribution to the molar heat capacity from the internal rotation.

(i) As in example 5.6(a), $f_{free} = 2{\cdot}64$ for ethane at 298 K, and $1/f_{free} = 0{\cdot}379$. The entropy decrease from the free rotation value is (235·8 − 229·5) J K^{-1} mol^{-1} = 0·75R. By interpolation of the values in Table 10.12, it is seen that this corresponds to $\mathbf{V}_m/RT = 6{\cdot}18$, hence:

$$\mathbf{V}_m = 15{\cdot}3 \times 10^3\,\mathrm{J\,mol^{-1}}$$

Note that this is not the same as the value quoted in Table 5.5; an error of 1 J K^{-1} mol^{-1} in the entropy *difference* produces, however, an error of 1·0 in $\mathbf{V}_m/RT$, and therefore an error of 2·5 × 10^3 J mol^{-1} in $\mathbf{V}_m$.

(ii) From Table 10.11, with $1/f_{free} = 0{\cdot}379$, and $\mathbf{V}_m/RT = 6{\cdot}18$, we see that the contribution to C_V from internal rotation is 0·91R = 7·6 J K^{-1} mol^{-1}.

□ □ □

10.7 Direct evaluation of thermodynamic properties from observed energy levels

If the internal molecular energy levels $\epsilon_{int(i)}$ and their degeneracies g_i are known, then f_{int} can be evaluated by the direct numerical summation of:

$$f_{int} = \sum_i g_i \mathrm{e}^{-\epsilon_{int(i)}/kT} \qquad (10.11)$$

Since the convention is adopted of measuring energies relative to the lowest level (cf. Section 5.1), the molar internal (in the mechanical, not the thermodynamical, sense) energy relative to the zero point energy is:

$$(\tilde{E} - \tilde{E}_0)_{\text{int}} = RT^2 \left(\frac{\partial \ln f_{\text{int}}}{\partial T}\right)_V = \frac{RT^2}{f_{\text{int}}} \left(\frac{\partial f_{\text{int}}}{\partial T}\right)_V$$

The molar heat capacity is obtained, as usual, by differentiating this expression, and the molar entropy is:

$$\tilde{S}^\circ_{\text{int}} = R\left[\ln f_{\text{int}} + \frac{T}{f_{\text{int}}}\left(\frac{\partial f_{\text{int}}}{\partial T}\right)_V\right]$$

The final results are most conveniently expressed in terms of the functions:

$$f' = \sum_i g_i \left(\frac{\epsilon_{\text{int}(i)}}{kT}\right) e^{-\epsilon_{\text{int}(i)}/kT} \tag{10.12}$$

and

$$f'' = \sum_i g_i \left(\frac{\epsilon_{\text{int}(i)}}{kT}\right)^2 e^{-\epsilon_{\text{int}(i)}/kT} \tag{10.13}$$

Then:

$$(\tilde{E} - \tilde{E}_0)_{\text{int}} = RT\left(\frac{f'}{f_{\text{int}}}\right) \tag{10.14}$$

$$\tilde{C}_{V(\text{int})} = R\left[\frac{f''}{f_{\text{int}}} - \left(\frac{f'}{f_{\text{int}}}\right)^2\right] \tag{10.15}$$

$$\tilde{S}_{\text{int}} = R\left(\ln f_{\text{int}} + \frac{f'}{f_{\text{int}}}\right) \tag{10.16}$$

and

$$\left(\frac{\tilde{G}^\circ - \tilde{E}^\circ_0}{T}\right)_{\text{int}} = -R\ln f_{\text{int}} \tag{10.17}$$

In the computation of a thermodynamic function, f_{int}, f', and f'' are calculated simultaneously, as the sums 10.11, 10.12, and 10.13, respectively.

Example 10.7

Calculate the electronic contribution to the entropy of Ni(g) at 298·2 K, from the energy levels and degeneracies given in Table 10.14. (This particular example, although of little physical interest is convenient for the purpose of illustration.)

Table 10.14

State	Electronic ϵ_i (joules per atom) $\times 10^{21}$	g_i	$\frac{\epsilon_i}{kT}$	$g_i e^{-\epsilon_i/kT}$	$g_i \frac{\epsilon_i}{kT} e^{-\epsilon_i/kT}$
$^3F_4$	0	9	0	9·000	0
$^3D_3$	4·0664	7	0·988	2·606	2·574
$^3D_2$	17·470	5	4·245	0·072	0·304
$^3F_3$	26·453	7	6·427	0·011	0·073
$^3D_1$	34·017	3	8·265	0·001	0·006
$^3F_2$	44·013	5	10·694	0·000	0·001
Sum				11·690	2·958

The scheme of computation is shown in Table 10.14. Finally:

$$f_{el} = 11{\cdot}690 \quad \text{and} \quad f'_{el} = 2{\cdot}958.$$

Therefore:

$$\tilde{S}_{el} = R\left(\ln 11{\cdot}690 + \frac{2{\cdot}958}{11{\cdot}690}\right)$$

$$= 22{\cdot}55 \text{ J K}^{-1}\text{ mol}^{-1}.$$

□ □ □

This method of calculating thermodynamic properties directly from the spectrum has been applied to the vibration–rotation spectra of simple molecules. It is more accurate [± 0·02% compared with ± 0·2%, in the case of Cl_2(g) at 300 K] than the method described in Section 10.3, but it involves much more computation than the latter; its most important application is to the evaluation of contributions from excited electronic states.

Problems

10.1 The first three lines in the pure rotation ($v' - v = 0$) spectrum of HI occur in the microwave region, and have the following frequencies (cm^{-1}): 128, 256, and 385. Calculate B for HI and the rotational entropy of HI at 300 K.

10.2 Czerny observed part of the pure rotation spectrum of HCl in the far infrared region. The following frequencies (cm^{-1}) were reported: 83·03, (104·1), 124·30, 145·03, 165·51, 185·86, 206·38, 226·50. From the separations between successive lines, determine an average value of B for HCl, and deduce the values of J for the upper levels involved in the transitions. Calculate the rotational entropy of HCl at 300 K.

10.3 Obtain Equation 10.9 and deduce that the frequency separation between the lines of maximum intensity in the P- and R-branches is $\sqrt{(8T/\Theta_{rot})}$. In early work, the separation was found to be 55 cm^{-1} in the case of CO. Estimate the moment of inertia of CO.

10.4 The following is the rotational structure of one of the vibration–rotation bands of $H^{35}Cl$; the frequencies are, in cm^{-1}:

P-branch	R-branch
2865·09	2906·25
43·56	25·78
21·49	44·89
2798·78	63·24
75·79	80·90
52·03	97·78
27·75	3014·29
03·06	29·96
2677·73	44·88
51·97	59·07
25·74	72·76

Plot the separations between successive lines against J or $J + 1$ as appropriate (see page 189) and determine the rotational constant B for $H^{35}Cl$. Use this value to calculate the rotational contribution to the free energy function at 300 K.

10.5 The translational entropy of a mixture of x_A mole fractions of A and x_B mole fractions of B is proportional to $(x_A \ln M_A + x_B \ln M_B)$. Show that the error introduced by replacing this term by $\ln M$, where M is the apparent (averaged) molecular weight $(x_A M_A + x_B M_B)$, is proportional to $\ln(\overline{M}/M)$, where $\overline{M}$ is the geometric mean $M_A{}^{x_A} M_B{}^{x_B}$. Deduce, that, if A and B differ only in isotopic constitution, the error is, in practice negligible. (N.B. this has nothing to do with the entropy of mixing referred to on page 153).

10.6 Calculate the standard molar spectroscopic entropy of NH_3(g) from the following data. The principal moments of inertia are $A = 4{\cdot}437 \times 10^{-47}$, $B = C = 2{\cdot}016 \times 10^{-47}\,\mathrm{kg\,m^2}$. The vibrational frequencies (in $\mathrm{cm^{-1}}$) are 950, 1628(2), 3336, and 3444. Compare with the calorimetric value $45{\cdot}98 \pm 0{\cdot}10\,\mathrm{cal\,K^{-1}\,mol^{-1}}$.

10.7 Calculate the contribution of the following molecular vibration frequencies to the value of $\tilde{C}_V$ at 200 K for solid benzene; the values are in $\mathrm{cm^{-1}}$:

406(2)	1008	1190
538	1030(2)	1480(2)
608(2)	1145	1520
783	1160	1645(2)
849·7(2)	1170(2)	1854
992·6		

There are six other frequencies with values greater than $3000\,\mathrm{cm^{-1}}$.

10.8 (a) The Debye characteristic temperature of the benzene crystal (derived from measured $\tilde{C}_V$ values below 50 K) is $\Theta_D = 150$ K. Calculate the contribution to $\tilde{C}_V$ from the *lattice* vibrations (see Section 6.4 and Table 6.3) to $\tilde{C}_V$ at 200 K. Combine this value with the result of the preceding problem to determine $\tilde{C}_V$ for the benzene crystal at 200 K.

(b) Calculate $\tilde{C}_P$ for benzene at 200 K from the formula for $\tilde{C}_P - \tilde{C}_V$ given in the footnote to page 105 (melting point of benzene = 278 K. Experimentally, $\tilde{C}_P = 10{\cdot}00R$ for benzene at 200 K.

10.9 Using the data and results of Examples 5.6 and Table 5.5, calculate the contributions from restricted internal rotation to the entropy, heat capacity, and free energy function of ethane and propane.

10.10 For methylamine CH_3NH_2(g) at 267 K:

$$\tilde{S}_{\mathrm{cal}} - \tilde{S}_{\mathrm{free}} = -0{\cdot}82R\,,$$

where $\tilde{S}_{\mathrm{free}}$ is the molar spectroscopic entropy calculated by assuming *free* internal rotation. Assuming that the difference between the two values of $\tilde{S}$ is due entirely to the non-zero value of $\mathbf{V}_{\mathrm{m}}$, calculate the rotational barrier in methylamine. [Assume that the geometry of the methyl group is the same as in ethane (problem 5.6). The lengths of the C–N and N–H bonds are respectively $1{\cdot}47 \times 10^{-10}$ m and $1{\cdot}01 \times 10^{-10}$ m; the CNH angle is 112°.]

10.11 From the infrared spectrum of methyl chloroform, CH_3CCl_3 it can be deduced that (i) the vibrational entropy is $2{\cdot}16$ cal K^{-1} mol^{-1} at 287 K, (ii) the barrier to internal rotation is 2967 cal mol^{-1}, and (iii) the product of the principal moments of inertia is $D = 6{\cdot}14 \times 10^{-134}$ kg m^2. Calculate $\tilde{S}_{\mathrm{spec}}$ for methyl chloroform at 287 K. The corresponding calorimetric value is $76{\cdot}22 \pm 0{\cdot}16$ cal K^{-1} mol^{-1}.

10.12 Calculate the free energy functions of CO, CO_2, H_2O, and H_2 for $T = 1200$ K. Note the data in Tables 5.2–5.4; the product of the principal moments of inertia of H_2O is evaluated in Example 5.5, and the moment of inertia of CO_2 (linear) is $71{\cdot}88 \times 10^{-47}$ kg m^2. The value of K_p for the reaction:

$$CO_2(g) + H_2(g) \rightleftharpoons CO(g) + H_2O(g)$$

is 1·37 at 1200 K. Calculate ΔE_0° for this reaction, and compare with the calorimetrically determined value, 40,330 J.

10.13 A suggested step in the reaction of hydrogen and oxygen is:

$$H_2 + OH\cdot \longrightarrow H_2O + H\cdot \qquad \text{(i)}$$

(a) The spectroscopically observed bond dissociation energies are:

$$H_2 \longrightarrow 2H\cdot \qquad \Delta\epsilon_0 = 4{\cdot}476\,\mathrm{eV}$$

$$H_2O \longrightarrow H\cdot + OH\cdot \qquad \Delta\epsilon_0 = 5{\cdot}113\,\mathrm{eV}.$$

Calculate ΔE_0° for reaction (i) in J mol^{-1}.

(b) For OH· and H·, $g_{\mathrm{el}} = 2$; the moment of inertia of OH· is $1{\cdot}511 \times 10^{-47}$ kg m^2, and the vibrational frequency is $11{\cdot}71 \times 10^{13}$ s^{-1}; the necessary data for H_2 and H_2O are given in problem 10.12. Calculate the change in free energy function associated with reaction (i) at 2000 K, and, by combining this value with the result of part (a), calculate K_p for that temperature.

10.14 From the data in problem 5.1, calculate the molar electronic energy and the electronic free energy function of gaseous silicon at 5000 K.

CHAPTER ELEVEN

Some Applications of Classical Dynamics

The theory discussed in the preceding chapters has been based upon quantum mechanics. It is fortunate that the quantal treatment is, in many ways, simpler than the classical, whilst at the same time being fundamentally more correct. In Section 2.1 it was remarked, however, that certain problems are more easily solved in terms of classical mechanics. An example already met with is the calculation of the rotational partition function of non-linear molecules; others will be encountered below.

The conditions under which it is permissible to replace quantum by classical mechanics are easily deduced from a general theorem known as the *limiting principle* (Fowler, pp. 18, 19). This states that quantum mechanical calculation of the properties of molecules in states *with large quantum numbers* yields the same results as the corresponding calculation by classical mechanics, or:

$$\underset{n\to\infty}{\mathrm{Lt}}\ (\text{quantized system}) = (\text{classical system})$$

This is so, for example, for the translational properties of molecules of a gas, since (see Section 4.7) the quantum numbers are very large, for the important translational energy levels.

In the first two sections of this chapter, we review the terminology of classical dynamics. For a fuller account of the equations of classical mechanics, the reader is referred to Chapter II of Tolman's book.

11.1 Generalized coordinates: degrees of freedom

The position of a single particle in space is determined when three coordinates, relative to fixed axes, are specified. The coordinates can be chosen in numerous ways: they may be, for example, rectangular Cartesian coordinates, x. y, z; or cylindrical polar coordinates, ρ, z, ψ; or spherical polar coordinates, r, θ, ψ, and so on. Similarly, the spatial configuration of a system of $\mathfrak{N}$ particles is determined when three coordinates are specified for each particle, i.e., when a total of $3\mathfrak{N}$ coordinates are given.

It often happens that, because of physical restrictions on possible positions, the spatial configuration can be defined by less than $3\mathfrak{N}$ coordinates. For example, the instantaneous position of a weight vibrating vertically at the end of a spring is determined if the instantaneous distance from the fixed end of the spring is given. In this case, since we state that the weight moves only *vertically*, say in the z-direction, we are effectively restricting the values of two (x and y) coordinates to those of the fixed end of the spring; it is then said that there are two *constraints*, i.e., previously determined conditions, in the system. In general, the minimum number of independent coordinates necessary for the specification of the spatial configuration is called the *number of degrees of freedom* (denoted by $\mathfrak{F}$). The difference $(3\mathfrak{N} + \mathfrak{F})$ is c, the number of constraints in the system, or

$$\mathfrak{F} = 3\mathfrak{N} - c$$

It is often convenient to specify a configuration in terms of variables which may not be coordinates in the simple sense described above. For example, the positions of the three nuclei in the H_2O molecule could be determined from the following nine quantities: (i) the three coordinates $x, y, z,$ of the oxygen nucleus, (ii) the HOH angle, (iii) the two H–O bond lengths, r_1 and r_2, (iv) the three angles made by the HOH plane with the X, Y, and Z axes. The configuration would equally be determined by nine independent combinations of these parameters; for example, instead of (iii), one might choose to specify the values of $q_1 = (r_1 + r_2)$ and $q_2 = (r_1 - r_2)$. Any such variables, which specify the configuration of a mechanical system, are called *generalized coordinates.*

11.2 Phase space

Superficial considerations suggest that the derivation of the Maxwell–Boltzmann law given in Section 2.2 could equally be applied to systems in which the energies are given by classical rather than quantum mechanics. There would, of course, be the obvious difference that the classical energies form a continuum, instead of being restricted to discrete values; but this would be allowed for by integrating, instead of summing, over the allowed energies. Thus, *apparently*, one could obtain equations similar to those given in Chapter 3 for thermodynamic functions, in which:

$$f = \sum_{\substack{\text{states}\\ j}} e^{-\epsilon_j/kT} \quad \text{is replaced by} \int_0^\infty e^{-\epsilon/kT}\, d\epsilon$$

In fact, the analogy is false, because the integral is over *energy values*, whereas the sum is over *quantum states*. The quantum sum over *energy levels* contains degeneracy factors:

$$f = \sum_{\substack{\text{levels}\\ i}} g_i\, e^{-\epsilon_i/kT};$$

and the more careful analysis, which we shall now make, shows that the simple integration of $e^{-\epsilon/kT}$ *does not, in general, provide the proper replacement for f.*

In classical, as in quantum, mechanics, a *state* of a molecule is one physically distinguishable condition of the molecule. In quantum mechanics, each state is characterized by a wave function (see Section 1.1), and observable properties associated with the state can be deduced when the wave function is known. In classical mechanics, on the other hand, the distinguishing properties of a dynamical state at any instant are (i) the positions, (ii) the velocities of the individual parts of the system. For example, a ball moving vertically passes through an infinite number of instantaneous dynamical states, each of which is characterized by a definite height h and velocity v of the ball. The energy associated with one state is:

$$\epsilon = mgh + \tfrac{1}{2} mv^2.$$

The corresponding 'partition function' is then:

$$\sum_{\text{all } h,\, v} e^{-\epsilon/kT} = \sum_{\text{all } h,\, v} e^{-(mgh + \frac{1}{2} mv^2)/kT}$$

and the sum would be evaluated as a double integral over h and v; however, we shall not pursue this example further.

In the general case, the position of a mechanical system with $\mathcal{F}$ degrees of freedom is specified by giving the values of $\mathcal{F}$ generalized coordinates $q_1, q_2, \ldots, q_{\mathfrak{F}}$. The dynamical state can correspondingly be defined by giving these values, together with the values of $\mathfrak{F}$ *generalized velocities* $\dot{q}_1, \dot{q}_2, \ldots, \dot{q}_{\mathfrak{F}}$. The generalized velocity denoted by $\dot{q}_i$ is the time-derivative of the corresponding generalized coordinate:

$$\dot{q}_i \equiv \frac{dq_i}{dt} \quad (11.1)$$

In practice, it is found to be more convenient to define the dynamical state in terms of the *generalized momenta*, $p_1\ p_2, \ldots, p_{\mathfrak{F}}$. For systems in which the potential energy depends only upon the coordinates (for example, an isolated molecule not absorbing or emitting radiation), the generalized momentum p_i is given by:

$$p_i = \frac{\partial \mathbf{T}}{\partial \dot{q}_i} \ (i = 1, 2, \ldots, \mathfrak{F}) \quad (11.2)$$

where $\mathbf{T}$ is the total kinetic energy (obtained, of course, by adding together the kinetic energies of the individual parts) of the system. p_i is said to be the *generalized momentum conjugate* to q_i. It is found possible to obtain relatively simple expressions for $\mathbf{T}$ in terms of the p_i; the great importance of this fact will become apparent in subsequent paragraphs. At the end of this section, two examples of the determination of $\mathbf{T}$ in terms of p_i are given.

In the statistical treatment of dynamical states, we shall find useful the notations of coordinate geometry. Let us consider firstly a particle with just one degree of freedom. The dynamical state of such a particle, which is determined by one position coordinate q and the conjugate momentum p, can be plotted as the *point* (p,q). The set of all possible values of p and q (i.e., points) is said to constitute the *phase space* of the particle. In the following discussion, it is convenient to imagine that the phase space is divided up into small, equal rectangles, or *cells*, with sides δp and δq, and which are numbered 1,2, ... j, ... (see Fig. 11.1).

Consider now a whole assembly of such particles. Suppose that, at some instant, m_1 particles have values of p and q corresponding to points inside cell 1, m_2 particles have values of p and q inside cell 2, and so on. If δ_p and δ_q are chosen to be small enough, the range of energies for points inside the j-th cell is sufficiently small for the m_j particles to be regarded as having the same energy. The

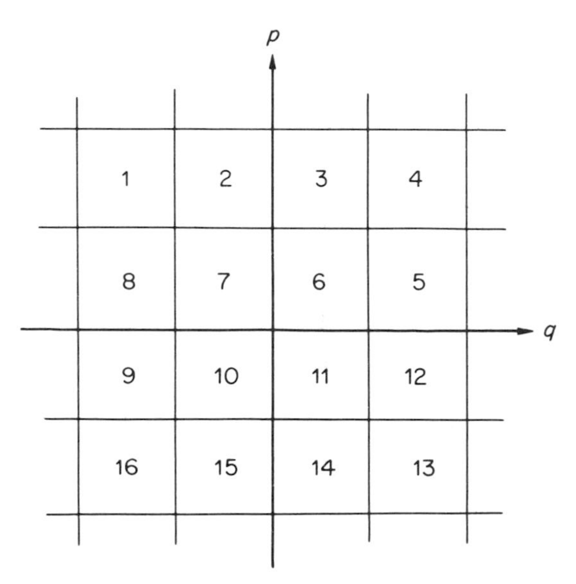

Fig. 11.1 Cells in phase space for one degree of freedom

most probable value of m_j is now obtained by just the same* arguments which led to equation 2.15:

$$\bar{m}_j = e^{-\alpha} e^{-\epsilon_j/kT} \tag{11.3}$$

the only difference being that, in Equation 2.15, $\bar{m}_j$ is the most probable number of molecules in the j-th quantum state, whereas, in Equation 11.3, $\bar{m}_j$ is the most probable number of molecules in the j-th cell in phase space.

As in Section 2.3, the constant $e^{-\alpha}$ is eliminated by summing over all the values of $\bar{m}_j$:

* The analogue of assumption (4) in Section 2.1 is the straightforward postulate that all elements of phase space compatible with a given energy of the system are equally likely. Historically, this point of view was not adopted by Maxwell and Boltzmann, who introduced another postulate which they respectively named the 'principle of continuity of path' and 'ergodic hypothesis'. An account and criticism of the ergodic hypothesis is given in Chapter III of Tolman's book.

$$N = \sum_{\substack{\text{cells}\\ j}} \overline{m}_j$$

$$= e^{-\alpha} \sum_j e^{-\epsilon_j/kT}$$

$$= \frac{e^{-\alpha}}{\delta p\, \delta q} \iint e^{-\epsilon/kT}\, dp\, dq \tag{11.4}$$

The last step is simply the evaluation of the sum as an integral, a device which we have already encountered (Section 4.5 and 5.4) in the case of a single variable. In the present case, summation is over all values of p and q, so a double integration is necessary. The reason for division by the product $\delta p\, \delta q$ on the right-hand side of Equation 11.4 is straightforward, and is discussed in Appendix 1.

The integral:

$$Q = \iint e^{-\epsilon/kT}\, dp\, dq$$

is called the *phase integral,* and is the classical analogue of the partition function f. From Equation 11.4. therefore:

$$e^{-\alpha} = \frac{N\, \delta p\, \delta q}{Q}$$

and

$$\overline{m}_j = \frac{N}{Q} e^{-\epsilon_j/kT}\, \delta p\, \delta q. \tag{11.5}$$

This equation shows that $\overline{m}_j$ is determined by the corresponding value of ϵ. It is no longer necessary to label the individual cells; moreover, since δp and δq can be chosen to be arbitrarily small, it is customary to write δn instead of $\overline{m}_j$. Hence, Equation 11.5 becomes:

$$\delta n = \frac{N}{Q} e^{-\epsilon/kT}\, \delta p\, \delta q \tag{11.6}$$

where δn is the most probable number of molecules which have values of p and q in some small, specified range, and ϵ is the energy corresponding to these values.

The above discussion is easily generalized to the case in which the particles have $\mathcal{F}$ degrees of freedom, i.e., in which a dynamical state is defined by $\mathcal{F}$ generalized coordinates q_i and $\mathcal{F}$ conjugate momenta p_i. Such a state can be represented as a point in a hypothetical phase space of 2 $\mathcal{F}$ dimensions, with 2 $\mathcal{F}$ rectangular axes. This hyperspace can be divided into cells having sides $\delta q_1, \delta q_2, \ldots, \delta q_{\mathcal{F}}, \delta p_1, \delta p_2, \ldots, \delta p_{\mathcal{F}}$. Equation 11.6 is thus generalized to:

$$\delta n = \frac{N}{Q} \mathrm{e}^{-\epsilon/kT} \, \delta q_1 \, \delta q_2 \, ... \, \delta q_{\mathcal{F}} \, \delta p_1 \, \delta p_2 \, ... \, \delta p_{\mathcal{F}} \qquad (11.7)$$

where

$$Q = \underset{(2\mathcal{F})}{\int ... \int} \mathrm{e}^{-\epsilon/kT} \, \mathrm{d}q_1 \, \mathrm{d}q_2 \, ... \, \mathrm{d}q_{\mathcal{F}} \, \mathrm{d}p_1 \, \mathrm{d}p_2 \, ... \, \mathrm{d}p_{\mathcal{F}}$$

is the phase integral. Equation 11.7 is the classical form of the Maxwell–Boltzmann law. From this law, the equations for thermodynamic functions can now be derived by the methods of Chapter 3, with Q replacing f.

We conclude this section with two examples of the application of Equation 11.2

Example 11.2(a)

For a single point particle:

$$\mathbf{T} = \tfrac{1}{2}m(\dot{x}^2 + \dot{y}^2 + \dot{z}^2) \qquad (11.8)$$

$$p_x = \frac{\partial \mathbf{T}}{\partial \dot{x}} = m\dot{x}$$

$$p_y = \frac{\partial \mathbf{T}}{\partial \dot{y}} = m\dot{y}$$

$$p_z = \frac{\partial \mathbf{T}}{\partial \dot{z}} = m\dot{z}$$

and consequently, substituting for $\dot{x}$, $\dot{y}$, $\dot{z}$, in Equation 11.8:

$$\mathbf{T} = \frac{1}{2m}(p_x^2 + p_y^2 + p_z^2)$$

□ □ □

Example 11.2(b)

A diatomic molecule may be regarded as consisting of two point masses m_1 and m_2 joined together. The kinetic energy can be expressed as the sum of terms in Cartesian coordinates, as in Equation 11.8, one term for each of the point masses:

$$\mathbf{T} = \tfrac{1}{2}m_1(\dot{x}_1^2 + \dot{y}_1^2 + \dot{z}_1^2) + \tfrac{1}{2}m_2(\dot{x}_2^2 + \dot{y}_2^2 + \dot{z}_2^2)$$

We choose the following (generalized) coordinates: r is the distance between the point masses; X,Y,Z are the coordinates of the centre of mass of m_1 and m_2; θ,φ are polar angles of the line joining the mass points, referred to the $x.y.z$ axes as shown In Fig. 11.2. It is then found that:

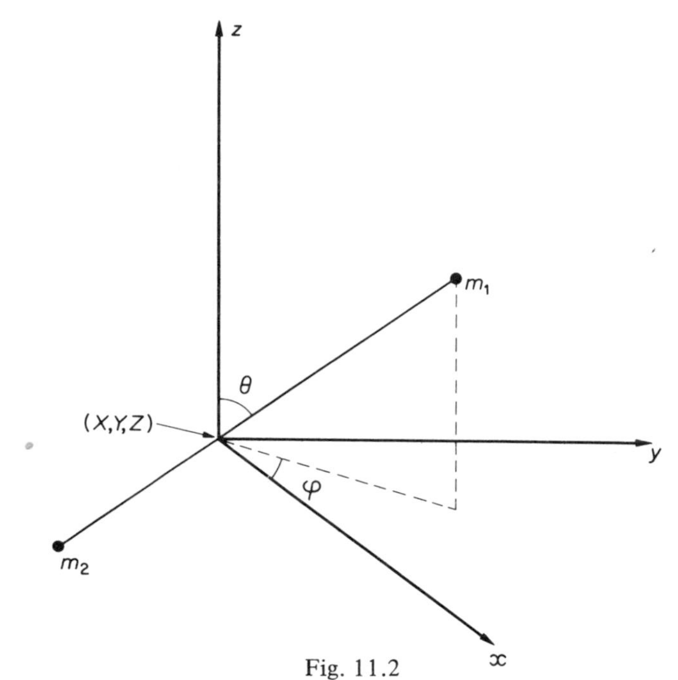

Fig. 11.2

$$\mathbf{T} = \frac{m_1 + m_2}{2}(\dot{X}^2 + \dot{Y}^2 + \dot{Z}^2) + \frac{r^2}{2}\frac{m_1 m_2}{m_1 + m_2}(\dot{\theta} + \sin^2\theta\,\dot{\varphi}^2)$$

$$+ \frac{1}{2}\frac{m_1 m_2}{m_1 + m_2}\dot{r}^2$$

This can be expressed in a simpler form, because $(m_1 + m_2)$ is the total mass m of the molecule, $m_1 m_2/(m_1 + m_2)$ is the reduced mass μ, and μr^2 is the moment of inertia I of the molecule. Thus:

$$\mathbf{T} = \frac{1}{2}m(\dot{X}^2 + \dot{Y}^2 + \dot{Z}^2) + \frac{1}{2}I(\dot{\theta}^2 + \sin^2\theta\,\dot{\varphi}^2) + \frac{1}{2}\mu\dot{r}^2 \qquad (11.9)$$

The three terms on the right-hand side of Equation 11.9 have simple physical meanings. The first is the kinetic energy of translation of the molecule as a whole, the second is the instantaneous kinetic energy of rotation, and the third is the instantaneous kinetic energy of vibration.

The velocities can be eliminated from Equation 11.9 by introducing the conjugate momenta. By differentiation of Equation 11.9 with respect to the various generalized velocities:

$$p_x = m\dot{X}; \quad p_y = m\dot{Y}; \qquad p_z = m\dot{Z};$$
$$p_\theta = I\dot{\theta}; \quad p_\varphi = I\sin^2\theta\,\dot{\varphi}; \quad p_r = \mu\dot{r}$$

and substitution in Equation 11.9 then yields:

$$\mathbf{T} = \frac{1}{2m}(p_x^2 + p_y^2 + p_z^2) + \frac{1}{2I}\left(p_\theta^2 + \frac{1}{\sin^2\theta}p_\varphi^2\right) + \frac{1}{2\mu}p_r^2 \qquad (11.10)$$

We shall refer again to Equation 11.10.

□ □ □

11.3 The relationship between Q and f

In Section 5.7, it was shown that a valid approximation of the partition function f can be obtained by replacing summation by integration, provided that the difference between successive energy levels is small in comparison with kT. By using this approximation, we obtained, for the translational partition function, the result:

$$f_{\text{trans}} = \frac{(2\pi mkT)^{3/2}\,V}{h^3}$$

It turns out that such approximations to the quantum mechanical partition function are closely related to the corresponding *phase integrals* of classical mechanics. Thus, as we shall subsequently show, the phase integral contains the translational factor:

$$Q_{\text{trans}} = (2\pi mkT)^{3/2}\,V$$

an expression which differs from the corresponding partition function only by the absence of the factor $1/h^3$. The relationship between f_{trans} and Q_{trans} is a special case of a general theorem: *if there are $\mathfrak{F}$ degrees of freedom, the approximation of replacing summation by integration in the evaluation of the partition function yields the same result as the phase integral divided by $h^{\mathfrak{F}}$*, i.e., in this approximation:

$$f = Q/h^{\mathfrak{F}}. \qquad (11.11)$$

For translation $\mathfrak{F} = 3$, and $f_{\text{trans}} = Q_{\text{trans}}/h^3$; for the rotation of a diatomic molecule $\mathfrak{F} = 2$ (there are two variables θ and φ), and $f_{\text{rot}} = Q_{\text{rot}}/h^2$.

Equation 11.11 can be regarded as a consequence of the *Heisenberg uncertainty principle*. The latter is summarized in the following statements: (i) quantum mechanics is concerned only with the calculation of quantities which can be related to experimental measurements; (ii) experimental observations carried out upon particles of molecular size, or less, inevitably disturb the dynamical state of the particles to a significant extent; (iii) calculation of the effects of this disturbance show that, for any generalized coordinate and its conjugate momentum:

$$\Delta p \, \Delta q \approx h \tag{11.12}$$

where $\Delta p \, \Delta q$ is the least possible value of the product of the errors, when p and q are simultaneously determined.

A rough demonstration (but see the comment made in the closing paragraph of this section) of how Equation 11.11 arises is as follows. Let us imagine that the phase space of a particle having one degree of freedom is divided, by lines parallel to the p and q axes, into rectangles, each rectangle having an area h. Now consider two phase points within a given rectangle, say (p,q) and $(p + \Delta p, q + \Delta q)$. By geometry, since both points lie within the rectangle:

$$\Delta p \, \Delta q < \text{area of rectangle, } h.$$

According to Heisenberg's principle, states corresponding to such pairs of points cannot be physically distinguished from each other, i.e., *all* the phase points inside a given rectangle correspond to only one physically distinct state. The element of area in phase space is $\mathrm{d}p \, \mathrm{d}q$, and the number of distinct states corresponding to this area is:

$$\frac{\mathrm{d}p \, \mathrm{d}q}{\text{area of 1 rectangle}} = \frac{\mathrm{d}p \, \mathrm{d}q}{h}$$

Similarly, the number of states corresponding to the element $\mathrm{d}p_1 \, \mathrm{d}p_2 \ldots \mathrm{d}p \quad \mathrm{d}q_1 \, \mathrm{d}q_2 \ldots \mathrm{d}q_{\mathfrak{F}}$ is:

$$\left(\frac{\mathrm{d}p_1 \, \mathrm{d}q_1}{h}\right) \left(\frac{\mathrm{d}p_2 \, \mathrm{d}q_2}{h}\right) \cdots \left(\frac{\mathrm{d}p_{\mathfrak{F}} \, \mathrm{d}q_{\mathfrak{F}}}{h}\right),$$

and Equation 11.11 follows.

The above argument contains several unsatisfactory features, the most important of which being the fact that Equation 11.12 is not an exact statement of the Uncertainty Principle. More rigorously (Margenau and Murphy, p. 348):

$$\Delta p_i \, \Delta q_i \geqslant h/4\pi,$$

where Δp_i and Δq_i are standard deviations. It does not seem to be possible to use this inequality to derive Equation 11.11. A general proof of 11.11 has, however, been given by Kirkwood (Hill, p. 462); alternatively, the equation can be verified by evaluating the partition function and the phase integral in individual cases (Fowler, p. 18).

11.4 Factorization of the phase integral

Just as the calculation of the partition function is simplified by expressing it as the product of simpler sums (Section 3.4), so the calculation of the phase integral is simplified by expressing it as the product of simpler integrals.

Quite generally, if $f_1(a)$ and $f_2(b)$ are two functions of *independent* variables a and b, then:

$$\int_{a_1}^{a_2}\int_{b_1}^{b_2} f_1(a)\, f_2(b)\, \mathrm{d}b\, \mathrm{d}a = \int_{a_1}^{a_2} f_1(a)\, \mathrm{d}a \int_{b_1}^{b_2} f_2(b)\, \mathrm{d}b$$

For example:

$$\int_{x_1}^{x_2}\int_{y_1}^{y_2} xy\, \mathrm{d}y\, \mathrm{d}x = \int_{x_1}^{x_2} x\, \mathrm{d}x \int_{y_1}^{y_2} y\, \mathrm{d}y = \tfrac{1}{4}(x_2^2 - x_1^2)(y_2^2 - y_1^2).$$

Multiple integrals can be factorized in the same way. It will be shown that it is often possible to factorize the integrand of the phase integral into independent functions, i.e.:

$$Q = \int_{(2\mathfrak{F})} \cdots \int f_1(q_1)\, f_2(q_2) \ldots f_{\mathfrak{F}}(q_{\mathfrak{F}})\, f_{\mathfrak{F}+1}(p_1)\, f_{\mathfrak{F}+2}(p_2) \ldots f_{2\mathfrak{F}}(p_{\mathfrak{F}})\, \mathrm{d}q_1\, \mathrm{d}q_2 \ldots \mathrm{d}q_{\mathfrak{F}}\, \mathrm{d}p_1\, \mathrm{d}p_2 \ldots \mathrm{d}p_{\mathfrak{F}}.$$

Hence:

$$Q = \int f_1(q_1)\, \mathrm{d}q_1 \int f_2(q_2)\, \mathrm{d}q_1 \ldots \int f_{\mathfrak{F}}(q_{\mathfrak{F}})\, \mathrm{d}q_{\mathfrak{F}} \times \int f_{\mathfrak{F}+1}(p_1)\, \mathrm{d}p_1 \int f_{\mathfrak{F}+2}(p_2)\, \mathrm{d}p_2 \ldots \int f_{2\mathfrak{F}}(p_{\mathfrak{F}})\, \mathrm{d}p_{\mathfrak{F}} \qquad (11.13)$$

and the problem of evaluating a $2\mathfrak{F}$-fold integral simplifies to one of evaluating $2\mathfrak{F}$ simple integrals.

Equation 11.10 shows that the instantaneous value of the kinetic energy of a diatomic molecule, (treated as a pair of point masses) is the sum of three terms:

$$\mathbf{T} = \epsilon_{\text{trans}} + \epsilon_{\text{rot}} + \epsilon_{\text{vib}(k)} \qquad (11.14\text{a})$$

where:

$$\epsilon_{\text{trans}} = \frac{1}{2m}(p_x^2 + p_y^2 + p_z^2) \text{ is the translational energy} \qquad (11.14\text{b})$$

$$\epsilon_{rot} = \frac{1}{2I}(p_\theta^2 + \frac{1}{\sin^2\theta} p_\varphi^2) \text{ is the rotational kinetic energy} \quad (11.14c)$$

$$\epsilon_{vib(k)} = \frac{1}{2\mu} p_r^2 \text{ is the kinetic energy of vibration} \quad (11.14d)$$

There is also a potential energy **V**. If effects of external fields are not considered (cf. Section 11.11), **V** does not change with the orientation of the molecule as a whole, i.e., it is independent of x, y, z, θ, and φ. **V** does, however, depend upon r, and, if Hooke's Law (Equation 5.7) is assumed:

$$\mathbf{V} = \tfrac{1}{2} k(r - r_0)^2 \quad (11.15)$$

where k is the force constant and **V** is the potential energy relative to that of the equilibrium configuration. Vibrations of small amplitude do not seriously alter the value of the moment of inertia or, therefore, the rotational energy term. Thus, the total energy is, to a good degree of approximation, the sum of three independent terms:

$$\epsilon = \epsilon_{trans} + \epsilon_{rot} + \epsilon_{vib}$$

where ϵ_{trans} and ϵ_{rot} are as defined above, and:

$$\epsilon_{vib} = \epsilon_{vib(k)} + \tfrac{1}{2} k\ (r - r_0)^2$$

Therefore:

$$\begin{aligned} e^{-\epsilon/kT} &= e^{-(\epsilon_{trans} + \epsilon_{rot} + \epsilon_{vib})/kT} \\ &= e^{-\epsilon_{trans}/kT}\ e^{-\epsilon_{rot}/kT}\ e^{-\epsilon_{vib}/kT}, \end{aligned}$$

i.e., $e^{-\epsilon/kT}$ factorizes into three independent terms. The phase integral can now be written in the form 11.13, i.e., as the product of three factors, corresponding respectively to translation, rotation, and vibration:

$$Q = Q_{trans}\, Q_{rot}\, Q_{vib},$$

where

$$Q_{trans} = \iiiint\!\!\iint e^{-\epsilon_{trans}/kT}\, dp_x\, dp_y\, dp_z\, dx\, dy\, dz,$$

$$Q_{rot} = \iiiint e^{-\epsilon_{rot}/kT}\, dp_\theta\, dp_\varphi\, d\theta\, d\varphi,$$

$$Q_{vib} = \iint e^{-\epsilon_{vib}/kT}\, dp_r\, dr,$$

and we shall find that each of these multiple integrals is itself expressible as the product of simpler factors.

11.5 Evaluation of the partition function of a diatomic molecule by classical mechanics

(i) The Translational Partition Function

If x, y, and z are the coordinates of the centre of gravity of the molecule, and m is the molecular mass, then, according to Equation 11.14b:

$$\epsilon_{\text{trans}} = \tfrac{1}{2} m (\dot{x}^2 + \dot{y}^2 + \dot{z}^2)$$
$$= \frac{1}{2m}(p_x^2 + p_y^2 + p_z^2)$$

The translational factor in the phase integral for a molecule moving in a rectangular box of dimensions a, b, and c, is therefore:

$$Q_{\text{trans}} =$$
$$\int_{-\infty}^{\infty}\int_{-\infty}^{\infty}\int_{-\infty}^{\infty}\int_{0}^{a}\int_{0}^{b}\int_{0}^{c} \exp[-(p_x^2 + p_y^2 + p_z^2)/2mkT]\mathrm{d}x\,\mathrm{d}y\,\mathrm{d}z\,\mathrm{d}p_x\,\mathrm{d}p_y\,\mathrm{d}p_z \qquad (11.16)$$

Since: $\exp[-(p_x^2 + p_y^2 + p_z^2)/2mkT$

$$= \exp(-p_x^2/2mkT)\exp(-p_y^2/2mkT)\exp(-p_z^2/2mkT)$$

Equation 11.16 can be written in the form 11.13:

$$Q_{\text{trans}} = \int_{-\infty}^{\infty} \mathrm{e}^{-p_x^2/2mkT}\,\mathrm{d}p_x \int_{-\infty}^{\infty} \mathrm{e}^{-p_y^2/2mkT}\,\mathrm{d}p_y \int_{-\infty}^{\infty} \mathrm{e}^{-p_z^2/2mkT}\,\mathrm{d}p_z$$
$$\times \int_0^a \mathrm{d}x \int_0^b \mathrm{d}y \int_0^c \mathrm{d}z$$

Each of the first three integrals is of the form:

$$\int_{-\infty}^{\infty} \mathrm{e}^{-ax^2}\,\mathrm{d}x = \left(\frac{\pi}{a}\right)^{1/2} \qquad (11.17)$$

i.e., $\mathrm{e}^{-p_x^2/2mkT}\,\mathrm{d}p_x = (2\pi mkT)^{1/2}$.

Hence:

$$Q_{\text{trans}} = (2\pi mkT)^{3/2} abc$$
$$= (2\pi mkT)^{3/2} V \qquad (11.18)$$

and the classical value of the partition function is, by Equation 11.11, since $\mathfrak{F} = 3$:

$$f_{\text{trans}} = Q_{\text{trans}}/h^3$$
$$= \frac{(2\pi mkT)^{3/2} V}{h^3},$$

which is identical with the result obtained by quantum mechanics;

It may be remarked that, for the space part of the integral Q, the result:

$$\iiint \mathrm{d}x\,\mathrm{d}y\,\mathrm{d}z = V$$

is true whatever the shape of the vessel containing the gas, so the above derivation is of general applicability.

(ii) The Vibrational Partition Function

According to Equations 11.14d and 11.15, the vibrational energy of a diatomic molecule is:

$$\epsilon_{\mathrm{vib}} = \frac{1}{2\mu} p_r^2 + \frac{1}{2} k(r - r_0)^2$$

Hence:*

$$\begin{aligned} Q_{\mathrm{vib}} &= \int_{-\infty}^{\infty}\int_{-\infty}^{\infty} \exp\left[-\frac{1}{kT}\left(\frac{p_r^2}{2\mu} + \frac{k(r-r_0)^2}{2}\right)\right] \mathrm{d}r\,\mathrm{d}p_r \\ &= \int_{-\infty}^{\infty} \exp\left(\frac{-p_r^2}{2\mu kT}\right) \mathrm{d}p_r \int_{-\infty}^{\infty} \exp\left(\frac{-k(r-r_0)^2}{2kT}\right) \mathrm{d}r \end{aligned} \tag{11.19}$$

Both of these integrals can be expressed in the form 11.17 (the second by putting $x = r - r_0$). Thus:

$$Q_{\mathrm{vib}} = (2\pi\mu kT)^{1/2} \left(\frac{2\pi kT}{k}\right)^{1/2} = 2\pi kT \left(\frac{\mu}{k}\right)^{1/2}$$

According to Equation 5.5:

$$\left(\frac{\mu}{k}\right)^{1/2} = \frac{1}{2\pi\omega}$$

where ω is the vibrational frequency. Thus:

$$Q_{\mathrm{vib}} = \frac{kT}{\omega}$$

Since ϵ_{vib} is a function of one generalized coordinate and its conjugate momentum, $\mathfrak{F} = 1$, and the corresponding classical partition function is:

* It might be objected that the limits of integration do not correspond to physical reality. If, however, r differs greatly from r_0, the potential energy, and hence ϵ_{vib} is large, and $\epsilon^{-\epsilon_{\mathrm{vib}}/kT}$ is small; consequently, the integrand is negligible for such values of r, and the limits can be chosen to be arbitrarily large, or small, respectively.

$$f_{\text{vib}} = \frac{kT}{h\omega} = \frac{T}{\Theta_{\text{vib}}}$$

This is *not* the same as the quantal result, but, as was discussed in Section 5.7, is an approximation to it, for high temperatures.

(iii) The Rotational Partition Function

From Equation 11.14c:

$$\epsilon_{\text{rot}} = \frac{1}{2I}\left(p_\theta^2 + \frac{p_\varphi^2}{\sin^2\theta}\right)$$

and

$$Q_{\text{rot}} = \int_0^{2\pi}\int_0^{\pi}\int_{-\infty}^{\infty}\int_{-\infty}^{\infty} \exp\left[-\frac{1}{2IkT}\left(p_\theta^2 + \frac{p_\varphi^2}{\sin^2\theta}\right)\right] \mathrm{d}p_\theta\ \mathrm{d}p_\varphi\ d\theta\ \mathrm{d}\varphi \qquad (11.20)$$

Examination of Fig. 11.2 shows that all spatial configurations are covered if θ runs from 0 to π and φ from 0 to 2π; hence the limits in Equation 11.20. Because of the term $p_\varphi^2/\sin^2\theta$, complete factorization of Equation 11.20 is not immediately possible. We obtain, however:

$$Q_{\text{rot}} = \int_0^{2\pi}\mathrm{d}\varphi\int_{-\infty}^{\infty}\exp\left(\frac{-p_\theta^2}{2IkT}\right)\mathrm{d}p_\theta \times$$
$$\int_0^{\pi}\int_{-\infty}^{\infty}\exp\left(\frac{-p_\varphi^2}{2IkT/\sin^2\theta}\right)\mathrm{d}p_\varphi\ \mathrm{d}\theta \qquad (11.21)$$

The first two integrals in this expression are immediately solved, giving, respectively, 2π and $(2\pi IkT)^{1/2}$. The double integral is most easily solved by integrating first with respect to p_φ and then with respect to θ:

$$\int_0^{\pi}\int_{-\infty}^{\infty}\exp\left(\frac{-p_\varphi^2}{2IkT\sin^2\theta}\right)\mathrm{d}p_\varphi\ \mathrm{d}\theta = \int_0^{\pi}\left[(2\pi IkT)^{1/2}\sin\theta\right]\mathrm{d}\theta$$
$$= (2\pi IkT)^{1/2} \times 2.$$

Hence, Equation 11.20 simplifies to:

$$Q_{\text{rot}} = 8\pi^2 IkT$$

Since $\mathfrak{F} = 2$:

$$f_{\text{rot}} = \frac{8\pi^2 IkT}{h^2} \qquad (11.22)$$

which is identical with Equation 5.22.

11.6 Rotation of polyatomic molecules

With the appropriate moment of inertia, the expression 11.14 for ϵ_{rot} applies to any linear polyatomic molecule, and therefore Equation 11.22 is again obtained for this case. The treatment of non-linear molecules is a little more complicated.

The orientation of a rigid body is uniquely defined when the positions of its principal axes (see p. 78) are given relative to x, y, and z directions fixed in space. We shall refer to the principal axes as the ζ-, ϵ-, and η-axes. The position of the ζ-axis is fixed by defining the polar angles θ and φ (see Fig. 11.3). Since the positions of the ϵ- and η-axes relative to ζ are maintained when these two axes are rotated simultaneously around the ζ-axis, a third angle χ is necessary to specify the orientation. This is numerically the angle through which the ξ-axis must be rotated about the ζ-axis, to bring it into the same plane as the z- and ζ-axes. The angles θ, φ, and χ are called the *Eulerian angles*.

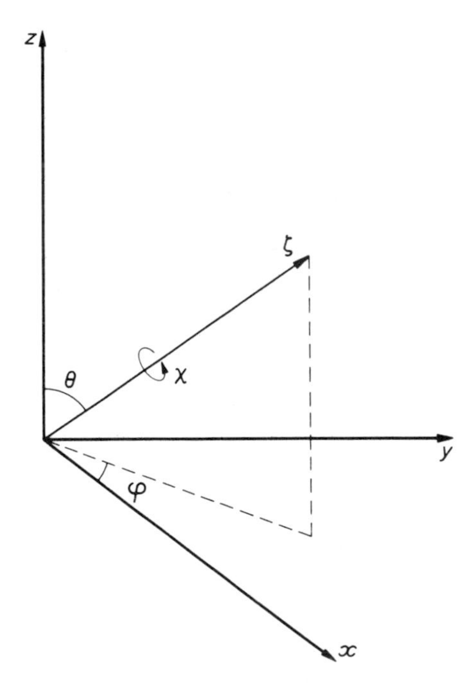

Fig. 11.3 The Eulerian angles

According to classical mechanics, the rotational kinetic energy is:

$$\frac{\gamma^2}{2A} + \frac{\lambda^2}{2B} + \frac{p_\chi^2}{2C} \qquad (11.23)$$

where A, B, and C are respectively the moments of inertia about the axes ξ, η, and ζ:

$$\gamma = \sin\chi\, p_\theta - \frac{\cos\chi}{\sin\theta}(p_\varphi - \cos\theta\, p_\chi)$$

$$\lambda = \cos\chi\, p_\theta + \frac{\sin\chi}{\sin\theta}(p_\varphi - \cos\theta\, p_\chi)$$

and p_θ, p_φ, p_χ are the momenta conjugate to the variables θ, φ, and χ.

The integral:

$$Q_{\text{rot}} = \int\int\int\int\int\int e^{-\epsilon_{\text{rot}}/kT}\, dp_\theta\, dp_\varphi\, d\theta\, d\varphi\, d\chi$$

is easily evaluated if, instead of integrating over p_0 and p_φ, one integrates over γ and λ. Noting that:

$$dp_\theta\, dp_\varphi = d\gamma\, d\lambda \sin\theta$$

(see Appendix 3), we have:

$$Q_{\text{rot}} = \int\int\int\int\int\int \exp\left[-\frac{1}{2kT}\left(\frac{\gamma^2}{A} + \frac{\lambda^2}{B} + \frac{p_\chi^2}{C}\right)\right] \times \sin\theta\, d\gamma\, d\lambda\, dp_\chi\, d\theta\, d\varphi\, d\chi \qquad (11.24)$$

$$= \int_{-\infty}^{\infty} e^{-\gamma^2/2AkT}\, d\gamma \int_{-\infty}^{\infty} e^{-\lambda^2/2BkT}\, d\lambda \int_{-\infty}^{\infty} e^{-p_\chi^2/2CkT}\, dp_\chi \times \int_0^{\pi} \sin\theta\, d\theta \int_0^{2\pi} d\varphi \int_0^{2\pi} d\chi$$

$$= (2\pi AkT)^{1/2}(2\pi BkT)^{1/2}(2\pi CkT)^{1/2} \times 2 \times 2\pi \times 2\pi$$

Since $\mathfrak{F} = 3$, the rotational partition function is:

$$f_{\text{rot}} = \left(\frac{8\pi^2 kT}{h^2}\right)^{3/2}(\pi ABC)^{1/2} \qquad (11.25)$$

11.7 Symmetry factors

It was shown in Section 5.4 that the value of f_{rot} given by Equation 11.22 or 11.25 must be reduced by the symmetry factor σ, because of restrictions placed by the Pauli principle upon the allowed quantum states. It is remarkable that a corresponding factor is also found to be necessary in the classical treatment. Consider the case of the diatomic molecule. If the nuclei constituting the masses m_1 and m_2 are identical, then, referring to Fig. 11.2, a spatial configuration defined by (θ,φ) is indistinguishable from that defined by $(\pi - \theta, \pi + \varphi)$; the two configurations differ only in the labelling of the nuclei. Since the phase integral is the sum over *physically distinct* dynamical states (points in phase space), we must therefore reduce the right-hand side of Equation 11.20 by half, because otherwise each state will have been counted twice in the evaluation of Q.

Similarly, if there are σ different sets of values of θ,φ,χ (within the respective ranges $0 < \theta \leqslant \pi, 0 < \varphi \leqslant 2\pi, 0 < \chi \leqslant 2\pi$) which produce the same orientation of a polyatomic molecule, then the right–hand side of Equation 11.25 must be divided by σ.

These are just the results obtained in Section 5.4.

11.8 The equipartition of energy

Another result which has already been mentioned (Section 5.7), and which we can now easily derive, is the principle of equipartition of energy. Inspection of Equations 11.14 and 11.23 shows that, in each of the examples discussed above, it happens that the part of the energy being considered depends upon the *squares* of the variables concerned. In each case, the contribution to the molecular energy associated with a particular variable w (w might be a momentum, e.g., p_θ or a coordinate, e.g., r) is of the form:

$$\epsilon_w = cw^2$$

where c is some parameter.

Correspondlngly, the phase integral, and hence the partition function, contains the factor:

$$\int_{-\infty}^{\infty} e^{-cw^2/kT}\, dw = \left(\frac{\pi kT}{c}\right)^{1/2}$$

Consequently, the contribution to the molar energy is:

$$\tilde{N}kT^2\left(\frac{\partial \ln f_w}{\partial T}\right) = \frac{\tilde{N}kT}{2} = \frac{RT}{2}$$

the same for all variables w. This is the equipartition principle.

Note that the equipartition principle is merely a consequence of the quadratic dependence of ϵ upon the variables, rather than a basic law of statistical mechanics. Tolman (Tolman, p. 95) obtained a more general equipartition principle, which, however, only reduces to the above simple result when the quadratic expression for ϵ applies.

11.9 Mean values in classical theory

Equation 2.23b for the average value of some molecular property $\boldsymbol{p}$ has a parallel in classical theory. By an obvious analogy:

$$\bar{\boldsymbol{p}} = \frac{\int_{2\mathcal{F}} \cdots \int e^{-\epsilon/kT} \, \boldsymbol{p} \, dq_1 \ldots dq_{\mathcal{F}} \, dp_1 \ldots dp_{\mathcal{F}}}{Q} \tag{11.26}$$

In general $\boldsymbol{p}$ *may* depend upon all the $2\mathcal{F}$ variables p_i and q_i, but, in practice, it will usually depend upon only *certain* of these. In the latter case, the numerator of Equation 11.26 factorizes into two parts, one dependent on $\boldsymbol{p}$ and the other independent of $\boldsymbol{p}$. If, for example, the property $\boldsymbol{p}$ of interest is the x-component of velocity $\dot{x}$, of a molecule, $\boldsymbol{p}$ will be *independent* of the angles θ, φ, and χ, and their conjugate momenta. Consequently, the numerator of Equation 11.26 contains as a factor the integral:

$$\int\!\!\int\!\!\int\!\!\int\!\!\int\!\!\int e^{-\epsilon_{rot}/kT} \, d\theta \, d\varphi \, d\chi \, dp_\theta \, dp_\varphi \, dp_\chi$$

which cancels with the corresponding factor Q_{rot} *in* Q. A similar cancellation will be found for all the other variables *except* p_x, the values of which, of course, depends upon $\dot{x}$. Therefore, we obtain finally:

$$\bar{\dot{x}} = \frac{\int_{\infty}^{-\infty} e^{-p_x^2/2mkT} \, \dot{x} \, dp_x}{\int_{\infty}^{-\infty} e^{-p_x^2/2mkT} \, dp_x} = \frac{\int_{\infty}^{-\infty} e^{-p_x^2/2mkT} \, (p_x/m) \, dp_x}{(2\pi mkT)^{1/2}}$$

A closely related problem is the following: calculate the number of molecules (in the most probable distribution) with a given value of one coordinate, irrespective of the values of other coordinates. The required number is obtained by integrating Equation 11.26 over all the coordinates *except the one whose value is specified.* Suppose, for example, that it is required to calculate the number of molecules for which the momentum in the x-direction has some value between p_x and $p_x + \delta p_x$. The required number is obtained by performing $(2\mathcal{F} - 1)$ integrations of Equation 11.7:

$$\delta_n = \frac{N}{Q} \int_{(2\mathcal{F}-1)} \cdots \int e^{-\epsilon/kT} \, dq_1 \ldots dq_{\mathcal{F}} \, dp_1 \ldots dp_{\mathcal{F}-1} \, \delta p_x. \tag{11.27}$$

As before, the integrals will cancel with corresponding factors in Q, and we will be left with:

$$\delta_n = \frac{N \, e^{-p_x^2/2mkT} \, \delta p_x}{\int e^{-p_x^2/2mkT} \, dp_x}. \tag{11.28}$$

Applications of the above methods are discussed in the following two sections.

11.10 Molecular speeds

(i) Distribution of molecular speeds

It is required to calculate, for the most probable distribution, the number of molecules which have speeds within some small specified range of values, say between c and $c + \delta c$. To do this, we first calculate the (most probable) number of molecules having translational momenta in specified ranges. By the argument which led to Equation 11.28, this number is:

$$\delta n = \frac{N \, e^{-(p_x^2 + p_y^2 + p_z^2)/2mkT} \, \delta p_x \, \delta p_y \, \delta p_z}{\iiint e^{-(p_x^2 + p_y^2 + p_z^2)/2mkT} \, dp_x \, dp_y \, dp_z} \tag{11.29}$$

Now $p_x = m\dot{x}$, $p_y = m\dot{y}$, $p_z = m\dot{z}$; moreover the denominator in Equation 11.29 is [cf. Section 11.5(i)]:

$$(2\pi mkT)^{3/2}$$

Hence:

$$\delta n = \frac{N \, e^{-m(\dot{x}^2 + \dot{y}^2 + \dot{z}^2)/2kT} \, m^3 \, \delta\dot{x} \, \delta\dot{y} \, \delta\dot{z}}{(2\pi mkT)^{3/2}}$$

is the number of molecules with components of velocity in the small specified ranges $\delta\dot{x}$, $\delta\dot{y}$, $\delta\dot{z}$. If the resultant velocity of a molecule is c, then:

$$c^2 = \dot{x}^2 + \dot{y}^2 + \dot{z}^2$$

If the direction of this velocity is given by the polar angles θ and φ, then it can be shown* that:

$$\delta\dot{x}\ \delta\dot{y}\ \delta\dot{z} = c^2 \sin\theta\ \delta\theta\ \delta\varphi\ \delta c \qquad (11.30)$$

and therefore:

$$\delta n = \left(\frac{m}{2\pi kT}\right)^{3/2} e^{-mc^2/2kT}\ c^2 \sin\theta\ \delta\theta\ \delta\varphi\ \delta c$$

The number of molecules having a *speed* c, irrespective of direction, is obtained by integrating over θ and φ [θ from 0 to π, φ from 0 to 2π; see Section 11.5(iii)] :

$$\delta n = \left(\frac{m}{2\pi kT}\right)^{3/2} e^{-mc^2/2kT}\ c^2\ \delta c \qquad (11.31)$$

Equation 11.31 is the Maxwell–Boltzmann law of distribution of molecular speeds.

If the range δc is arbitrarily assumed to be unity, the function:

$$n(c) = 4\pi \left(\frac{m}{2\pi kT}\right)^{3/2} e^{-mc^2/2kT}\ c^2 \qquad (11.32)$$

gives the fractional number† of molecules with speeds between c and $c + 1$. Equation 11.32 can be expressed in the form:

$$\left(\frac{n(c)}{\lambda}\right) = \frac{4}{\pi^{1/2}} (\lambda c)^2\ e^{-(\lambda c)^2}$$

where:

$$\lambda = \left(\frac{M}{2kT}\right)^{1/2} = \left(\frac{M}{2RT}\right)^{1/2}$$

Hence the graph of $n(c)/\lambda$ versus λc is a curve common to all gases at all temperatures (Fig. 11.4).

* If r is a line representing the vector c, the projection of r on axes (say) X, Y, and Z are the components $\dot{x}$, $\dot{y}$, and $\dot{z}$. It can be shown (see Appendix 3) that:

$$\delta X\ \delta Y\ \delta Z = r^2 \sin\theta\ \delta\theta\ \delta\varphi\ \delta r$$

Equation 11.30 follows.

† Note that $n(c)$ defined in this way is not a pure number: it has the dimensions of $(\text{speed})^{-1}$. The product of $n(c)$ with δc is a pure number, $\delta n/N$, and this has the same *numerical* value as $n(c)$ if $\delta c = 1$.

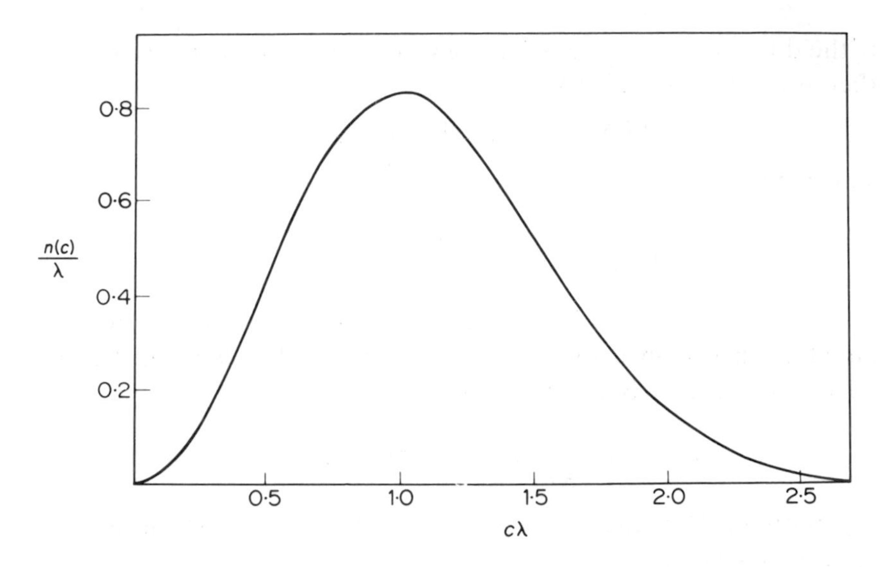

Fig. 11.4 Distribution of molecular speeds in a perfect gas

(ii) Averages of molecular speed

The average speed $\overline{c}$ is given, according to Section 11.9 by:

$$\overline{c} = \frac{\iiiint\!\!\iint e^{-mc^2/2kT}\ c\ dx\ dy\ dz\ dp_x\ dp_y\ dp_z}{Q_{\text{trans}}}$$

$$= \frac{m^3 V}{(2\pi mkT)^{3/2}\ V} \int_{-\infty}^{\infty}\int_{-\infty}^{\infty}\int_{-\infty}^{\infty} e^{-mc^2/2kT}\ c\ d\dot{x}\ d\dot{y}\ d\dot{z}$$

According to Equation 11.30:

$$d\dot{x}\ d\dot{y}\ d\dot{z} = c^2 \sin\theta\ d\theta\ d\varphi\ dc$$

and therefore:

$$\overline{c} = \left(\frac{m}{2\pi kT}\right)^{3/2} \int_0^{\infty}\int_0^{2\pi}\int_0^{\pi} e^{-mc^2/2kT}\ c^3\ \sin\theta\ d\theta\,\gamma\, d\varphi\ dc$$

$$= \left(\frac{m}{2\pi kT}\right)^{3/2} 4\pi \int_0^{\infty} e^{-mc^2/2kT}\ c^3\ dc \qquad (11.33)$$

Since (see Appendix 4):

$$\int_0^{\infty} e^{-ax^2}\ x^3\ dx = \frac{1}{2a^2}\,,$$

we obtain:

$$\bar{c} = \left(\frac{8kT}{m\pi}\right)^{1/2} = 14{\cdot}5 \left(\frac{T}{M}\right)^{1/2} \text{ m s}^{-1} \qquad (11.34)$$

The *mean square speed* $\overline{c^2}$ is obtained in a similar way. Thus, we obtain, instead of Equation 11.33:

$$\overline{c^2} = \left(\frac{m}{2\pi kT}\right)^{3/2} 4\pi \int_0^\infty e^{-mc^2/2kT} c^4 \, dc$$

and since:

$$\int_0^\infty e^{-ax^2} x^4 \, dx = \frac{3}{8a^2}\left(\frac{\pi}{a}\right)^{1/2},$$

there results:

$$\overline{c^2} = \frac{3kT}{m}$$

The square root of this value is called the *root mean square* speed:

$$\sqrt{\overline{c^2}} = \left(\frac{3kT}{m}\right)^{1/2} = 15{\cdot}8 \left(\frac{T}{M}\right)^{1/2} \text{ m s}^{-1} \qquad (11.35)$$

11.11 Electric and magnetic fields

Further illustration of the procedures of Section 11.9 is provided by consideration of the behaviour of molecules in externally applied fields. We here consider the classical theory of the effect of a uniform electric or magnetic field upon a gas.

The potential energy of interaction of a (neutral) molecule with an electric field of strength F is:

$$\epsilon' = \mu F \cos\theta$$

where μ is the *dipole moment* of the molecule and θ is the angle between the direction of the dipole and the direction of the field. (The term μF is the potential energy of the dipole when its direction coincides with that of the field, and $-\mu F \cos\theta$ is the work done in turning the dipole through an angle θ.) Since, for a given field, μF is constant, it can be included in the zero-point energy (cf Section 5.1). The other part of ϵ' is angle dependent and, therefore, constitutes the *potential energy of rotation* ϵ of the molecule in the field:

$$\epsilon = -\mu F \cos\theta \qquad (11.36)$$

Let us suppose that the direction of the dipole coincides with a principal axis; this would be the case, for example, for a linear molecule. In fact, the result which we shall finally obtain is valid whether or not this assumption be true. (The derivation for the general case is given by Rushbrooke, p. 159.) If it is true, however, θ in Equation 11.36 is identical with the Eulerian angle θ in Equation 11.23. The total rotational energy (kinetic + potential) is therefore:

$$\epsilon_{rot} = \frac{\delta^2}{2A} + \frac{\lambda^2}{2B} + \frac{p_x^2}{2C} - \mu F \cos\theta \qquad (11.37)$$

for non-linear molecules. There is a corresponding expression, based upon Equation 11.14c for diatomic or linear polyatomic molecules, but we shall not consider this explicitly because the treatment is so similar to that of the general case (problem 11.6).

The rotational partition function for a molecule in an electric field becomes (cf. Equation 11.24):

$$f_{rot} = \frac{1}{h^3} \int \cdots \int e^{-\epsilon_{rot}/kT} \sin\theta \, d\delta \, d\lambda \, dp_x \, d\theta \, d\varphi \, d\chi \, ,$$

where ϵ_{rot} is given by Equation 11.37. Integration over coordinates other than θ is unaffected by the presence of the $-\mu F \cos\theta$ term, and we have immediately:

$$f_{rot} = \frac{(4\pi^2 kT)^{3/2}}{h^3} (2\pi ABC)^{1/2} \int_0^{\pi} e^{\mu F \cos\theta/kT} \sin\theta \, d\theta$$

The dipole moment of any one molecule can be resolved into components:

$\mu \sin\theta$ perpendicular to the field F

$\mu \cos\theta$ parallel to the field F.

It is easily shown that the average value of the component perpendicular to the field is zero (see problem 11.7). The average value, $\bar{\mu}$, of $\mu \cos\theta$, i.e., *the average value of the component of* μ *parallel to the field* is, according to Section 11.9:

$$\bar{\mu} = \frac{(4\pi^2 kT)^{3/2}(2\pi ABC)^{1/2} \int_0^{\pi} e^{\mu F \cos\theta/kT} (\mu \cos\theta) \sin\theta \, d\theta}{(4\pi^2 kT)^{3/2}(2\pi ABC)^{1/2} \int_0^{\pi} e^{\mu F \cos\theta/kT} \sin\theta \, d\theta}$$

$$= \frac{\int_0^{\pi} \mu \, e^{\mu F \cos\theta/kT} \cos\theta \sin\theta \, d\theta}{\int_0^{\pi} e^{\mu F \cos\theta/kT} \sin\theta \, d\theta} \qquad (11.38)$$

Let $\cos\theta = x$, so that $\sin\theta\, d\theta = -\,dx$. Then:

$$\int_0^{\pi} e^{\mu F\cos\theta/kT}\sin\theta\, d\theta = -\int_1^{-1} e^{-\mu F x/kT}\, dx$$

$$= \frac{kT}{\mu F}\left(e^{\mu F/kT} - e^{-\mu F/kT}\right).$$

Now, $(e^y - e^{-y})/2$ is the function $\sinh y$ so:

$$\int_0^{\pi} e^{\mu F\cos\theta/kT}\sin\theta\, d\theta = \frac{\sinh y}{y} \qquad (11.39)$$

where $y = \mu F/kT$. The numerator of Equation 11.38 is easily evaluated since:

$$\frac{d}{dy}\int_0^{\pi} e^{y\cos\theta}\sin\theta\, d\theta = \int_0^{\pi} e^{y\cos\theta}\cos\theta\sin\theta\, d\theta$$

and therefore:

$$\int_0^{\pi} \mu\, e^{\mu F\cos\theta/kT}\cos\theta\sin\theta\, d\theta = \mu\frac{d}{dy}\left(\frac{\sinh y}{y}\right)$$

$$= \mu\left(\frac{\cosh y}{y} - \frac{\sinh y}{y^2}\right) \qquad (11.40)$$

Substituting Equations 11.39 and 11.40 into 11.38:

$$\mu\mu = \mu\left(\frac{\cosh y}{\sinh y} - \frac{1}{y}\right) = \mu\left(\coth y - \frac{1}{y}\right)$$

Numerical considerations show that, for values of F of practical importance, y is small.

For example, a diatomic molecule may be regarded as having nett charges $+z$ and $-z$ at the positions of the nuclei. The dipole moment is then:

$$\mu = zr$$

where r is the internuclear distance.

Let us suppose that z is the electronic charge:

$$z = e = 1{\cdot}6 \times 10^{-19}\, C$$

and that

$$r = 1Å = 10^{-10}\, m$$

Then:

$$\mu = 1{\cdot}6 \times 10^{-19} \times 10^{-10} = 1{\cdot}6 \times 10^{-29}\, Cm$$

Suppose that:

$$F = 10^5 \text{ volts per metre}$$

and that:

$$T = 300 \text{ K}.$$

Then:

$$\mu F = 1{\cdot}6 \times 10^{-29} \times 10^{5} = 1{\cdot}6 \times 10^{-24} \text{ J}$$

and

$$kT = 1{\cdot}38 \times 10^{-23} \times 300 = 4{\cdot}14 \times 10^{-21} \text{ J}$$

Therefore:

$$y = \frac{\mu F}{kT} = 4 \times 10^{-4}.$$

For small values of y, the expansion:

$$\coth y = \frac{1}{y} + \frac{1}{3} y + \ldots ,$$

can be used, i.e.:

$$\overline{\mu} = \mu \left(\frac{y}{3}\right) = \frac{\mu^2 F}{3kT} . \qquad (11.41)$$

In deriving* Equation 11.41, we have neglected the effect of the electric field upon the charge distribution within the molecule. Actually, the molecule will be *polarized* by the field, i.e., it will acquire an additional (induced) dipole moment μ_i *in the direction of the field,* given by:

$$\mu_i = \alpha F$$

where α is called the *polarizability* of the molecule in the direction of the field. As an approximation (which in fact gives, to the first power of $1/T$, a result identical with the exact answer) we take a value of α averaged over all orientations; that is, we assume that α is the mean polarizability of the molecule. The overall average component of dipole moment in the direction of the field is then:

$$\overline{\mu} + \mu_i = F \left(\frac{\mu^2}{3kT} + \alpha\right)$$

The *electrical susceptibility* κ of a gas is defined to be

$$\kappa = \frac{1}{F}\frac{N}{V}(\overline{\mu} + \mu_i)$$

* Equation 11.41 can also be obtained by a quantum-mechanical treatment. This value of $\overline{\mu}$ comes entirely from the molecules of the $J = 0$ level, and the rest do not contribute. In this case, therefore, the equivalence of the quantal and classical treatments is *not* a consequence of the limiting principle (page 213). See Fowler and Guggenheim, p. 634.

i.e.,
$$\kappa = \frac{N}{V}\left(\frac{\mu^2}{3kT} + \alpha.\right) \tag{11.42}$$

The electrical susceptibility is related to the dielectric constant D of the gas by:

$$D = 1 + 4\pi\kappa$$

and finally, we obtain the important equation:

$$\frac{N}{V}\left(\frac{D-1}{4\pi}\right) = \frac{\mu^2}{3kT} + \alpha\,. \tag{11.43}$$

Since D can be measured experimentally, this equation permits the empirical determination of μ. Thus a graph of $\frac{V}{N}\left(\frac{D-1}{4\pi}\right)$ *versus* $\frac{1}{T}$ is linear, and μ is found from the gradient. The intercept on the OY axis gives α the polarizability.

The theory described above is due to Debye. It is, however, closely related to the earlier work of Langevin on *magnetic* susceptibilities. Langevin's theory holds for gases but was developed for treating the more important problem of the magnetic susceptibilities of *solids.* The magnetic susceptibility of a crystal such as one of potassium ferric alum arises from the magnetic moments of individual atoms of atomic ions (in the example cited from Fe^{+++} ions). If μ_m is the *magnetic* moment of an ion and H is the applied magnetic field, there is a potential energy analogous to that given by Equation 11.36:

$$\epsilon = -\mu_m H \cos\theta\,,$$

and the mean component of magnetic moment in the direction of the field is given by an equation corresponding to 11.41:

$$\bar{\mu}_m = \frac{\mu_m^2 H}{3kT}$$

An equation for the magnetic susceptibility, analogous to 11.43 now follows. The first term in it:

$$\frac{1}{H}\frac{N}{V}\bar{\mu}_m = \frac{N}{V}\frac{\mu_m^2}{3kT}$$

is called the *paramagnetic susceptibility* of the substance. The inverse proportionality between paramagnetic susceptibility and temperature, indicated by the theory, had previously been discovered experimentally by P. Curie, and is known as *Curie's law.*

The magnetic field also *induces* a magnetic moment in an ion or molecule, and this gives rise to a term analogous to α in Equation 11.43. The corresponding contribution to the magnetic susceptibility is called the *diamagnetic* susceptibility. There are, however, two differences between the electric and magnetic susceptibilities. Firstly, the induced *magnetic* moment is in the *opposite* direction to that of the field, i.e., the diamagnetic susceptibility is negative; on the other hand, the induced electric moment is always in the direction of the field, and α is positive. This difference is a consequence of electrodynamics. Secondly, the diamagnetic susceptibility is much smaller (by a factor of $\sim 10^{-3}$ to 10^{-2}) than the paramagnetic term, and (in contrast to the case of the electric field) may be neglected unless the paramagnetic susceptibility is zero.

Problems

11.1 Show that the root mean square of the angular momentum P_θ is:

$$\sqrt{\overline{p_\theta^2}} = (2IkT)^{\frac{1}{2}}$$

11.2 Obtain an expression for the root mean square momentum of a classical harmonic oscillator.

11.3 The following is a description of a very simple molecular beam apparatus:

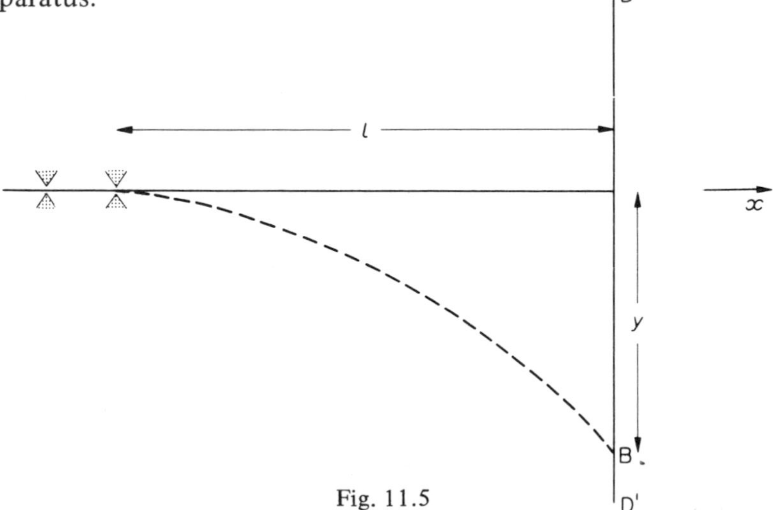

Fig. 11.5

A collimated beam of atoms energes from a heated vessel containing a vapour. If the speed of an atom emerging is v_x, it will travel a distance l in l/v_x seconds. Under the influence of gravity, it will fail a distance $y = \frac{1}{2}g(l/v_x)^2$ in this time. Therefore, if y and l are measured, v_x is determined. The number of atoms reaching B is measured by a detector (a heated tungsten filament) which can be moved along DD′. The variation of this number with the position of B gives the distribution of speeds in the x-direction.

(a) Calculate the average value $\overline{v_x}$ and the root mean square value $\sqrt{\overline{v_x^2}}$ for caesium atoms emerging from a furnace at 1000 K.

(b) What proportion of the atoms would have values of v_x between $0{\cdot}99\sqrt{\overline{v_x^2}}$ and $1{\cdot}01\sqrt{\overline{v_x^2}}$?

(c) How could such a method be used to determine the equilibrium constant for the reaction:

$$Bi_2 \rightleftharpoons 2Bi$$

at 2000 K?

11.4 The potential energy due to gravity of a molecule at height z is mgz.

(a) What is the partition function for molecules in a gravitational field?

(b) Derive the barometric formula for the variation of atmospheric pressure with height:

$$P/P' = e^{(z' - z)mg/kT}.$$

11.5 The potential energy of an ion having charge q in an electric field F which acts in the z-direction is qFz. Obtain the quasi-classical translational partition function for ions in an electric field.

Calculate the contribution from an electric field of 1 V m^{-1} to the free energy change of the reaction:

$$LiCl_{(g)} \rightarrow Li^+_{(g)} + Cl^-_{(g)}$$

when the $LiCl_{(g)}$ is contained in a cube of side 1 m at 1000 K ($q = 1{\cdot}6 \times 10^{-19}$ C). Why is such a calculation physically unrealistic?

11.6 Using Equation 11.14c, write down an expression for the rotational energy of a diatomic molecule, the axis of which makes an angle θ to an electric field F. Obtain the partition function, and derive an expression for the fractional number of molecules with values of θ between 45° and 46°.

11.7 Referring to Fig. 11.2, show that the components of the molecule $m_1\ m_2$ (or, for that matter, any vector represented by the line $m_1\ m_2$) are:

$$\mu_z = \mu \cos\theta$$
$$\mu_x = \mu \sin\theta \cos\varphi$$

Show that averaging μ_x over φ gives zero.

11.8 Döppler broadening of spectral lines
(a) For a monatomic gas show that the number of atoms with x-components of velocity between v_x and $v_x + \delta v_x$ is:

$$n(v_x) = n_0\, e^{-mv_x^2/2kT}$$

where n_0 is the number with x-components between 0 and δv_x.
(b) If an atom emits light of wavelength λ_0 whilst moving away from the observer with a velocity v_x, it will *appear* to the observer to be emitting light of wavelength:

$$\lambda = \lambda_0 \left(1 + \frac{v_x}{c}\right)$$

(the Döppler effect), where c is the speed of light. Express v_x in terms of λ, λ_0, and c, and hence obtain Rayleigh's formula for the variation of intensity of a spectral line with wavelength:

$$I(\lambda) = I(\lambda_0)\, e^{-mc^2(\lambda-\lambda_0)^2/2\lambda_0^2 kT}.$$

11.9 The choice of coordinates q_i in the expression for the phase integral of a molecule is arbitrary. If Cartesian coordinates of the nuclei are used, the conjugate momenta are all of the form $m_i\dot{x}_i$.
(a) Show that the phase integral can always be expressed as

$$Q = \left(\prod_{\alpha} (2\pi m_\alpha kT)\right)^{3/2} \int_{(3n)} \cdots \int e^{-U/kT}\, dq_1 \ldots dq_{3n} ,$$

where U is a potential function of the $3n$ Cartesian coordinates, and the product is over all the nuclei.
(b) Deduce that if A and B differ only in isotopic constitution, so that the potential function U is the same for both, then, for a given T and V:

$$\frac{Q_A}{Q_B} = \frac{\sigma_B}{\sigma_A} \prod \left(\frac{m_{\alpha A}}{m_{\alpha B}}\right)^{3/2}$$

where σ_A and σ_B are the symmetry numbers of A and B.

(c) Noting that the quasi-classical vibrational partition function $\frac{h\omega_i}{kT}$ is replaced by $\frac{e^{-h\omega_i/2kT}}{1 - e^{-h\omega_i/kT}}$ (including the zero-point vibrational energy), deduce that (cf problem 9.8):

$$\frac{f_A}{f_B} = \frac{\sigma_B}{\sigma_A} \prod \left(\frac{m_{\alpha A}}{m_{\alpha B}}\right)^{3/2} \cdot \prod_i \frac{\omega_{iA}}{\omega_{iB}} \cdot \frac{1 - e^{-h\omega_{iB}/kT}}{1 - e^{-h\omega_{iA}/kT}} \cdot \frac{e^{-h\omega_{iA}/2kT}}{e^{-h\omega_{iB}/2kT}}$$

CHAPTER TWELVE

The Transition State Theory of Reaction Rates

Statistical mechanics has wide and important application in the theories of chemical reaction kinetics. It would be inappropriate to include much discussion of this topic in a book on statistical *thermodynamics,* a book concerned with equilibrium states of a system, rather than the rates at which such states may be achieved. However, the formulation of one theory of chemical kinetics is so closely related to the methods developed in the previous chapters, that we include a brief* account of it; this is the *theory of absolute reaction rates* (H. Eyring) or *transition state theory* (M.G. Evans). We shall not discuss the alternative, *collision theory* of reaction rates, which, after a period of eclipse by the transition state theory, is becoming increasingly important in present-day physical chemistry.

12.1 Potential energy surfaces: the activated complex

The transition state theory is most easily introduced by consideration of the atom–diatomic molecule reaction:

$$A + BC \longrightarrow AB + C \qquad (12.1)$$

and, in the first three sections of this chapter, discussion is confined to this reaction. The simplest case of reaction 12.1 is that in which

* Accounts of the current status of reaction rate theories are given by H. S. Johnston, *Gas Phase Reaction Rate Theory,* Ronald, New York (1966), and K. J. Laidler, *Theories of Chemical Reaction Rates,* McGraw-Hill, New York (1969). For an account of the earlier work on transition state theory, see S. Glasstone, K. J. Laidler, and H. Eyring. *The Theory of Rate Processes,* McGraw-Hill, New York (1941).

A, B, and C are all hydrogen atoms:

$$H + HH \longrightarrow HH + H.$$

This reaction occurs, for example, in the conversion of para-hydrogen into orthohydrogen, and its rate constant has been deduced from the observed kinetics of such conversion.

Let us consider reaction 12.1 as a process of classical mechanics; the modifications necessitated by quantum mechanics will be discussed subsequently. Suppose that, initially, the atom A and the molecule BC are isolated, and that BC is in the equilibrium configuration. They approach, exerting electrostatic forces upon each other, and reaction ultimately occurs. Finally, the products of reaction, AB and C, separate, and AB attains its equilibrium internuclear distance. In such a process, the *total* energy of the three-atom system remains constant (it is assumed that energy is neither absorbed from, nor lost to, the surroundings, in the form of radiation or by a collision with another molecule). Because of the varying nature of the forces of interaction, however, the potential energy **V** changes during the process; any increase in potential energy occurs at the expense of kinetic energy, and *vice versa*.

The potential energy at any stage of the reaction can, in principle, be calculated when the position coordinates of each atom are specified. Actually, the potential energy can be expressed in terms of a smaller number of variables than this, since it is determined by the positions of the atoms *relative* to one another (see page 71). Clearly, in the present case, the potential energy is determined if the A–B, B–C, and A–C distances are specified. To simplify the matter further, let us consider only reactions during which the three atoms remain collinear; the justification for this is given later, in Section 12.3. If the atoms are collinear, only two of the interatomic distances are independent; for example, the A–C distance can be calculated if the A–B and B–C distances are known. Thus, the potential energy of the three-atom systems can now be expressed in terms of two independent variables, say r_{AB} and r_{BC}. Quantum-mechanical calculation of **V** for various values of r_{AB} and r_{BC} has been carried out for the $(H + H_2)$ system. The calculated dependence of **V** upon r_{AB} and r_{BC} can be represented by constructing a surface in three dimensions, in which r_{AB} and r_{BC} are respectively plotted along the x- and y-axes, and the potential energy along the z-axis.

Use of such a three-dimensional representation is not, in practice, convenient. There are, however, two methods by means of which the potential energy surface can be represented in two-dimensional diagrams.

(i) *Contour Diagrams*

The contour diagram for the linear (H H H) system is shown in Fig. 12.1. The numbers on the contours are the potential energies of the points (r_{AB}, r_{BC}) connected by the contour.

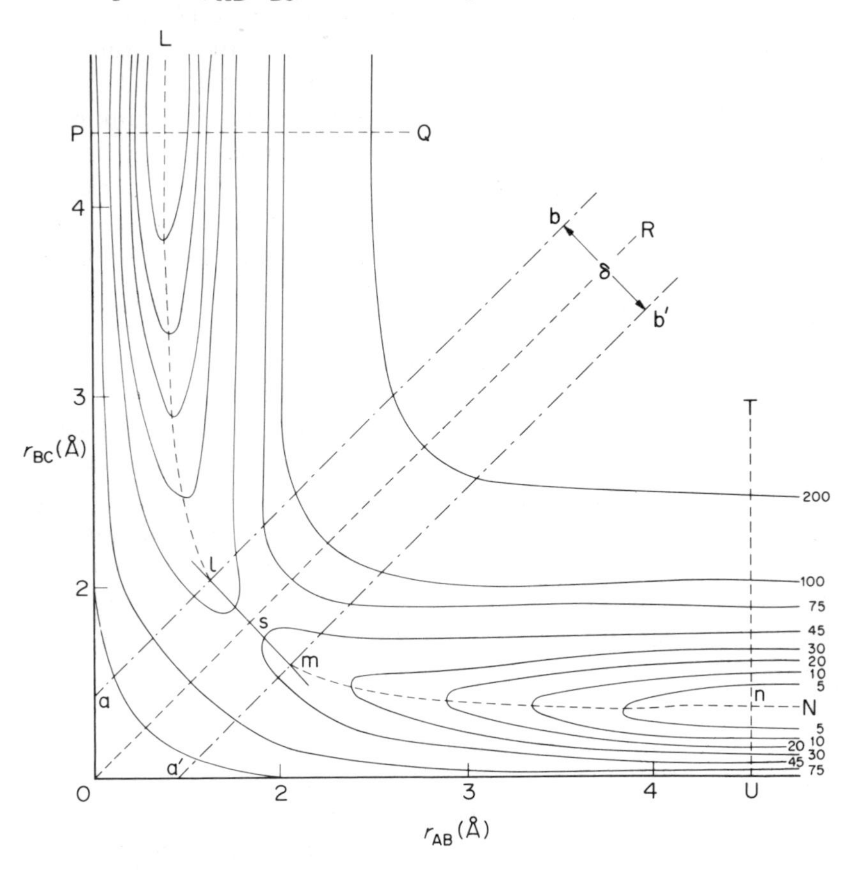

Fig. 12.1 Potential energy contours for the linear (H H H) system. This is based upon the calculations of I. Shavitt, R. M. Stevens, F. L. Minn and M. Karplus, J. Chemical Physics, 1968, **48**,6. Energies are expressed as kJ mol^{-1}

(ii) *Sectional Diagrams*

Sections, or cuts, may be taken through the three-dimensional figure. Suppose that r_{BC} is given the constant value P, and the value of **V** is plotted for various values of r_{AB}. This corresponds to the cut PQ in the contour diagram, and the plot is shown in Fig. 12.2(a) (see problem 12.1). A similar result is obtained for the cut TU, in which r_{AB} is held constant while r_{BC} varies; the corresponding plot is shown in Fig. 12.2(b).

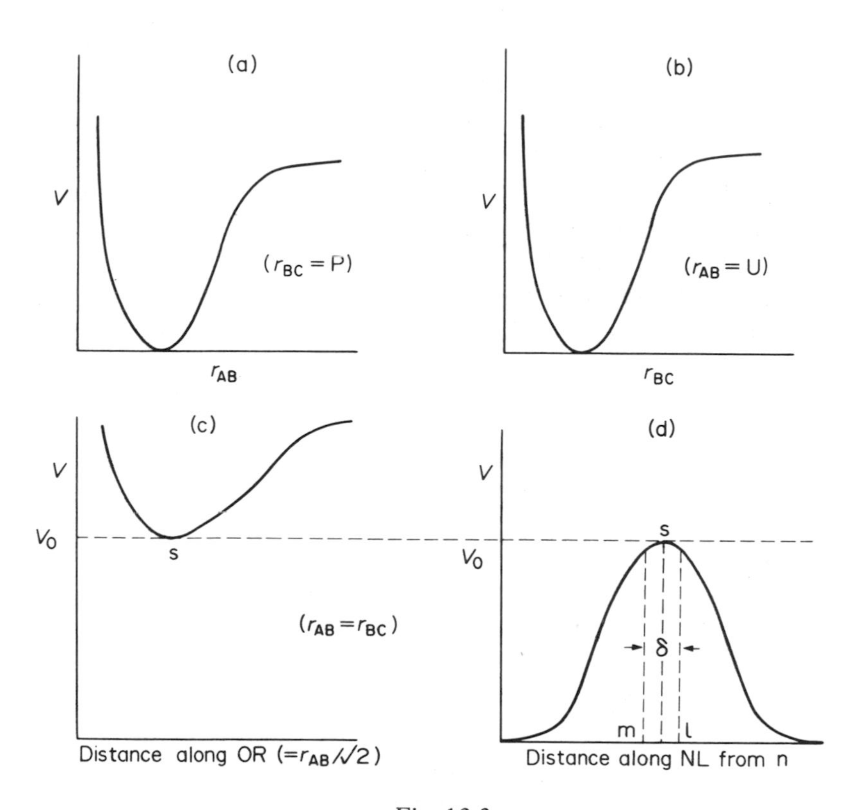

Fig. 12.2

Fig. 12.2(c) shows the section OR. in this case, $r_{AB} = r_{BC}$ for all the points, and the distance along OR from the origin is r_{AB} sec 45 = $r_{AB}\sqrt{2}$.

Sections can also be taken along curves. For example, by taking the section along the line NL, we obtain the curve shown in Fig 12.2(d). In the figure, the distance from some arbitrary point (say n) on this line is denoted by q_*. Note that NL coincides with the direction of r_{AB} at the begining, and with the direction of r_{BC} at the end.

The *maximum* in Fig. 12.2(d) corresponds to the same point s as the *minimum* in Fig. 12.2(c). Thus, at point s, the curvature is ∪ in the direction of OR and ∩ in the direction of NL. A point such as s, at which a surface has different kinds of curvature in different directions, is called a *saddle point* of the surface.

Let:

$$q_1 = (r_{BC} + r_{AB})/\sqrt{2}$$

and (12.2)

$$q_2 = (r_{BC} - r_{AB})/\sqrt{2}$$

The position of a point on the contour diagram can now be specified by giving the values of q_1 and q_2, instead of r_{AB} and r_{BC}.

It is easily shown that q_1 and q_2 are the Cartesian coordinates of the point (r_{AB}, r_{BC}) relative to the axes obtained by rotating OY and OX through 45° about OZ. Since the new position of OX coincides with OR, *q_1 is the distance from* 0 *in the direction* OR, *and q_2 the distance from* s *in the direction perpendicular to* OR.

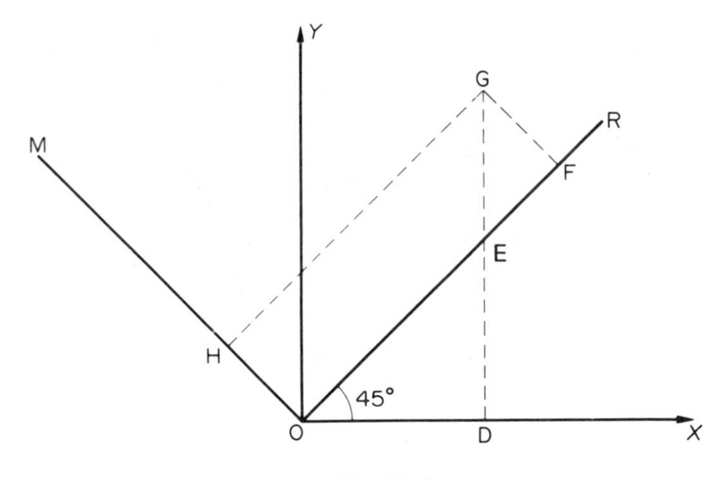

Fig. 12.3

In Fig. 12.3:

$$r_{AB} = \text{OD and } r_{BC} = \text{GD}$$

The coordinates relative to OR and OM are OF and GF. From the figure:

$$\text{GE} = \text{GD} - \text{ED} = \text{GD} - \text{OD} = r_{BC} - r_{AB}$$

Therefore:

$$\text{GF} = \text{GE} \sin 45^\circ = \frac{1}{\sqrt{2}}(r_{\text{BC}} - r_{\text{AB}})$$

$$= q_2$$

$$\text{OF} = \text{OE} + \text{EF} = \text{OD} \sec 45^\circ + \text{GE} \cos 45^\circ$$

$$= r_{\text{AB}}\sqrt{2} + (r_{\text{BC}} - r_{\text{AB}})/\sqrt{2}$$

$$= \frac{1}{\sqrt{2}}(r_{\text{BC}} + r_{\text{AB}})$$

$$= q_1$$

For small displacements from the point s, the increase $\mathbf{V} = \mathbf{V}' - \mathbf{V}_0'$ in potential energy is given by 'Hooke's Law' expressions (since $\partial\mathbf{V}/\partial q_1 = \partial\mathbf{V}/\partial q_2 = 0$, at this point; see the note on page 68):

for displacement of q_1 : $\mathbf{V} = \mathbf{V}' - \mathbf{V}_0' = \frac{1}{2}k_1(q_1 - q_1^0)^2$ (12.3a)

for displacement of q_2 : $\mathbf{V} = \mathbf{V}' - \mathbf{V}_0' = \frac{1}{2}k_2(q_2 - q_2^0)^2$ (12.3b)

where q_1^0 and q_2^0 are the values of the coordinates at s, k_1 and k_2 are the corresponding force constants, and $\mathbf{V}_0'$ is the potential energy at s. It can then be shown (see the note below) that, *if* k_1 *and* k_2 *were positive,* q_1 and q_2 would vary periodically with time, with frequencies:

$$\omega_1 = \frac{1}{2\pi}\sqrt{\frac{k_1}{m}} \quad \text{and} \quad \omega_2 = \frac{1}{2\pi}\sqrt{\frac{3k_2}{m}} \qquad (12.4)$$

where m is the mass of the hydrogen atom. The corresponding vibrations (normal modes) of the system are represented in Fig. 12.4

Fig. 12.4,

It can be seen, however, from Fig. 12.2(c) and (d) that, whereas displacement of q_1 from q_1^0 increases V, displacement of q_2 from

q_2^0 *decreases* **V**; i.e., the force constant k_2 in Equation 12.4 is *negative.* The corresponding vibrational frequency ω_2 of the second mode, calculated from Equation 12.4, would therefore be imaginary. This means, physically, that motion of this kind, when once started, will continue without changing direction, i.e., that r_{AB} will continue to become smaller, and r_{BC} to become larger, until ultimately the products of the reaction are formed. For this reason, the coordinate q_2 is called the *reaction coordinate** of the system.

We have defined q_* as the distance along the curve NL in Fig. 12.1, and from the figure it is clear that, *within some small range* δ:

$$q_2 \quad \text{is the same as} \quad q_* - q_*^0,$$

where q_*^0 is the value of q^* at s. A system of three hydrogen atoms which has values of r_{AB} and r_{BC} represented by points inside the area aa′b′b, i.e., *with values of* q_* *within the range* δ, is said to be in the *transition state;* the group of the three atoms in the transition state is called an *activated complex.* When reaction occurs, the system (A + BC) passes over to the system (AB + C), and must therefore at some stage achieve values of r_{AB} and r_{BC} corresponding to points within the rectangle aa′b′b; i.e., it must pass through the transition state [but see Section 12.3(iii)].

After this rather long preamble, we now introduce statistical mechanics into the problem by calculating (a) the average concentration of activated complexes, (b) the average rate at which they decompose. The product of these two expressions gives the rate of reaction.

Note on the vibrations of the (H H H) activated complex

The atoms are collinear; let x'_A, x'_B, x'_C be the coordinates of A, B, and C. The centre of mass is at x, defined by

$$m_A x'_A + m_B x'_B + m_C x'_C = (m_A + m_B + m_C)x$$

i.e., $x'_A + x'_B + x'_C = 3x$ for the three hydrogen atom system. Let:

$$x_A = x'_A - x, \text{ etc., so that:}$$

$$x_A + x_B + x_C = 0 \tag{12.5}$$

* Thus defined, the term ‘reaction coordinate’ has a precise meaning. It is also, however, used in an obscure way to mean some (unspecified) measure of the extent to which a rate process has occurred. See Johnston, *op. cit.*, Ch. 16.

Let:

$$q_1 = \frac{1}{\sqrt{2}}(r_{BC} + r_{AB}) = \frac{1}{\sqrt{2}}[(x_C - x_B) + (x_B - x_A)] = \frac{1}{\sqrt{2}}[x_C - x_A] \tag{12.6}$$

and

$$q_2 = \frac{1}{\sqrt{2}}(r_{BC} - r_{AB}) = \frac{1}{\sqrt{2}}[(x_C - x_B) - (x_B - x_A)] = \frac{1}{\sqrt{2}}[x_C - 2x_B + x_A] \tag{12.7}$$

From Equations 12.5, 12.6 and 12.7:

$$x_A = \frac{1}{\sqrt{2}}\left(\frac{1}{3}q_2 - q_1\right)$$

$$x_B = -\frac{\sqrt{2}}{3}q_2$$

$$x_C = \frac{1}{\sqrt{2}}\left(\frac{1}{3}q_2 + q_1\right)$$

The vibrational kinetic energy (i.e., the kinetic energy when the x-axis and the centre of mass are fixed) is:

$$\begin{aligned} \mathbf{T} &= \frac{1}{2}m(\dot{x}_A^2 + \dot{x}_B^2 + \dot{x}_C^2) \\ &= \frac{1}{2}m(\dot{q}_1^2 + \frac{1}{3}\dot{q}_2^2) \end{aligned} \tag{12.8}$$

The potential energy is, according to Equation 12.3:

$$\mathbf{V} = \mathbf{V}_0 + \frac{1}{2}k_1(q_1 - q_1^0)^2 + \frac{1}{2}k_2(q_2 - q_2^0)^2.$$

The total vibrational energy is therefore:

$$\mathrm{E} = \mathbf{T} + \mathbf{V} = \mathbf{V}_0 + \left[\frac{1}{2}m\dot{q}_1^2 + \frac{1}{2}k_1(q_1 - q_1^0)^2\right] + \left[\frac{1}{2}\frac{m}{3}\dot{q}_2^2 + \frac{1}{2}k_2(q_2 - q_2^0)^2\right] \tag{12.9}$$

Since the total energy does not vary with time, $\mathrm{d}E/\mathrm{d}t = 0$, and hence, differentiating Equation 12.9:

$$\left[m\dot{q}_1\ddot{q}_1 + k_1\dot{q}_1(q_1 - q_1^0)\right] + \left[\frac{m}{3}\dot{q}_2\ddot{q}_2 + k\dot{q}_2(q_2 - q_2^0)\right] = 0 \tag{12.10}$$

where $$\ddot{q}_1 = \mathrm{d}\dot{q}_1/\mathrm{d}t = \mathrm{d}^2q/\mathrm{d}t^2.$$

Now q_1 and q_2 are *independent* variables, and therefore each bracket in Equation 12.10 must separately be zero. For the first bracket:

$$m\dot{q}_1\ddot{q}_1 + k_1\dot{q}_1(q_1 - q_1^0) = 0$$

or

$$\ddot{q}_1 + \frac{k_1}{m}(q_1 - q_1^0) = 0$$

But this is the differential equation for simple harmonic motion; q_1 therefore varies periodically with a frequency:

$$\omega_1 = \frac{1}{2\pi}\sqrt{\frac{k_1}{m}}$$

Similarly, q_2 varies periodically with a frequency:

$$\omega_2 = \frac{1}{2\pi}\sqrt{\frac{3k_2}{m}}$$

12.2 Calculation of rate of reaction

Before discussing the transition state theory, we digress to mention an alternative way of using the calculated potential energy surface to determine the rate of reaction. This is the method of *reaction dynamics.* Suppose that, by *classical* dynamics, the instantaneous positions and momenta of the three atoms in the (H H H) system are specified, and correspond to a point in the region of N in Fig. 12.1. The instantaneous positions and momenta at any future time can now be calculated from a knowledge of the potential surface and the classical equations of motion. Similarly, the *average* behaviour of many (H H H) systems can be investigated by repeating the calculations for many different initial conditions; from this average, the overall rate of reaction is computed. This is currently an active field of research; the most serious drawbacks to it are: firstly, the necessity of using classical, rather than quantal equations of motion; secondly, the calculations involve the whole of the potential energy surface, which must be calculated with some accuracy; and thirdly, the fact that to obtain the average over a sufficient number of calculations, an electronic computer is necessary.

In contrast with the above method, the transition state theory concerns itself with only two parts of the potential energy surface, namely, one corresponding to the initial state, the other to the transition state. This enormous simplification is achieved by making the basic supposition that *the activated complexes can be regarded as being in equilibrium with the reactants.* Since equilibrium is assumed, the dynamical changes occurring during the formation of the activated complex from the reactants need not be considered; correspondingly, a knowledge of the nature of the potential energy surface for intermediate configurations is unnecessary.

The most convincing justification of the basic assumption seems to be the following. Suppose that the reaction:

$$A + B \rightleftharpoons C$$

has proceeded to equilibrium, so that the rates of the forward and reverse reactions are equal. Under this condition, there is also equilibrium with intermediate states; that is, A and B *are* in equilibrium with the activated complex. The rate of formation of activated complexes *from* A and B, and their subsequent decomposition *to* C is independent of the concentration of C and consequently is the same *whether or not* the reaction has *actually* proceeded to equilibrium.

The atom–diatomic molecule reaction can be written as:

$$A + BC \rightleftharpoons X^{\ddagger} \longrightarrow AB + C$$

where $X^{\ddagger}$ denotes an activated complex. It is conventional to use the symbol ‡ to denote properties accociated with the transition state. The equilibrium constant is:

$$K^{\ddagger} = \frac{[X^{\ddagger}]}{[A]\,[BC]},$$

and therefore the concentration of activated complexes is:

$$[X^{\ddagger}] = K^{\ddagger}[A]\,[BC]$$

The rate of reaction is equal to this concentration multiplied by the frequency ν with which the activated complexes break up to give products:

$$\text{rate} = \nu K^{\ddagger}[A]\,[B] \qquad (12.12.)$$

If k is the experimentally determined bimolecular rate constant for the overall reaction:

$$\text{rate} = k[A]\,[B] \qquad (12.13)$$

and hence

$$k = \nu K^{\ddagger} \qquad (12.14)$$

The problem is therefore to calculate $K^{\ddagger}$ and ν.

(i) Calculation of $K^{\ddagger}$

The unit of concentration in Equation 12.13 will be taken as mole litre^{-1}, and this unit must therefore be used in Equations 12.11 and 12.12. Thus, the required equilibrium constant is of type K_c, for which the statistical mechanical expression is given in Equation 9.17:

$$K^{\ddagger} = e^{-\Delta\epsilon_0^{\ddagger}/kT} \frac{f_s^{\ddagger}}{f'_A f'_B} \left(\frac{1}{\tilde{\tilde{V}}^{\circ}\tilde{N}}\right)^{-1} \text{ l mol}^{-1} \qquad (12.15)$$

where $\Delta\epsilon_0^{\ddagger}$ is the increase in potential energy when the system moves from the equilibrium configuration of the reactants to the configuration corresponding to the saddle point s on the potential energy surface. $f_s^{\ddagger}$ is the partition function of the activated complex, i.e., the sum (or integral) of $e^{-\epsilon/kT}$ over all those states (or phase points) of the system which correspond to points within the area aa′b′b in Fig. 12.1.

The partition function $f_s^{\ddagger}$ can be factorized into electronic, translational, vibrational, and rotational terms in the usual way. One of the vibrational factors corresponds, however, to the energy associated with the displacement of q_2 (or q_*) from the value at s. We shall denote this factor by f_*. It differs from ordinary partition function factors in two ways: (i) it cannot be evaluated as an ordinary vibrational factor because the corresponding value of ω, given by Equation 12.4, is imaginary; (ii) only terms corresponding to values of q_* in the range δ contribute to f_*.

Actually, we shall find it unnecessary to calculate f_*, because it cancels with a factor in $K^{\ddagger}$ when k is evaluated from Equation 12.14.

(ii) Calculation of ν

Since q_*^0 is the mid-point of the range δ, the transition state begins at:

$$q_{*(1)} = q_*^0 - \tfrac{1}{2}\delta$$

and ends at:

$$q_{*(2)} = q_*^0 + \tfrac{1}{2}\delta$$

The time τ taken for the system to pass through the transition state is therefore the time taken for q_* to change from the value $q_{*(1)}$ to the value $q_{*(2)}$. It can be seen from Fig. 12.1 that the directions of the contours through the rectangle aa′b′b are essentially the same as the direction of the line lm, i.e., the direction of q_*. Consequently, for any value of q_1, the straight line joining $q_{*(1)}$ and $q_{*(2)}$ is almost coincident with some contour, and therefore the *potential energy of the activated complex remains essentially constant* when q_* changes from $q_{*(1)}$ to $q_{*(2)}$. (This is equivalent to the statement that k_2 is small.) Since the total energy is constant, this means that the kinetic energy associated with q_* does not change during this process. The kinetic energy can be shown to be (see Equation 12.8):

$$\epsilon_* = \frac{1}{2}\left(\frac{m}{3}\right)\dot{q}_*^2, \tag{12.16}$$

and therefore:

$$\dot{q}_* = \frac{\mathrm{d}q_*}{\mathrm{d}t}$$

has a constant value during the process. The time τ is, in consequence:

$$\frac{q_{*(2)} - q_{*(1)}}{\dot{q}_*} = \frac{\delta}{\dot{q}_*}$$

The momentum conjugate to q_* is, according to Equation 11.2:

$$p_* = \frac{\partial}{\partial \dot{q}_*}\left[\frac{1}{2}\left(\frac{m}{3}\right)\dot{q}_*^2\right] = \frac{m}{3}\dot{q}_* \tag{12.17}$$

Hence:

$$\tau = \frac{m}{3}\frac{\delta}{p} \tag{12.18}$$

Suppose that $n_1, n_2, \ldots, n_i, \ldots$ activated complexes have momenta $p_{*(1)}, p_{*(2)}, \ldots, p_{*(i)}, \ldots$. If $n_1, n_2, \ldots, n_i, \ldots$ are maintained at their equilibrium values, the rate of reaction is:

$$\text{rate} = \frac{n_1}{\tau_1} + \frac{n_2}{\tau_2} + \ldots + \frac{n_i}{\tau_i} + \ldots, \tag{12.19}$$

where $\tau_1, \tau_2, \ldots, \tau_i, \ldots$ are the time intervals corresponding to $p_{*(1)}, p_{*(2)}, \ldots, p_{*(i)}, \ldots$ calculated according to Equation 12.18. Equation 12.19 simply expresses the fact that the rate of reaction is the total rate with which activated complexes pass through the transition state.

The mean value of $1/\tau$ is, by definition:

$$\overline{\left(\frac{1}{\tau}\right)} = \frac{1}{n}\left[n_1\left(\frac{1}{\tau_1}\right) + n_2\left(\frac{1}{\tau_2}\right) + \ldots + n_i\left(\frac{1}{\tau_i}\right) + \ldots\right]$$

where n is the total number of activated complexes, and so:

$$\text{rate} = \overline{\left(\frac{1}{\tau}\right)}.$$

Thus, the frequency ν in Equations 12.2 and 12.14 is the same as $\overline{1/\tau}$, i.e., the average value of $3p_*/m\delta$. From Equations 12.16 and 12.17:

$$\epsilon_* = \frac{3}{2}\frac{p_*^2}{m}$$

Using the averaging procedure of Section 11.9, we obtain immediately:

$$\begin{aligned}\overline{\left(\frac{1}{\tau}\right)} &= \frac{1}{Q_*}\int_0^{\infty}\int_{-\delta/2}^{\delta/2} e^{-3p_*^2/2mkT}\left(\frac{3p_*}{m\delta}\right) dq_*\, dp_* \\ &= \frac{1}{Q_*}\left(\frac{mkT}{3}\right)\left(\frac{3}{m\delta}\right)\delta \\ &= \frac{kT}{Q_*}\end{aligned}$$

where Q_* is the factor corresponding to p_* and q_* in the phase integral. According to the general rule (Equation 11.11):

$$Q_* = f_*\, h$$

so that:

$$\nu = \overline{\left(\frac{1}{\tau}\right)} = \frac{kT}{f_*\, h}$$

Substituting for $K^{\ddagger}$ and ν in Equation 12.14:

$$\begin{aligned}k &= \frac{kT}{hf_*}K^{\ddagger} \\ &= \frac{kT}{hf_*}\left(\frac{f_s^{\ddagger}}{f_A'f_{BC}'}\right)\left(\frac{1}{\tilde{N}\tilde{\tilde{V}}^{\circ}}\right)^{-1} e^{-\Delta\epsilon_0^{\ddagger}/kT} \qquad (12.20)\\ &= \frac{kT}{h}\left(\frac{f^{\ddagger}}{f_A'f_B'}\right)\left(\frac{1}{\tilde{N}\tilde{\tilde{V}}^{\circ}}\right)^{-1} e^{-\Delta\epsilon_0^{\ddagger}/kT}\end{aligned}$$

where:

$$f^{\ddagger} = \frac{f_s^{\ddagger}}{f_*} \qquad (12.21)$$

i.e., $f^{\ddagger}$ is the ordinary partition function for $X^{\ddagger}$ with the factor corresponding to the coordinate q_* omitted. Equation 12.20 permits the calculation of the rate of the ($H + H_2$) reaction, from structural data; we shall generalize this equation in Section 12.4.

12.3 Further discussion of the reaction rate calculation

(i) In the derivation of Equation 12.20, only *linear* configurations of the three atoms have been considered. This is because, for a given pair of values of r_{AB} and r_{BC}, non-linear configurations have a higher energy than the linear one. Essentially, for given AB and BC bond lengths, $1/r_{AC}$ (and hence the repulsion between atoms A and C) has the lowest value with a linear configuration: this qualitative conclusion is supported by detailed calculations.* Non-linear configurations would contribute to the partition function, or phase integral, throught the Boltzmann factor $e^{-\epsilon/kT}$, corresponding to these configurations, and if ϵ is sufficiently large (i.e., large distortion from linearity), this term is small enough to be neglected. *Small* displacements from linearity will be significant; these will be associated with out-of-line vibrations. Since the 'equilibrium' configuration of the H–H–H complex is linear, these are $(3 \times 3 - 5) = 4$ normal modes of vibration. The two 'in line' vibrations we have considered previously; the remaining two are represented in Fig. 12.5. Neither of these leads to decomposition of the activated complex (because, in both cases, the

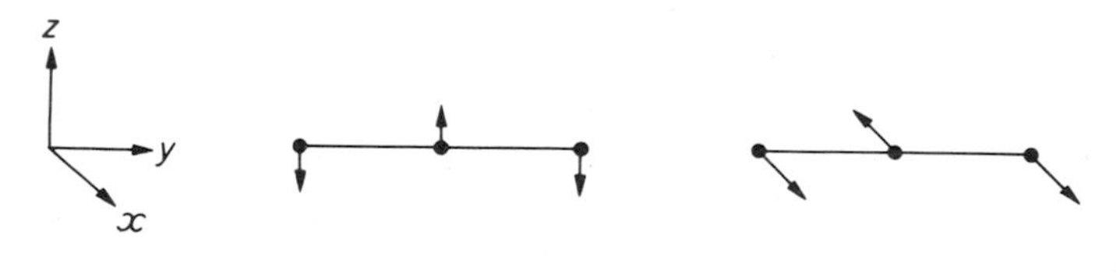

Fig. 12.5

potential energy increases with displacement), and it would have been an unnecessary complication to consider them explicitly. The corresponding factors f_{vib} must, however, be included in the calculation of the partition function of the activated complex.

* S.F. Boys and I. Shavitt, 'A fundamental calculation of the energy surfaces for the system of three hydrogen atoms', Technical Report WIS-AF-13 (1959).

(ii) In the classical derivation of Equation 12.20, $\Delta\epsilon_0^{\ddagger}$ is the difference between the potential energy of the initial equilibrium configuration of (A + BC) and that of the configuration s in Fig. 12.1. In quantum mechanics, however, $\Delta\epsilon_0^{\ddagger}$ is the difference between the energy of (A + BC) when BC is the lowest vibrational state and the energy of the activated complex in *its* lowest vibrational state. Thus, the classical calculation gives a value of $\Delta\epsilon_0^{\ddagger}$ which is too low by the energy difference:

$$(\text{Zero-point energy of } X^{\ddagger}) - (\text{zero-point energy of BC}).$$

In the application of Equation 12.20, therefore, it should be understood that this quantity is included in the value of $\Delta\epsilon_0^{\ddagger}$ used.

(iii) In classical mechanics, the system must, if it completes the reaction, at some stage achieve potential energies and configurations corresponding to points within the rectangle aa′b′b in Fig. 12.1. Therefore, if it is to react, the system must have a kinetic energy $\mathbf{T}$, convertible in c to potential energy, where $\mathbf{T}$ is at least equal to $\Delta\epsilon_0^{\ddagger}$.

In quantum mechanics (cf. Section 11.3), the exact potential energy and configuration at any instant are not definitely known, and one can only speak of the *probability* that a system with a given energy will react. This probability is:

(a) greater than 0 even if $\mathbf{T} < \Delta\epsilon_0^{\ddagger}$
(b) less than 1 even if $\mathbf{T} > \Delta\epsilon_0^{\ddagger}$.

According to (a), reaction *can* occur when $\mathbf{T} < \Delta\epsilon_0^{\ddagger}$; such a process is known as *quantum tunnelling*. According to (b), reaction *might* not occur when $\mathbf{T} > \Delta\epsilon_0^{\ddagger}$, i.e., the activated complex might disintegrate to re-form the reactants; this process is known as reflection. The effects associated with (a) and (b) together constitute what is called the 'tunnel effect'. Wigner (Johnston, *op. cit.*, Ch. 2) showed that, to a first approximation (valid for high temperatures), the tunnel effect is allowed for by multiplying the right-hand side of Equation 12.20 by Γ:

$$\Gamma = 1 + \frac{1}{24}\left(\frac{h|\omega_2|}{kT}\right)^2 + \dots \qquad (12.22)$$

where $|\omega_2|$ is the value of ω_2 given by Equation 12.4, divided by $\sqrt{-1}$. For lower temperatures, other estimates of the tunnelling correction must be used (Johnston, *op. cit.*, Ch. 2).

Example 12.3

Calculate the rate constant of the reaction

$$H + H\text{–}H \longrightarrow H\text{–}H + H$$

at 1000 K, from the following data:

Hydrogen: $r_{BC} = 0{\cdot}74 \times 10^{-10}$ m

$\omega = 4395\ \text{cm}^{-1}$.

H–H–H *complex* (derived from the calculated potential energy surface):

$$r_{AB} = r_{BC} = 0{\cdot}94 \times 10^{-10}\ \text{m}$$
$$\omega_1 = 1943\ \text{cm}^{-1}$$
$$\omega_2 = 1360\sqrt{-1}\ \text{cm}^{-1}$$
$$\omega_3 = \omega_4 = 950\ \text{cm}^{-1}$$

(potential energy at the saddle point) – (minimum potential energy of reactants) = $1{\cdot}07 \times 10^{-19}$ J.

The zero-point energy change in the formation of the activated complex is (cf. Table 10.1):

$$\tfrac{1}{2}(1{\cdot}986 \times 10^{-23}) \times (1943 + 950 + 950 - 4395) = -0{\cdot}054 \times 10^{-19}\ \text{J}$$

Hence:

$$\Delta\epsilon_0^{\ddagger} = 1{\cdot}02 \times 10^{-19}\ \text{J}$$

and (Table 10.7):

$$\Delta\epsilon_0^{\ddagger}/kT = 7{\cdot}39$$
$$\exp(-\Delta\epsilon_0^{\ddagger}/kT) = 6{\cdot}19 \times 10^{-4}$$

The procedure for evaluating the terms in Equation 12.20 is that used in Section 9.4. Considering the factors in turn:

electronic: It is assumed that $f_{el}^{\ddagger} = 2$, since there is one unpaired electron. This term cancels with:

$$f_{el\,H} = 2.$$

rotational:

$$\frac{f_{rot}^{\ddagger}}{f_{rot\,H_2}} = \frac{I^{\ddagger} \times \sigma_{H_2}}{I_{H_2} \times \sigma^{\ddagger}} = \frac{(r_{AB}^2 + r_{BC}^2)^{\ddagger} \times 2}{2(r_{BC}/2)^2 \times 2}$$
$$= \frac{4(0{\cdot}94)^2}{(0{\cdot}74)^2} = 6{\cdot}454,$$

translational:

$$\frac{f^{\ddagger}_{\text{trans}}}{f^{\circ}_{\text{trans H}} f^{\circ}_{\text{trans H}_2}} \left(\frac{1}{\tilde{N}}\right)^{-1} = (0{\cdot}02559\text{T}^{5/2}) \left(\frac{3}{1 \times 2}\right)^{3/2},$$

(by Equation 4.27)

$$= 2{\cdot}271 \times 10^{-6}$$

vibrational:

$$\frac{f^{\ddagger}_{\text{vib}}}{f_{\text{vib H}_2}} = \frac{1 - e^{-U_{\text{H}_2}}}{(1 - e^{-U_1})(1 - e^{-U_3})(1 - e^{-U_4})}$$

where:

$$U_x = \frac{h\omega_x}{kT} = \frac{1{\cdot}439}{1000}\omega_x, \text{ when } \omega_x \text{ is in cm}^{-1}.$$

$$U_{\text{H}_2} = 6{\cdot}325;\ U_1 = 2{\cdot}797;\ U_3 = U_4 = 1{\cdot}386.$$

Hence the vibrational factor is 1·91.

$$\tilde{V}^{\circ} = 82{\cdot}06 \text{ l mol}^{-1} \text{ at } 1000 \text{ K}$$

$$\frac{kT}{h} = \frac{1{\cdot}381 \times 10^{-23} \times 1000}{6{\cdot}625 \times 10^{-34}} = 2{\cdot}084 \times 10^{13} \text{ s}^{-1}$$

The tunnelling correction is, according to Equation 12.22:

$$\Gamma = 1 + \frac{1}{24}\left(1{\cdot}439 \times \frac{1360}{1000}\right)$$

$$= 1{\cdot}16.$$

Finally, therefore:

$$k = (6{\cdot}19 \times 10^{-4}) \times (6{\cdot}45) \times (2{\cdot}27 \times 10^{-7}) \times (1{\cdot}91)$$

$$\times (82{\cdot}06) \times (2{\cdot}08 \times 10^{13}) \times (1{\cdot}16)$$

$$= 3{\cdot}39 \times 10^{7} \text{ l mol}^{-1} \text{ s}^{-1}$$

□ □ □

The experimentally determined value is about 10^9 l mol^{-1} s^{-1}. Much better agreement between theoretical and experimental values of k can be obtained if empirical elements are introduced into the calculation. A rather successful approach has been to determine the properties of the (H H H) complex from the known relationships between bond energy, force constants, and bond order n [$n = 1$ for H_2, n = 0·5 for H–H in the (H H H) complex].

12.4 Generalization of the theory

In this section, an attempt is made to generalize the theory developed for the (H H H) system.

As before, the potential energy at any stage of the reaction can, in principle, be calculated when the relative positions of the atoms are known. The corresponding potential energy 'diagram' would be a surface in a conceptual space of many dimensions, i.e., **V** will depend upon many (say η) variables, or coordinates.

Let us firstly consider the reaction as a process of *classical* mechanics. The initial equilibrium configuration of the reactant molecules is a configuration of *minimum* potential energy, i.e., any distortion from this configuration must lead to an *increase* in potential energy (at the expense of kinetic energy). Similarly, any distortion of the product molecules from their equilibrium configuration would result in an increase in potential energy, or conversely, the potential energy must *decrease* as the equilibrium state of the products is approached.

The change from the reactant configuration (say r) to the product configuration (say p) can be made in an infinite number of ways, each of which may be referred to as a reaction path. In any one reaction path, each of the η coordinates will, in general, change, continuously or otherwise. Let us select one of these coordinates, say q (q might be, for example, an internuclear distance, or a bond angle, or, more generally, an algebraic function of internuclear distances and bond angles), and plot the change of potential energy against it. The simplest type of result is that shown in Fig. 12.6(a). Much more complicated curves might, of course, be found, depending upon the reaction path and the choice of q.

The potential energy curve must necessarily possess at least one maximum m *between the points* r *and* p; this is a geometrical necessity. In any reaction path, therefore, there will be some point of maximum potential energy. Suppose that the height of the maximum above r is $\Delta\epsilon_0$. Since the total energy is constant, the potential energy at m can never be greater than the initial total energy, and only molecular pairs with sufficient energy can react. *The most favoured reaction path is therefore the one for which* $\Delta\epsilon_0$ *has the least possible value* [Fig. 12.6(b)]. We will denote this particular value of $\Delta\epsilon_0$ by $\Delta\epsilon_0^{\ddagger}$, and refer to it as the potential energy barrier of the reaction.

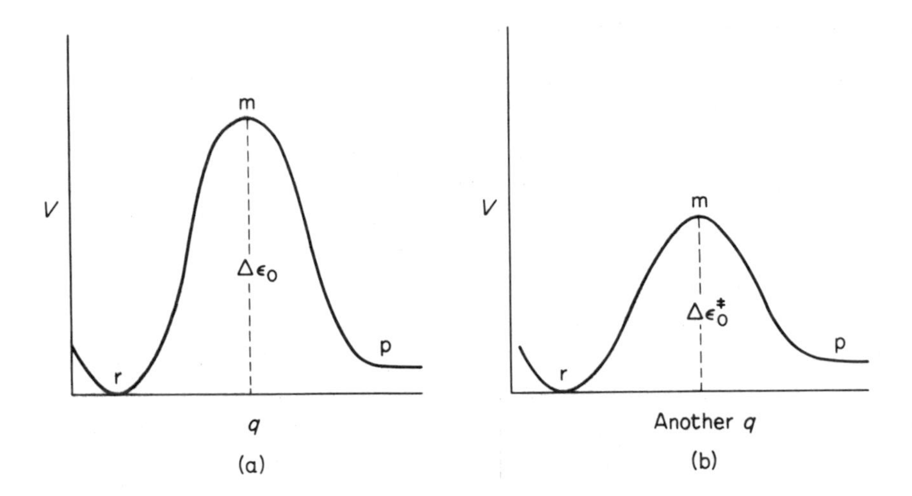

Fig. 12.6

Let us suppose that the reacting system has a configuration corresponding to point m on the potential energy curve for the most favoured reaction path. Any displacement along this path will lead to a decrease in potential energy. Any *other* displacement will lead to an *increase* in potential energy, *because* m *is the lowest maximum.* There exists, therefore, a situation analogous to that found for the (H H H) system, namely, m is a saddle point on the potential energy surface. For small displacements from m, coordinates $q_1, q_2, \ldots$ can be found, such that:

$$\mathbf{V} - \mathbf{V}_0 = \tfrac{1}{2} k_1(q_1 - q_1^0)^2 + \tfrac{1}{2} k_2(q_2 - q_2^0)^2 + \ldots ,$$

where $k_1, k_2, \ldots$ are the corresponding force constants, and $\mathbf{V}_0$ is the potential energy at m. Since in one direction $(\mathbf{V} - \mathbf{V}_0)$ is negative, *at least one* of the force constants must be negative. It is assumed that only one, say k_* is negative. The corresponding coordinate q_* is called the reaction coordinate, and is treated in exactly the same way as was q_2 in the (H H H) system. Thus, Equation 12.20 applied to the general bimolecular process:

$$A + B \rightleftharpoons X^{\ddagger} \longrightarrow \text{products},$$

$$k = \left(\frac{kT}{h}\right) \Gamma\left(\frac{f^{\ddagger}}{f_A' f_B'}\right) \left(\frac{1}{\tilde{\tilde{V}}{}^{\circ}\tilde{N}}\right)^{-1} e^{-\Delta\epsilon_0^{\ddagger}/kT}$$

The same analysis can be applied to unimolecular reactions:

$$A \rightleftharpoons X^{\ddagger} \longrightarrow \text{products.}$$

$X^{\ddagger}$ is here a grossly distorted form of molecule A. Then:

$$k = \left(\frac{kT}{h}\right) \Gamma \left(\frac{f^{\ddagger}}{f_A'}\right) e^{-\Delta\epsilon_0^{\ddagger}/kT} \qquad (12.23b)$$

Γ, the correction factor for the tunnel effect, is often assumed to be unity, in the application of these equations.

Evaluation of $f^{\ddagger}$ in Equation 12.23a is more difficult than the corresponding calculation for the (H H H) system because the structural details of $X^{\ddagger}$ are not known with certainty. Nevertheless, qualitative information and correlation of rate constants of different reactions can be obtained by using the following approximate method.

(i) By use of chemical intuition a geometrical structure is postulated for the activated complex. From this, all the factors in $f^{\ddagger}$ except those arising from vibration can be calculated, or at least (in the case of the electronic and internal rotation factors) estimated.

(ii) The vibrational factors are estimated by deducing approximate vibrational frequencies, using the concept of bond and group vibrations referred to briefly on p. 72. The reaction coordinate is identified with the set of atomic displacements which lead towards the formation of the products; detailed knowledge of this coordinate is unnecessary because it does not enter explicitly into the rate-constant calculation.

An example of this type of calculation is given below (Section 12.6 and problem 12.7).

12.5 Alternative forms of the rate constant equation

In the literature on reaction kinetics, the following alternative method of expressing Equations 12.23, which has the advantage of conceptual simplicity, is often used. It is based upon the generation of thermodynamic functions from the equilibrium constant for the formation of activated complexes.

$K^{\ddagger}$ is expressed in terms of concentrations, and the corresponding constant in terms of pressure is (Equation 9.5):

$$K_p = K^{\ddagger}(\tilde{V}^{\circ})^{\Delta\nu}$$

Therefore:

$$k = \frac{kT}{hf_*}\Gamma K^{\ddagger} = \left(\frac{kT\Gamma}{h}\right)\frac{K_p}{f_*}(\tilde{\tilde{V}}^{\circ})^{-\Delta\nu} \tag{12.24}$$

Now:

$$\Delta G^{\circ} = -RT\ln K_p \tag{12.25}$$

is the standard free energy change associated with the formation of one mole of activated complex. Let:

$$\Delta G^{\ddagger} = -RT(\ln K_p - \ln f_*)$$

i.e., $\Delta G^{\ddagger}$ is the standard molar free energy of formation of the activated complex *less* the contribution $-RT \ln f_*$ from f_*; $\Delta G^{\ddagger}$ is therefore calculated in the usual way, *but with the omission of the factor* f_* from the partition function. The corresponding molar enthalpy of formation is:

$$\Delta H^{\ddagger} = \Delta G^{\ddagger} - T\left(\frac{\partial \Delta G^{\ddagger}}{\partial T}\right)_P$$

the entropy is:

$$\Delta S^{\ddagger} = \frac{\Delta H^{\ddagger} - \Delta G^{\ddagger}}{T},$$

and the energy is:

$$\Delta E^{\ddagger} = \Delta H^{\ddagger} - \Delta\nu RT,$$

all these quantities being calculated with the omission of f_*. Note that the energy of any one activated complex is the *same* as the energy of the reacting molecules, but the *average* energy of the activated complexes is greater than the average energy of the reactants, because only molecules with energies greater than the average react. $\Delta E^{\ddagger}$ is equal to the difference between these averages, multiplied by the Avogadro number; this is essentially the interpretation of activation energy (defined below) given by Tolman.

Substitution of Equation 12.25 in 12.24 gives:

$$k = \frac{kT}{h}\Gamma\, e^{-\Delta G^{\ddagger}/RT}\, (\tilde{\tilde{V}}^{\circ})^{-\Delta\nu}. \tag{12.26a}$$

Therefore:

$$k = \frac{kT}{h}\Gamma\, e^{\Delta S^{\ddagger}/R}\, e^{-\Delta H^{\ddagger}/RT}\, (\tilde{\tilde{V}}^{\circ})^{-\Delta\nu} \tag{12.26b}$$

and

$$k = \frac{kT}{h}\Gamma\, e^{(\Delta S/R - \Delta\nu)}\, e^{-\Delta E^{\ddagger}/RT}\, (\tilde{\tilde{V}}^{\circ})^{-\Delta\nu} \tag{12.26c}$$

Equations 12.26 are alternative forms of Equation 12.23.

Experimentally determined values of log k vary almost exactly linearly with $1/T$, i.e., these values satisfy the empirical *Arrhenius equation:*

$$k = A\,e^{-E_a/RT} \tag{12.27}$$

A and E_a are constants determined from the linear plot of log k *versus* $1/T$, called, respectively, the frequency factor and the activation energy of the reaction. The linearity of this plot is merely a consequence of the dominance of the exponential term in Equation 12.23. The temperature coefficient of ln k at a given pressure is therefore:

$$\left(\frac{\partial \ln k}{\partial T}\right)_P = \frac{E_a}{RT^2} \tag{12.28}$$

From Equation 12.24:

$$\ln k = \ln\left(\frac{\Gamma kT}{h}\right) + \ln(\tilde{\tilde{V}}^\circ)^{-\Delta\nu} + \ln K_p - \ln f_*$$

Noting that:

$$\tilde{\tilde{V}}^\circ = \frac{RT}{P^\circ}$$

and that:

$$\left(\frac{\partial}{\partial T}(\ln K_p - \ln f_*)\right)_P = \frac{\Delta G^\ddagger}{RT^2} - \frac{1}{RT}\left(\frac{\partial \Delta G}{\partial T}\right)_P = \frac{\Delta H^\ddagger}{RT^2},$$

we obtain:

$$\left(\frac{\partial \ln k}{\partial T}\right)_P = \frac{1}{T} - \frac{\Delta\nu}{T} + \frac{\Delta H^\ddagger}{RT^2}$$

Comparing this with Equation 12.28:

$$E_a = \Delta H^\ddagger + (1 - \Delta\nu)RT \tag{12.29}$$

Since:

$$\ln A = \ln k + \frac{E_a}{RT},$$

substitution of Equation 12.26b for k gives:

$$A = \left(\frac{kT\Gamma}{h}\right)(\tilde{\tilde{V}}^\circ)^{-\Delta\nu}\,e^{(1-\Delta\nu)}\,e^{\Delta S^\ddagger/R} \tag{12.30}$$

where $\Delta\nu = 0$ for unimolecular reactions, and -1 for bimolecular reactions.

12.6 The interpretation of experimental data

Equation 12.30 permits the quantitative interpretation of the observed values of A for various reactions, because it is often possible to decide, on physical grounds, the likely order of magnitude of $\Delta S^{\ddagger}$. In unimolecular reactions, $f^{\ddagger}_{\text{trans}}$ and $f_{\text{trans A}}$ are the same, and $f^{\ddagger}_{\text{rot}}$ and $f_{\text{rot A}}$ nearly so; consequently, $\Delta S^{\ddagger}$ is governed by the ratio of $f^{\ddagger}_{\text{vib}}$ to $f_{\text{vib A}}$. If, therefore, it is assumed that both $f^{\ddagger}_{\text{vib}}$ and $f_{\text{vib A}}$ are about unity, $\Delta S^{\ddagger} \approx 0$, and (assuming that $\Gamma \approx 1$):

$$A \approx \frac{kT}{h}\,\mathrm{e} \approx 10^{13}\ \mathrm{s}^{-1}$$

the observed value for many unimolecular reactions. In bimolecular reactions, since two molecules combine to form one $X^{\ddagger}$, there is a loss of translational entropy. The translational entropy of a gas is of the order of 100 J K^{-1} mol^{-1}, at ordinary temperatures, and *if* the changes in the totational and vibrational entropies can be neglected:

$$\Delta S^{\ddagger} \approx -100\ \mathrm{J\,K^{-1}\,mol^{-1}} \approx 14R$$

Hence:

$$A \approx \frac{kT}{h}\,\mathrm{e}^{-12}(0{\cdot}082T) \approx 10^{9}\ \mathrm{l\,mol^{-1}\,s^{-1}}$$

the observed value for many bimolecular reactions.

More detailed calculation of $\Delta S^{\ddagger}$ is required if anything more than such qualitative results is sought. A typical example of such calculation is provided by the paper of H.E. O'Neal and S.W. Benson (*J. Phys. Chem.*, 1967, **71**, 2903). These authors considered unimolecular reactions of the type:

$$\underset{\text{(I)}}{CH_3CH_2Cl} \longrightarrow \underset{\text{(II)}}{CH_2{=}CH_2} + HCl$$

which presumably proceed via the structure

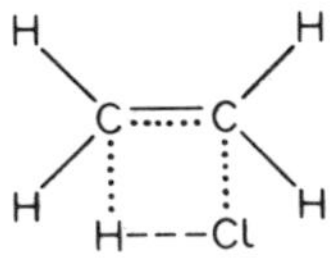

(III)

The broken line in structure (III) is meant to indicate a weak interaction between the H and Cl atoms, which ultimately leads to the H–Cl bond; the dotted lines are meant to indicate bond properties (in particular, interatomic distances and electronic charge distributions) intermediate between those in (I) and (II).

The value of $\Delta S^{\ddagger}$, in this reaction, is governed by changes in the vibrational frequencies (see above; it was also assumed that $g_{el}^{\ddagger} = g_{elI} = 1$). The vibrational frequencies of (I) were obtained as those of the component groups. The vibrational frequencies of (III) were obtained by assuming that the dotted lines in structure (III) represent bonds of *half* order; the frequencies could then be obtained from empirical rules (Pauling's rule connecting bond order and bond length, and Badger's rule connecting bond length and force constant). The following values were thus obtained:

$$S_{vib\,I} = 35{\cdot}6\ \text{J K}^{-1}\ \text{mol}^{-1};\quad S_{vib}^{\ddagger} = 45{\cdot}1\ \text{J K}^{-1}\ \text{mol}^{-1}.$$

There is also an important contribution to $\Delta S^{\ddagger}$ arising from the replacement of an internal rotation in (I) by a vibration in (III). The entropy of internal rotation about the C–C bond in (I) was estimated by using Equation 10.10 for f_{free}, and assuming a rotational barrier $\mathbf{V}_m = 15{,}000\ \text{J mol}^{-1}$. Taking $\sigma_{int} = 3$ leads to the value $S_{int\,rot} = 15\ \text{J K}^{-1}\ \text{mol}^{-1}$. Hence:

$$\Delta S^{\ddagger} = 45{\cdot}1 - (35{\cdot}6 + 15) = -5{\cdot}5\ \text{J K}^{-1}\ \text{mol}^{-1}$$

and

$$A = 10^{13{\cdot}3}\ \text{s}^{-1}\ \text{at 600 K}.$$

The observed values of A for this reaction at 600K are $10^{13{\cdot}6 \pm 0{\cdot}5}\ \text{s}^{-1}$. Quantitatively satisfactory values of the frequence factors of some one hundred unimolecular reactions were obtained by this method. These calculations are rather insensitive to errors in the assigned values of vibrational frequencies. For example, a 1500 cm^{-1} vibration would contribute $0{\cdot}13R$, and a 1200 cm^{-1} vibration $0{\cdot}22R$, to the molar entropy. A corresponding discrepancy in the frequencies used in the calculation of A would therefore lead to an error of $0{\cdot}09R$ in $\Delta S^{\ddagger}$, but this leads to an error of only 0·04 in $\log A$.

Problems

12.1 To illustrate the construction of contour diagrams, consider the potential energy function:

$$\mathbf{V} = x^2 - y^2.$$

Evaluate **V** systematically for all integer values of x and y from 0 to 5; values for negative x and y follow by symmetry. Represent pairs of (x,y) values as points on squared graph paper (take the centre of the page of origin), and label each by the calculated value of **V**.
(a) Sketch the contours of $\mathbf{V} = \pm 1, \pm 4, \pm 9, \pm 16$, and ± 25 by interpolating between plotted points.
(b) Draw sections through the potential energy surface corresponding to (i) $y = 0$, (ii) $x = 0$, (iii) $y = \frac{1}{2}x$.

12.2 Apply the transition state theory to show that the rate constant of the two atom combination reaction A + B $\longrightarrow$ C is:

$$k = \pi R^2 \left(\frac{8kT}{\pi\mu}\right)^{1/2} e^{-\Delta E_0^{\ddagger}/RT}$$

where μ is the reduced mass, $m_A m_B/(m_A + m_B)$, and R is the internuclear distance in the transition state. Compare this result with that obtained according to the collision theory (Fowler and Guggenheim, p. 491).

12.3 Calculate the rate constant of the reaction:

$$D + D_2 \longrightarrow D_2 + D$$

at 1000 K, using the dated of Example 12.1.

12.4 Repeat the calculation of Example 12.1 by first evaluating $\Delta S^{\ddagger}$ using the tables in Chapter 10 (this is a very useful method for more complicated reactions).

12.5 It is not too unrealistic to make rough estimates of frequency factors by assuming the following values to be universally applicable:

$$f_{trans} \approx 10^{32}$$

$$f_{rot} \approx 10^2 \text{ (linear molecules);}$$

$$f_{int} \approx 10 \text{ for each internal rotation}$$

$$f_{vib(i)} \approx 10^{0\cdot 3} \text{ for each normal mode}$$

$$kT/h \approx 10^{13}\ s^{-1}$$

Estimate $\Delta S^{\ddagger}$ and A ($\mathrm{l\,mol^{-1}\,s^{-1}}$) for the following reactions (the experimental values of A are given):

Reaction	$\log_{10}A$
$H_2 + Cl\cdot \rightarrow HCl + H\cdot$	10·9
$H\cdot + C_2H_4 \rightarrow C_2H_5\cdot$	10·4
$OH\cdot + CO \rightarrow CO_2 + H\cdot$	8·6
$NO + O_3 \rightarrow NO_2 + O_2$	9·0
$CH_3CH_2Cl \rightarrow C_2H_4 + HCl$	13·5–14·6

12.6 Calculate $\Delta S^{\ddagger}$ and A for the reaction:

$$Cl\cdot + H_2 \rightarrow HCl + H$$

at 500K, from the following data:

Initial State

Data for H_2 are given in Tables 5.2 and 5.4. The electronic partition function of the chlorine atom is disuussed on page 67.

Transition State (estimated) Cl .... H .... Cl linear.

The H....H and Cl....H bond lengths are $0{\cdot}92 \times 10^{-10}$m and $1{\cdot}45 \times 10^{-10}$m. The vibrational frequencies are 560(2) and 1460 cm^{-1}. The electronic partition function is assumed to be 2. The experimental value of A is $8 \times 10^{10}\ \mathrm{l\,mol^{-1}\,s^{-1}}$.

12.7 Simple arguments (cf. pp. 72, 263) show that some of the vibrational frequencies of the transition state may be essentially the same as those of the reactants. In evaluating $\Delta S^{\ddagger}$, it is necessary to consider only those modes which are changed. Consider the reaction

$$\underset{\displaystyle H}{\overset{}{H_2\underset{|}{C}}} - \underset{\displaystyle Cl}{\underset{|}{CH_2}} \rightarrow \underset{\displaystyle H}{H_2\underset{\vdots}{C}} \cdots \underset{\displaystyle Cl}{\underset{\vdots}{CH_2}} \rightarrow H_2C = CH_2 + HCl \qquad \text{(i)}$$

The vibrations which change are (see the paper by O'Neal and Benson cited in the text):-

Initial	ω (cm^{-1})	Transition State	ω (cm^{-1})
H–C–H	1450 (2)	H···C···C	800
		H···C–H	1000
C–C–Cl	400	C···C···Cl	280
H–C–Cl	700	H–C···Cl	700
C–C	1000	C···C	1300
C–H	3000	C······H	2200
Internal rotation	——	C···C	400
C–Cl	650	Reaction co-ordinate	——

It is assumed that the C ... Cl displacement is the reaction co-ordinate. It is a satisfactory approximation (because of the large mass of the Cl atom) to replace I_{red} for the internal rotation by I, the moment of inertia of the methyl group along [see Example 5.6(a)]. For the rotational barrier, take $\mathbf{V}_m = 1{\cdot}75 \times 10^3\, R$. Calculate $\Delta S^{\ddagger}$ and $\log_{10} A$ for reaction (i) at 600 K. Reported experimental values of log A are in the range 13·5–14·6.

CHAPTER THIRTEEN

Quantum Effects

The treatment of statistical mechanics given in the preceding chapters is based upon the classical methods used by Maxwell and Boltzmann; it has involved quantum mechanics only in so far as the values of the allowed energy levels and their degeneracies are required in the evaluation of partition functions. As we have discussed in Chapter 11, most of the results could have been obtained without the use of quantum theory.* However, one case, that of the vibrational partition function, has been encountered in which the classical method of evaluation is inadequate. There are a few other, rather specialized cases, in which quantum theory plays a vital role; we have gathered them together under the title 'quantum effects'. They fall into two classes: (i) effects modifying the distribution law (Section 13.1–13.5) and (ii) effects modifying the partition function (Section 13.7). In Section 13.6, the opportunity is taken of discussing, in a somewhat more formal way than previously, the meaning of the term 'complexion'.

13.1 Bose–Einstein statistics

It was pointed out in Section 4.1 that, because in the gaseous phase molecules of a given type are indistinguishable from one another, one of the basic assumptions made in the derivation of

* But note that the quantal and quasi-classical expressions contain h, and that this term can never appear in a truly classical result; a little consideration shows, however, that powers of h always cancel when *thermodynamic functions* are evaluated.

the Maxwell–Boltzmann law is untrue. We showed that, as an approximation, the resulting error could be compensated for by removing the factor $N!$ from the Maxwell–Boltzmann Ω_D. Now, however, we derive a new distribution law, known as the *Bose–Einstein distribution law,* in which the indistinguishability of the molecules is taken into account in a rigorous way.

To illustrate the principle of evaluating Ω_D consider, as an example, the number of different ways in which two identical particles can be placed in one three-fold degenerate level. The possibilities are:

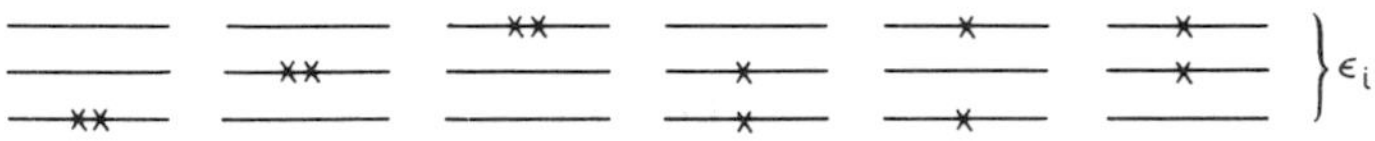

or, using an alternative and equivalent notation:

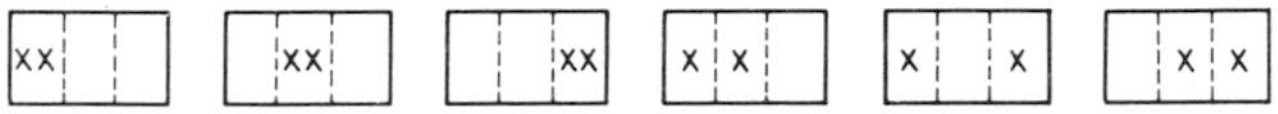

The second set of diagrams can be regarded in a different way, namely, as arising from *permutations of the partitions (broken lines) and the particles* (X). There are two partitions and two particles, and the number of permutations of these four things is $4! = 24$. However, half of these permutations differ only in the order of identical particles, and half again differ only in the order of identical partitions; so the number of different arrangements is $24/4 = 16$. This result may be contrasted with the number $3^2 = 9$, obtained (page 8) for distinguishable molecules.

In general, if a level is g_i-fold degenerate, there will be $(g_i - 1)$ partitions. If the distribution number is n_i, the partitions and particles can be permuted in $(n_i + g_i - 1)!$ ways. The number of ways in which the n_i particles can be placed in the i-th level is, therefore:

$$\frac{(n_i + g_i - 1)!}{n_i!\,(g_i - 1)!}$$

the factors $n_i!$ and $(g_i - 1)!$ allowing for, respectively, the equivalence of the particles and the equivalence of the partitions.

The number of complexions associated with a given distribution is the product of such terms over all the energy levels:

$$\Omega_D = \prod_i \frac{(n_i + g_i - 1)!}{n_i!\,(g_i - 1)!} \tag{13.1}$$

Therefore:

$$\ln \Omega_D = \sum_i \ln(n_i + g_i - 1)! - \sum_i \ln n_i! - \sum_i \ln(g_i - 1)!$$

As was discussed in Section 4.7, Stirling's theorem can be applied to $\ln(g_i - 1)!$ and to $\ln n_i!$, and so:

$$\begin{aligned}\ln \Omega_D &= \sum_i \,[(n_i + g_i - 1)\ln(n_i + g_i - 1) - (n_i + g_i - 1) - n_i \ln n_i \\ &\quad + n_i - (g_i - 1)\ln(g_i - 1) + (g_i - 1)] \\ &= \sum_i \,[(n_i + g_i - 1)\ln(n_i + g_i - 1) \\ &\quad - n_i \ln n_i - (g_i - 1)\ln(g_i - 1)]\end{aligned}$$

Therefore:

$$\begin{aligned}\frac{\partial \ln \Omega_D}{\partial n_i} &= 1 + \ln(n_i + g_i - 1) - 1 - \ln n_i \\ &= \ln\left(\frac{n_i + g_i}{n_i}\right) \qquad (13.2)\end{aligned}$$

if 1 is negligible compared with $(n_i + g_i)$. Writing α instead of γ in Equation 2.12, and substituting 13.2:

$$\ln\left(\frac{n_i + g_i}{n_i}\right) - \alpha - \beta\epsilon_i = 0$$

Hence:

$$e^{\alpha}\, e^{\beta\epsilon_i} = \frac{n_i + g_i}{n_i} = 1 + \frac{g_i}{n_i}$$

and therefore:

$$\bar{n}_i = \frac{g_i}{e^{\alpha}\, e^{\beta\epsilon_i} - 1} \tag{13.3}$$

where, as usual, $\bar{n}_i$ is the value of n_i for the most probable distribution. This is the *Bose–Einstein* distribution law.

13.2 Fermi–Dirac statistics

Let ν be the sum of the numbers of protons, neutrons, and electrons of which one molecule, or other particle, is composed; for example, $\nu = 4$ for H_2, $\nu = 5$ for ^3He, $\nu = 3$ for H^-, etc. All known experimental facts are consistent with the following hypothesis (the *exclusion principle*): *In any system of indistinguishable molecules (or other particles) for which ν has an odd value, no molecular quantum state can be occupied by more than one molecule at a time.* This is a generalization of the familiar *Pauli principle,* which refers specifically to electrons ($\nu = 1$) (see also page 289).

Now it is obvious that, in view of this principle, Bose–Einstein statistics *overestimates* the number of complexions of a system of particles for which ν is odd. In the example on page 273, the first three allocations of the particles to molecular quantum states would not be complexions (i.e., correspond to accessible quantum states) of such a system.

An expression for the number of complexions for this case is, however, easily obtained; it leads to yet another distribution law, known as the *Fermi–Dirac distribution law.*

Consider the i-th energy level, and again assume that it is g_i-fold degenerate. In the case of particles with an odd value of ν, no two of the n_i particles which occupy this level can occupy the same quantum state. Thus, n_i quantum states will each be occupied by one particle, and therefore $(g_i - n_i)$ states will be unoccupied. The number of ways in which this can be achieved is given by the combinatory rule, according to the following scheme:

$$g_i \begin{cases} \nearrow\ n_i \text{ occupied} \\ \searrow\ (g_i - n_i) \text{ unoccupied} \end{cases}$$

clearly, it is:

$$\frac{g_i!}{n_i!(g_i - n_i)!}$$

Hence the number of complexions corresponding to a particular distribution is:

$$\Omega_D = \prod_i \frac{g_i!}{n_i!(g_i - n_i)!} \cdot$$

Taking logarithms and applying Stirling's theorem:

$$\ln \Omega_D = \sum_i \ln g_i! - \sum_i \ln n_i! - \sum_i \ln(g_i - n_i)!$$

$$= \sum_i [g_i \ln g_i - n_i \ln n_i - (g_i - n_i) \ln(g_i - n_i)] .$$

Hence:

$$\frac{\partial \ln \Omega_D}{\partial n_i} = \ln(g_i - n_i) - \ln n_i$$

$$= \ln \frac{g_i - n_i}{n_i} .$$

Proceeding as in the previous section, we now obtain:

$$\overline{n}_i = \frac{g_i}{e^{\alpha} e^{\beta \epsilon_i} + 1} . \tag{13.4}$$

This is the *Fermi–Dirac* distribution law.

13.3 The classical limit

The Maxwell–Boltzmann distribution law can be written as:

$$\overline{n}_i = \frac{g_i}{e^{\alpha} e^{\beta \epsilon_i}}$$

If $e^{\alpha} e^{\beta \epsilon_i}$ happens to be much greater than 1, then this expression differs negligibly from:

$$\overline{n}_i = \frac{g_i}{e^{\alpha} e^{\beta \epsilon_i} + 1}$$

and from:

$$\overline{n}_i = \frac{g_i}{e^{\alpha} e^{\beta \epsilon_i} - 1}$$

Consequently, when $e^{\alpha} e^{\beta \epsilon_i}$ is large enough, the three distribution laws become identical. It is said, therefore, that the Maxwell–Boltzmann law is the *classical limit* of the Bose–Einstein and Fermi–Dirac distribution laws.

In Maxwell–Boltzmann statistics, e^{α} is given by Equation 2.16; thus:

$$e^{\alpha} = \frac{f}{N}$$

Since we are concerned with gaseous assemblies, the largest factor in f is the translational partition function:

$$f_{\text{trans}} = \frac{(2\pi mkT)^{3/2} V}{h^3}$$

It is thus found (Mayer and Mayer, p. 122):

for H_2 at its b.p., 20·3 K, $e^{\alpha} = 1{\cdot}4 \times 10^2$

for N_e at its b.p., 27·2 K, $e^{\alpha} = 9{\cdot}3 \times 10^3$

The lowest value of $e^{\beta\epsilon_i}$ is $e^0 = 1$. Hence, in both of the above cases:

$$e^{\alpha} e^{\beta\epsilon_i} >> 1,$$

and the Maxwell–Boltzmann law can be used without significant error. For gases with larger values of m. T, and V, the error becomes even less. There are, however, two systems of interest in physical chemistry for which e^{α} is not large, and for which, therefore, the Maxwell–Boltzmann law cannot be used. These are (i) electrons in metals, and (ii) liquid helium.

13.4 Fermi–Dirac gas: electrons in metals

A useful method for the study of metallic properties is to regard the conduction electrons as constituting a perfect gas, the particles of which can move freely between any points within the metal. The validity of such a model might be questioned, because the electrons are negatively charged particles, between which strong Coulomb repulsion must occur. However, a partial justification (see Mayer and Mayer, p. 387, for a fuller discussion) is that, on average, the Coulomb repulsion by the other electrons is roughly balanced by the Coulomb attraction of the nuclei; each electron effectively moves among essentially neutral atoms. It is therefore assumed that the *potential* energy U of the electrons does not differ greatly from one point in the metal to another, and can thus be treated as a constant; moreover, it is assumed that U is independent of temperature. Hence, the contribution to the total energy which is of interest statistically is the *kinetic* energy, and this can be evaluated by means of the procedures used in Chapter 4 for the perfect gas.

First, let us estimate the order of magnitude of the corresponding partition function. To do this, we take $T = 300$K and

N/V corresponding to $10^{-5}\,\text{m}^3\,\text{mol}^{-1}$ for a metal with one free electron per metal atom. Then:

$$\frac{f_{\text{trans}}}{N} = 2{\cdot}1 \times 10^{-4}$$

which is *small* rather than large. Consequently Maxwell–Boltzmann statistics cannot be applied to such an electron gas, and the Fermi–Dirac distribution law must be used.

The translational kinetic energy is given by (cf. Equation 4.25):

$$\epsilon_r = \frac{r^2h^2}{8m}$$

where r is a quantum number (not, in general, an integer). The average number of electrons with this energy is, according to Equation 13.4:

$$\bar{n}_r = \frac{g_r}{\mathrm{e}^{\alpha}\,\mathrm{e}^{\epsilon_r/kT} + 1} \tag{13.5}$$

where now g_r is the degeneracy in the sense described in Section 4.7. The value of g_r is, however, twice that given by Equation 4.27, because of the two alternatives for electron spin, i.e.:

$$g_r = V\pi r^2 \Delta r$$

The total number of electrons is therefore:

$$N = \sum_r n_r = \sum_r \frac{V\pi r^2 \Delta r}{\mathrm{e}^{\alpha}\,\mathrm{e}^{\epsilon_r/kT} + 1}$$

$$= \int_0^\infty \frac{V\pi r^2}{\mathrm{e}^{\alpha}\,\mathrm{e}^{\epsilon_r/kT} + 1}\,\mathrm{d}r \tag{13.6}$$

when summation is replaced by integration, in the usual way. Now:

$$r^2 = \frac{8m\epsilon}{h^2} \quad \text{and} \quad \mathrm{d}r = \frac{8m\,\mathrm{d}\epsilon}{h^2\,2r}$$

Hence Equation 13.6 can be written as:

$$N = \frac{1}{2}\left(\frac{8m}{h^2}\right)^{3/2} \pi\, V \int_0^\infty \frac{\epsilon^{1/2}}{\mathrm{e}^{\alpha}\,\mathrm{e}^{\epsilon/kT} + 1}\,\mathrm{d}\epsilon\,. \tag{13.7}$$

The integral has not been evaluated in closed form. An approximation is, however, obtained from the following considerations. If $(\alpha + \epsilon/kT)$ has a large negative value, $\mathrm{e}^{\alpha}\,\mathrm{e}^{\epsilon/kT}$ is small. Thus, if:

$$\epsilon/kT << -\alpha, \text{ i.e., if } \epsilon << -\alpha kT,$$

the integrand is effectively $\epsilon^{1/2}$. On the other hand, if:

$$\epsilon >> -\alpha kT$$

the integrand is very small, and may be neglected. Hence, as a rough approximation:

$$\int_0^\infty \frac{\epsilon^{1/2}}{e^\alpha e^{\epsilon/kT} + 1} d\epsilon \approx \int_0^{-\alpha kT} \epsilon^{1/2} d\epsilon \qquad (13.8)$$

$$= \tfrac{2}{3}(-\alpha kT)^{3/2}$$

Substituting this value into Equation 13.7, one obtains an approximation for α:

$$\alpha = -\frac{h^2}{8mkT}\left(\frac{3N}{\pi V}\right)^{2/3} = -\frac{\mu_0}{kT}$$

where:

$$\mu_0 = \frac{h^2}{8m}\left(\frac{3N}{\pi V}\right)^{2/3} \qquad (13.9)$$

is called the *Fermi energy*.

Substitution of numerical values into Equation 13.9 shows that μ_0/k is of the order of 10^5 K.

From Equation 13.5, the *average* number of electrons in any one quantum state j, for which the energy is ϵ_j, is:

$$\overline{m}_j = \frac{\overline{n}_r}{g_r} = \frac{1}{e^\alpha e^{\epsilon_j/kT} + 1}$$

Substituting for α from Equation 13.9, therefore:

$$\overline{m}_j = \frac{1}{e^{(\epsilon_j - \mu_0)/kT} + 1} \qquad (13.10)$$

When $T \to 0$:

$$e^{(\epsilon_j-\mu_0)/kT} \to \begin{cases} \infty \text{ if } \epsilon_j > \mu_0 \\ 0 \text{ if } \epsilon_j < \mu_0 \end{cases}$$

Hence, at the absolute zero of temperature, according to Equation 13.10:

$$\overline{m}_j = 0 \text{ if } \epsilon_j > \mu_0$$

$$\overline{m}_j = 1 \text{ if } \epsilon_j < \mu_0$$

in other words, at the absolute zero, quantum states with energies up to the Fermi energy μ_0 are singly occupied, and all other quantum states are empty.

The kinetic energy of the electrons is:

$$\begin{aligned} E &= \sum_r \bar{n}_r \epsilon_r \\ &= \int_0^\infty \frac{V\pi r^2 \epsilon}{e^\alpha e^{\epsilon/kT} + 1} dr \qquad (13.11) \\ &= \frac{1}{2}\left(\frac{8m}{h^2}\right)^{3/2} \pi V \int_0^\infty \frac{\epsilon^{3/2}}{e^\alpha e^{\epsilon/kT} + 1} d\epsilon \end{aligned}$$

Use of the same approximation which led to Equation 13.8 gives:

$$\begin{aligned} E &= \frac{1}{2}\left(\frac{8m}{h^2}\right)^{3/2} \pi V \int_0^{\mu_0} \epsilon^{3/2} \, d\epsilon \\ &= \tfrac{3}{5} N\mu_0 \end{aligned}$$

Thus, *in so far as the above approximation is valid, E* is *independent* of temperature. Actually, a more rigorous evaluation of the integral in Equation 13.11 shows that E *is* dependent upon temperature, but that the temperature-dependence is small, at ordinary temperatures, Thus, it is found (Fowler and Guggenheim, p. 457):

$$E = \tfrac{3}{5} N\mu_0 \left[1 + \frac{5\pi^2}{12}\left(\frac{kT}{\mu_0}\right)^2 + \ldots\right]$$

The additional terms are small at ordinary temperatures because k/μ_0 is of the order of $10^{-5}\,K^{-1}$. The electronic contribution to the heat capacity of a metal is:

$$\begin{aligned} C_V &= \left(\frac{\partial E}{\partial T}\right)_V = Nk\frac{\pi^2}{2}\frac{kT}{\mu_0} + \ldots \\ &\approx 10^{-4} T \, J\,K^{-1}\,mol^{-1} \end{aligned}$$

This is negligible at ordinary temperatures. Thus, the electrons do not contribute significantly to the specific heat of a metal; this is in complete contrast with the classical prediction that their contribution should be (3/2)R (cf. Section 5.7). The linear dependence of C_V upon T has been detected at very low temperatures (at which the vibrational C_V goes rapidly to zero, in accordance with the Debye

T^3 law). For example, for silver, $C_V = (6 \times 10^{-4})T\,\text{J K}^{-1}\,\text{mol}^{-1}$, between 1·7 K and 4 K.

The electronic contribution to the entropy of a metal is correspondingly small. When $T = 0$, the contribution of the free electrons to the entropy is zero, because there is only one way ($\Omega = 1$) of putting the N electrons into the N states of lowest energy. At other temperatures:

$$\tilde{S} = \int_0^T \frac{\tilde{C}_V\,\mathrm{d}T}{T} \approx 10^{-4}T\ \text{J K}^{-1}\,\text{mol}^{-1}$$

$$\approx 0{\cdot}03\ \text{J K}^{-1}\,\text{mol}^{-1} \text{ at } 300\,\text{K}.$$

13.5 Bose–Einstein gas: liquid helium

The low boiling point of helium indicates that the intermolecular forces in this system are unusually weak. Hence, there is some justification for neglecting such forces, in theoretical studies of the liquid phase.

For the abundant nuclide ^{4}He, the normal boiling point is 4·2 K, and, at this temperature e^{α} *calculated from* f_{trans} by using the molar volume of the *liquid,* is 0·67 for ^{4}He. This is too small for Maxwell–Boltzmann statistics to be applied to the system, and since ν, the sum of the numbers of protons, neutrons, in ^{4}He is 6 (an even number), the Bose–Einstein statistics must be used.

According to the Bose–Einstein distribution law 13.3, the number of molecules in the lowest ($\epsilon = 0$) level is:

$$\bar{n}_0 = \frac{1}{e^{\alpha} - 1} \quad (g_0 = 1) \tag{13.12}$$

Rearrangement of this equation gives:

$$e^{-\alpha} = \frac{\bar{n}_0}{\bar{n}_0 + 1}$$

Now, $\bar{n}_0$ cannot be a negative number, and so the right-hand expression must have a value which is positive and less than 1, i.e.:

$$0 \leqslant e^{-\alpha} < 1 \tag{13.13}$$

As in Section 13.4, the value of e^{α} is determined from the condition:

$$N = \sum \frac{g_r}{e^{\alpha} e^{\epsilon_r/kT} - 1} \tag{13.14}$$

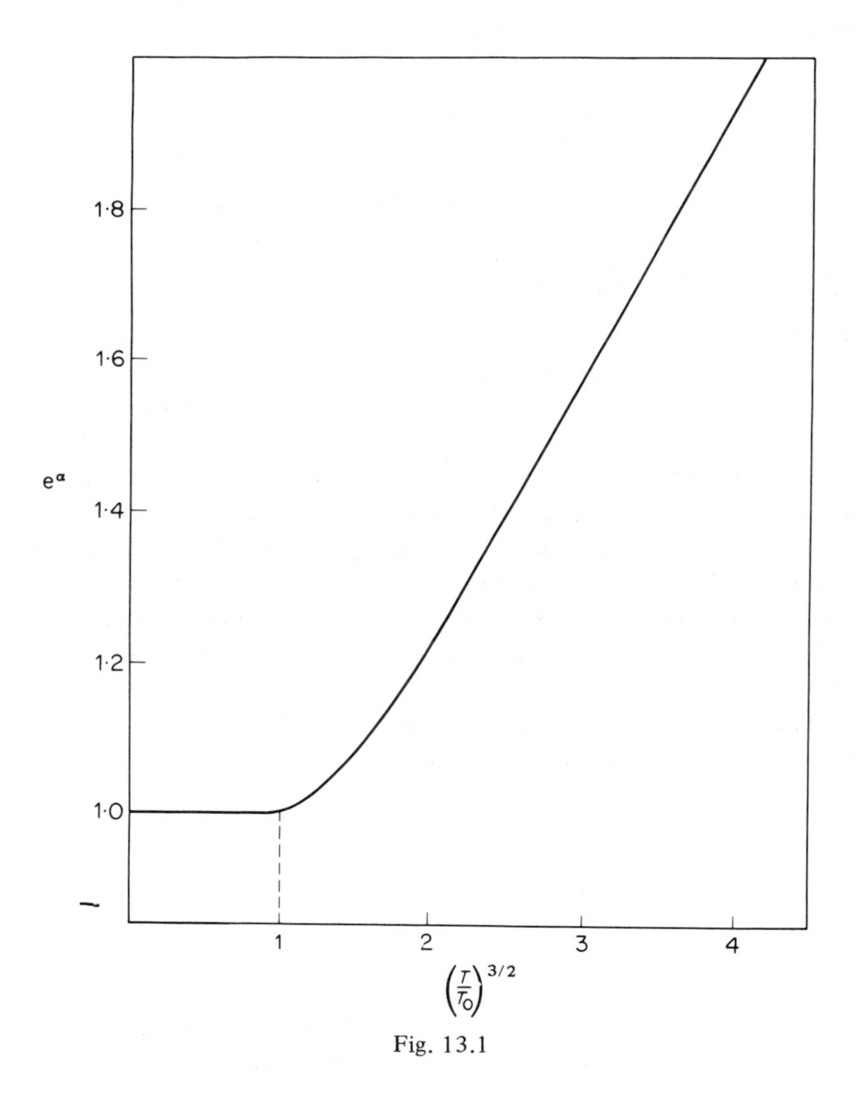

Fig. 13.1

Details are given at the end of the present section. The results are shown in Fig. 13.1. As T decreases, e^{α} decreases steadily until, at a temperature T_0, it reaches a value differing negligibly from 1; at temperatures lower than T_0, it remains at essentially the same value. We show below that:

$$T_0 = \left(\frac{N}{2{\cdot}612V}\right)^{2/3} \frac{h^2}{2\pi mk} \qquad (13.15)$$

For liquid ^{4}He, the value of T_0 is thus calculated to be 3.13K.

Now T_0 has a remarkable physical significance. According to Equation 13.12, the *fractional* number of molecules in the lowest translational energy level is:

$$\frac{\overline{n}_0}{N} = \frac{1}{N}\left(\frac{1}{e^{\alpha} - 1}\right)$$

If e^{α} is appreciably greater than 1, this is very small. For example, if $e^{\alpha} = (1 + 10^{-10})$, the right-hand expression becomes $10^{10}/N \approx 10^{-14}$. When, however, $e^{\alpha} \rightarrow 1$ the expression tends to unity (see the note below), and an appreciable fraction of the atoms are then in the lowest translational energy level. To summarize:

$$\text{at } T \approx T_0, \quad n_0 \approx N$$

and therefore, $n_i \approx 0$ $(i \neq 0)$.

Thus, it is predicted that liquid ^{4}He will display a phenomenon which in in some respects analogous to the condensation of a gas; above the temperature T_0, a negligible proportion of molecules are in the lowest energy level, whereas below T_0 a significant number of them are in this level. This 'condensation', however, has its origin in the Bose–Einstein distribution law, and not the presence of intermolecular forces.

The observed properties of liquid ^{4}He are, qualitatively, in keeping with these predictions. When liquid helium is cooled from its boiling point at atmospheric pressure, it appears to undergo a phase transition at 2·19 K, to a species called liquid helium II. Thus, the heat capacity curve (Fig. 13.2) exhibits a singularity (known as a λ-point) at 2·19 K. Above the λ-point, liquid helium is not greatly different from the other liquids, whereas below this temperature its physical properties are unique. For example, the viscosity of liquid helium II appears to be vanishingly small; we cannot go into the explanation of this superfluidity here, but it arises because of the difficulty of converting the kinetic energy of the flowing liquid into the thermal energy associated with the random motion of individual molecules.

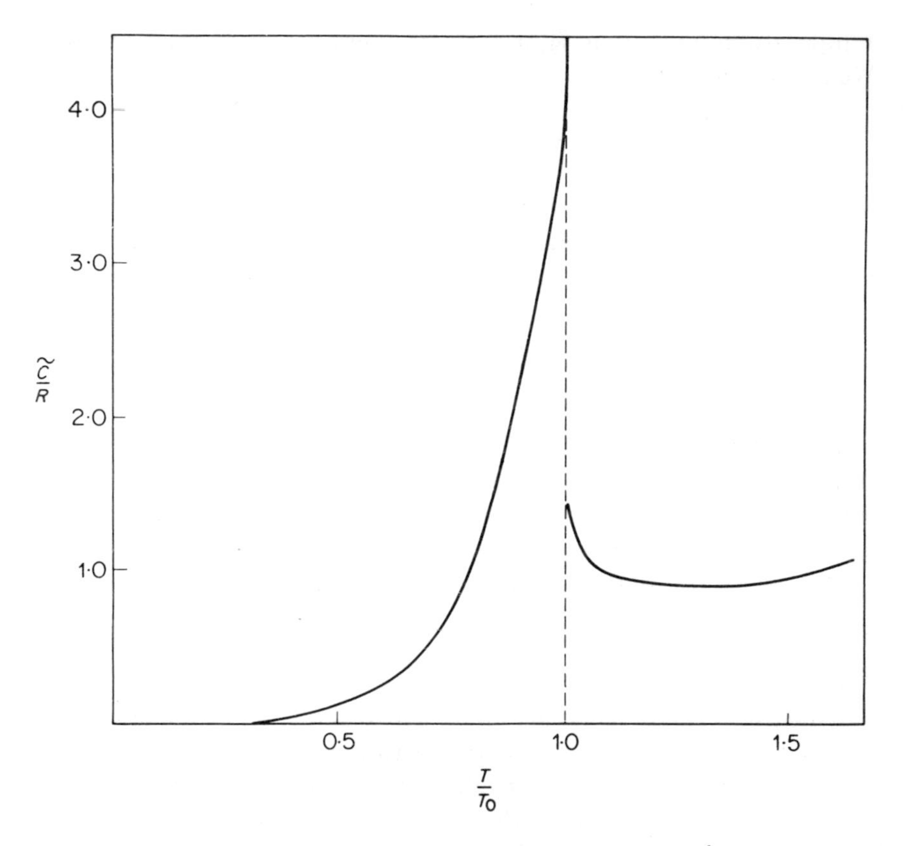

Fig. 13.2 Experimentally determined heat capacities of ^4He at low temperatures ($T_0 = 2{\cdot}19$ K)
(After F. London, *J. Phys. Chem.*, 1939, **43**, 49. Fig. 6)

The *theoretical* heat capacity curve, calculated by the above methods, is shown in Fig. 13.3; this also exhibits a singularity, at $T_0(3{\cdot}13\,\text{K})$. The shape of the experimental C_V *versus* T curve must, of course, be influenced by the presence of intermolecular forces, which have been neglected in the above treatment; nevertheless, the transition at 2·19 K is undoubtedly closely related to the quantum effects which we have described.

An equation analogous to 13.7 is obtained by approximating the sum 13.4 as an integral. However, two modifications are made. Firstly, the factor 2, arising from electron spin in the previous case, is now omitted. Secondly, it is easily seen (by putting $\epsilon = 0$ in the integrand) that the integral omits the contribution (given by Equation 13.12) to N arising from the lowest level. Since the latter becomes important at low temperatures, we include it as a separate term. Therefore:

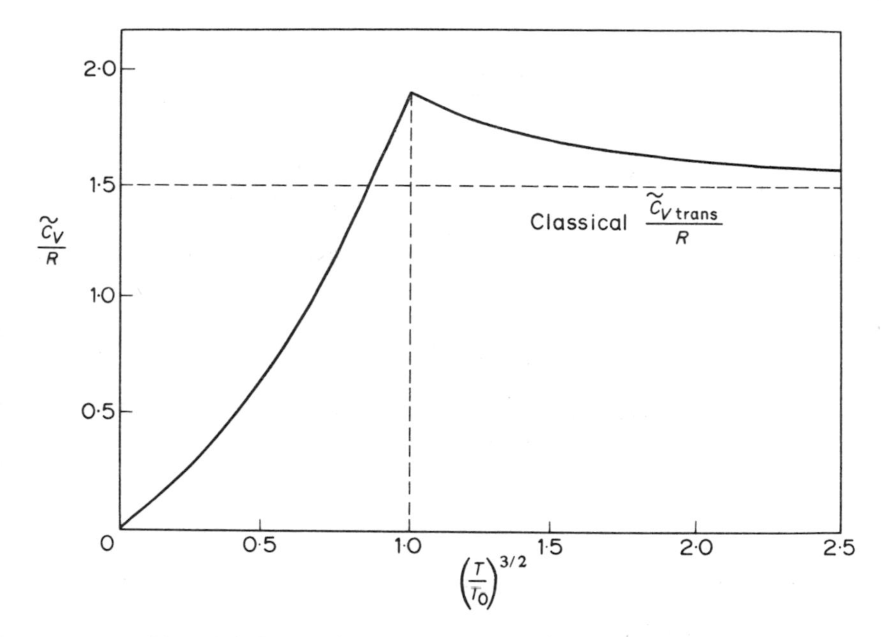

Fig. 13.3 Theoretical heat capacity of a Bose–Einstein gas

$$N = \frac{1}{e^{\alpha}-1} + \frac{1}{4}\left(\frac{8m}{h^2}\right)^{3/2} \pi V \int_0^\infty \frac{\epsilon^{1/2}}{e^{\alpha}\, e^{\epsilon/kT} - 1}\, d\epsilon \tag{13.16}$$

Now:

$$\int_0^\infty \frac{\epsilon^{1/2}}{e^{\alpha}\, e^{\epsilon/kT} - 1}\, d\epsilon = (kT)^{3/2} F'$$

where $F' = \int_0^\infty x^{\frac{1}{2}}/(e^{\alpha}e^{x} - 1)\, dx$, with $x = \epsilon/kT$. Expanding the integrand of F':

$$F' = \int_0^\infty \frac{x^{1/2}}{e^{\alpha}\, e^{x}} \left(\frac{1}{1 - e^{-x}\, e^{-\alpha}}\right)$$

$$= \int_0^\infty x^{1/2}\, (e^{-x}\, e^{-\alpha} + e^{-2x}\, e^{-2\alpha} + e^{-3x}\, e^{-3\alpha} + \ldots)\, dx$$

Integrating term by term, and using the result (Appendix 3) that:

$$\int_0^\infty e^{-ax}\, x^{1/2}\, dx = \tfrac{1}{2}\pi^{1/2}\, a^{-3/2}, \tag{13.17}$$

$$F' = \frac{\pi^{1/2}}{2}(e^{-\alpha} + 2^{-3/2}\, e^{-2\alpha} + 3^{-3/2}\, e^{-3\alpha} + \ldots)$$

$$= \frac{\pi^{1/2}}{2} F, \text{ say.}$$

Therefore:

$$N = \frac{(2\pi mkT)^{3/2} VF}{h^3} + \frac{1}{e^{\alpha} - 1} \tag{13.18}$$

or

$$e^{-\alpha} + 2^{-3/2} e^{-2\alpha} + 3^{-3/2} e^{-3\alpha} + \ldots = \frac{Nh^3}{(2\pi mkT)^{3/2} V} \tag{13.19}$$

unless e^{α} is very close to 1. From this, $e^{-\alpha}$ can be evaluated numerically. Rearranging Equation 13.18:

$$T = \frac{h^2}{2\pi mk} \left[\frac{1}{VF} \left(N - \frac{1}{e^{\alpha} - 1} \right) \right]^{2/3} \tag{13.20}$$

According to the inequalities 13.13, the largest value of $e^{-\alpha}$ is 1. Thus, according to Equation 13.17, the largest value of F is:

$$\begin{aligned} F_0 &= 1 + 2^{-3/2} + 3^{-3/2} + \ldots \\ &= 2{\cdot}612 + \ldots \end{aligned}$$

Three ranges of values of e^{α} in Equation 13.20 can be considered:

(i) $\infty > e^{\alpha} > 1{\cdot}000001$

(ii) $1{\cdot}000001 > e^{\alpha} > 1 + 10^{-18}$

(iii) $1 + 10^{-18} > e^{\alpha} > 1 + 10^{-24}$

In the first range, e^{α} is sufficiently large for $1/(e^{\alpha} - 1)$ to be negligible compared with N. Therefore:

$$T = \frac{h^2}{2\pi mk} \left(\frac{N}{VF} \right)^{2/3}$$

In the second range, e^{α} is sufficiently close to 1 for F to be approximated by F_0, but is sufficiently different from 1 for $1/(e^{\alpha} - 1)$ (which now $\approx 1/\alpha$) to be negligible compared with N. Therefore:

$$\begin{aligned} T &= \frac{h^2}{2\pi mk} \left(\frac{N}{VF_0} \right)^{2/3} \\ &= T_0, \text{ say.} \end{aligned}$$

In the third range, e^{α} is again sufficiently close to 1 for F to be approximated by F_0, and now, replacing $1/(e^{\alpha} - 1)$ by $1/\alpha$:

$$N = \frac{(2\pi mkT)^{3/2} VF_0}{h^3} + \frac{1}{\alpha}$$

or

$$\frac{1}{\alpha} = N \left[1 - \left(\frac{T}{T_0} \right)^{3/2} \right]$$

Since α is not negative, this means that this range of values of α must correspond to $T < T_0$.

It is concluded that if T is measurably greater than T_0, e^{α} must be in the range (i), whereas if T is measurably less than T_0, e^{α} must be in the range (iii). It is this fact which gives rise to the 'condensation' phenomenon.

13.6 Further discussion of the complexions of a system of independent particles

Suppose that the wave functions of a single molecule are:

$$\psi_1, \psi_2, \psi_3, \ldots, \psi_j, \ldots,$$

and the corresponding energies:

$$\epsilon_1, \epsilon_2, \epsilon_3, \ldots, \epsilon_j, \ldots$$

Consider two such molecules, say a and b, isolated from each other. Each will be in a quantum state characterized by one of the wave functions, and will have the corresponding energy; for molecule a, let us suppose that the energy is ϵ_r, and the wave function $\psi_r(\mathrm{a})$; for molecule b, the energy ϵ_s and the wave function $\psi_s(\mathrm{b})$.

Now suppose that the two molecules are brought together to form part of the same system, whilst still being distinguishable from each other (e.g., both might occupy sites in one crystal). If intermolecular forces can be neglected, the energy of a and b together is:

$$E = \epsilon_r + \epsilon_s,$$

and it can be shown [see note (a) below] that the product function:

$$\Psi_1 = \psi_r(\mathrm{a})\, \psi_s(\mathrm{b})$$

is an eigenfunction of the Schrödinger equation for the system of two molecules. This wave function characterizes a complexion of the system.

Since the two molecules are of the same type, the same wave functions ψ and energies ϵ are available to both molecules. Therefore, another complexion of the system is the one characterized by:

$$\Psi_2 = \psi_r(\mathrm{b})\, \psi_s(\mathrm{a})$$

in which molecule b happens to be in the quantum state associated with ψ_r and a in the state associated with ψ_s. The corresponding energy is again:

$$E = \epsilon_r + \epsilon_s$$

the same as that associated with Ψ_1.

Quite generally, for a system of N molecules a,b,c,d, ... , with wave functions $\psi_r(a)$, $\psi_s(b)$, $\psi_t(c)$, $\psi_u(d)$, ... , the total energy is:

$$E = \epsilon_r + \epsilon_s + \epsilon_t + \epsilon_u + \dots ,$$

and the wave function for the whole system is the product:

$$\Psi_1 = \psi_r(a)\, \psi_s(b)\, \psi_t(c)\, \psi_u(d) \dots .$$

Other wave functions Ψ_2, Ψ_3, ... , etc. having the same associated energy E are obtained by interchanging the quantum states of the individual molecules. Clearly, to each different way of assigning distinguishable molecules to the quantum states Ψ and energy levels ϵ there corresponds one product Ψ, i.e., one complexion; this was the assumption made in Section 1.2.

Identical molecules of a *gas* are physically indistinguishable from one another. This is allowed for in quantum mechanics by requiring that the wave function of the whole system must be such that Ψ *is everywhere unchanged in magnitude if the labels* a *and* b *of any two identical molecules are interchanged.* Now, clearly:

$$\Psi_1 = \psi_r(a)\ \psi_s(b)$$

has not this property, because interchanging the labels gives:

$$\Psi_2 = \psi_r(b)\, \psi_s(a)$$

a completely different wave function.

But the following linear combinations of Ψ_1 and Ψ_2 also satisfy the Schrödinger equation [see note (b) below]:

$$\begin{aligned} \Psi_S &= C(\Psi_1 + \Psi_2) = C[\psi_r(a)\, \psi_s(b) + \psi_r(b)\, \psi_s(a)] \\ \Psi_A &= C(\Psi_1 - \Psi_2) = C[\psi_r(a)\, \psi_s(b) - \psi_r(b)\, \psi_s(a)] \end{aligned} \qquad (13.21)$$

where C is a normalization constant (here equal to $1/\sqrt{2}$). Interchanging the labels a and b clearly leaves Ψ_S unchanged, but changes Ψ_A to:

$$C[\psi_r(b)\, \psi_s(a) - \psi_r(a)\, \psi_s(b)] = -\Psi_A ,$$

i.e., changes the sign but not the magnitude of Ψ_A at every point. Ψ_S is said to be *symmetric* in the molecules, and Ψ_A *antisymmetric* in the molecules.

Symmetric and antisymmetric wave functions can be constructed for the general system of N molecules. $N!$ products, $\Psi_1, \Psi_2, \Psi_3, \Psi_4, \ldots$ can be formed by permuting the molecules among the occupied quantum states, $\psi_r, \psi_s, \ldots$: Then:

$$\Psi_S = (1/\sqrt{N!})\,(\Psi_1 + \Psi_2 + \Psi_3 + \Psi_4 + \ldots)$$

$$\Psi_A = (1/\sqrt{N!})\,(\Psi_1 - \Psi_2 + \Psi_3 - \Psi_4 + \ldots)$$

the order of $\Psi_1, \Psi_2, \ldots$ in the latter expression being chosen so that each Ψ differs from the previous Ψ by the interchange of just two labels in the product. For example, considering a system of three molecules, and denoting $\psi_r(\mathrm{a})\,\psi_s(\mathrm{b})\,\psi_t(\mathrm{c})$ by *abc*, $\psi_r(\mathrm{b})\,\psi_s(\mathrm{a})\,\psi_t(\mathrm{c})$ by *bac*, etc.:

$$\Psi_A = \frac{1}{\sqrt{6}}\,(abc - bac + bca - acb + cab - cba) \tag{13.22}$$

In Chapter 4, the basic supposition was made that only *one complexion* is associated with each way of assigning N identical gas molecules to N different molecular quantum states. We have now found, however, that there are *two* possible wave functions Ψ_S and Ψ_A (i.e., quantum states of the *system*) associated with such an assignment. The difficulty is only removed by introducing a new assumption, which we call the *general exclusion principle,* and which, as we subsequently show, contains, as a special case, the exclusion principle referred to in Section 13.2.

As in Section 13.2, let ν be the sum of the numbers of protons, neutrons, and electrons in a molecule, or other particle. Then the general exclusion principle states that *for any (molecular or macroscopic) system of identical particles for which ν has an odd value, only wave functions of type Ψ_A are allowed; for systems of particles for which ν has an even value, only wave functions of type Ψ_S are allowed.* The spectroscopic and chemical evidence for the exclusion principle applied to systems of electrons is well known. Evidence for the *general* exclusion principle is chiefly spectroscopic, namely, the alternation of intensities in band spectra (see Section 13.7). The principle is further supported by the fact that certain experimental data (e.g., the heat capacities of

H_2 and D_2) are satisfactorily interpreted in terms of it.

Suppose that, in a system of three molecules, the quantum state ψ_r is doubly occupied, i.e., ψ_r appears twice in the product functions. The product functions are now:

$$\Psi_1 = \psi_r(\mathrm{a})\,\psi_r(\mathrm{b})\,\psi_t(\mathrm{c})$$
$$\Psi_2 = \psi_r(\mathrm{b})\,\psi_r(\mathrm{a})\,\psi_t(\mathrm{c})$$
$$\Psi_3 = \psi_r(\mathrm{b})\,\psi_r(\mathrm{c})\,\psi_t(\mathrm{a})$$

etc.

Obviously, Ψ_1 and Ψ_2 differ only in the order in which the terms in the product have been written down, i.e.:

$$\Psi_1 \equiv \Psi_2\,;$$

hence they cancel out in the expression 13.22 for Ψ_A. The same is true for the pairs Ψ_3,Ψ_4 and Ψ_5,Ψ_6. Therefore:

$$\Psi_A \equiv 0$$

The argument is easily generalized (Mayer and Mayer, p.64) to show that Ψ_A for a system of any number of molecules vanishes if any ψ_r occurs more than once, i.e., if any molecular quantum state ψ_r is occupied by more than one molecule.

The vanishing of a wave function means, physically, that such states do not occur. Therefore, in systems for which only wave functions of type Ψ_A are allowed (i.e., for which ν is odd), no more than one particle can occupy any particular quantum state ψ_r. This is just the exclusion principle of Section 13.2.

To conclude: we have shown in this section that our intuitive description of a complexion as a particular way of assigning individual molecules to energy levels is in keeping with the formal definition of the term (page 3).

Note (a)

For the isolated molecules, the Schrödinger equations are:

$$H_{\mathrm{a}}\,\psi_r(\mathrm{a}) = \epsilon_r\psi_r(\mathrm{a}) \tag{13.23}$$

$$H_{\mathrm{b}}\,\psi_s(\mathrm{b}) = \epsilon_s\psi_s(\mathrm{b}) \tag{13.24}$$

Since H_{a} operates only upon the coordinates of a, we can multiply both sides of Equation 13.23 by $\psi_s(\mathrm{b})$:

$$H_{\mathrm{a}}\psi_r(\mathrm{a})\,\psi_s(\mathrm{b}) = \epsilon_r\psi_r(\mathrm{a})\,\psi_s(\mathrm{b}) \tag{13.25}$$

and similarly, from Equation 13.24:

$$H_b\ \psi_s(b)\ \psi_r(a) = \epsilon_r\ \psi_s(b)\ \psi_r(a) \tag{13.26}$$

Adding Equation 13.25 and 13.26:

$$(H_a + H_b)\ [\psi_r(a)\ \psi_s(b)] = (\epsilon_r + \epsilon_s)\ [\psi_r(a)\ \psi_s(b)]$$

If there are no intermoleulcar forces, $(H_a + H_b)$ is just H, the Hamiltonian operator for the whole system; moreover, $(\epsilon_r + \epsilon_s)$ is E, the total energy. Thus:

$$H[\psi_r(a)\ \psi_s(b)] = E[\psi_r(a)\ \psi_s(b)]$$

and comparing this with the Schrödinger equation for the system:

$$H\Psi = E\Psi$$

we see that:

$$\Psi = \psi_r(a)\ \psi_s(b)$$

Note (b)

If Ψ_1 and Ψ_2 are two wave functions with the same associated energy, then:

$$H\Psi_1 = E\Psi_1$$

$$H\Psi_2 = E\Psi_2$$

Adding:

$$H(\Psi_1 + \Psi_2) = E(\Psi_1 + \Psi_2)$$

i.e., $(\Psi_1 + \Psi_2)$ is also an (un-normalized) eigenfunction of H.

13.7 Effect of symmetry in homonuclear diatomic molecules

It is possible, as an approximation, to break down the Schrödinger equation for a molecule into simpler parts, each of which can be treated separately (Pauling and Wilson, p.113; see also p.355); this is really the justification for our use of Equation 5.1:

$$\epsilon_{int} = \epsilon'_{vib} + \epsilon'_{rot} + \epsilon'_{el} + \epsilon'_{nuc}$$

the energy of a molecule not undergoing translation. By applying the argument given in note (a) at the end of the preceding section, it is easily shown that the molecular wave function ψ can then be expressed as the product of the corresponding wave functions:

$$\psi = \phi_{vib}\ \phi_{rot}\ \phi_{el}\ \phi_{nuc}.$$

If the two nuclei of a diatomic molecule are identical (e.g., H_2 or D_2, but not HD), a situation occurs which is entirely analogous to that discussed in the preceding section. The wave function ψ of an isolated homonuclear diatomic molecule must be such that interchanging the nuclei leaves the magnitude of ψ everywhere unchanged; moreover, the general exclusion principle applied: if ν (the sum of the numbers of protons and neutrons in the nucleus) is even, ψ must be symmetric in the nuclei; if $\nu$is odd, ψ must be antisymmetric in the nuclei.

The symmetry properties of ψ depend, of course, upon the properties of the four functions ϕ of which it is composed, i.e., upon how the four functions ϕ_{rot}, ϕ_{vib}, ϕ_{el}, and ϕ_{nuc} individually change, when the nuclei are interchanged (by rotating the molecule through 180° about a principle axis). The following can be established.

(i) ϕ_{rot} *is symmetric* if $J = 0,2,4,...,$ *and antisymmetric* if $J = 1,3,5,....$ ϕ_{rot} is a function of the polar angles θ and φ known as a spherical harmonic. Interchanging the nuclei is equivalent to changing θ to $(\pi - \theta)$ and φ to $(\varphi + \pi)$. The effect of such a displacement is to leave ϕ_{rot} unchanged if J is even, and to alter it to $-\phi_{rot}$ if J is odd.

(ii) ϕ_{vib} is always sy*mmetric* because ϕ_{vib} is a function only of the internuclear distance, the value of which is unaffected by interchanging the nuclei.

(iii) ϕ_{el} *is usually symmetric in the ground state;* this may be predicted by the molecular orbital theory.

(iv) The only nuclear property which enters into statistical thermodynamics is the nuclear spin. Consequently, ϕ_{nuc} will be regarded as a *nuclear spin wave function,* the behaviour of which determines the observable properties associated with nuclear spin. *The number of such functions* ϕ_{nuc} *depends upon the value of the quantum number* $\boldsymbol{I}$ *of the nuclei, defined in Section* 5.2. *There can be* $(I + 1)(2I + 1)$ *symmetric (ortho) spin states, and* $I(2I + 1)$ *antisymmetric (para) spin states.*

Let $\theta_r(a)$ be one of the $2I + 1$ nuclear spin functions of the nucleus a, and $\theta_s(b)$ be one of the $2I + 1$ spin functions of nucleus b. As in Equation 13.21, nuclear spin functions for the whole molecule that have the correct symmetry property are:

$$\phi_{\text{nuc}(S)} = \frac{1}{\sqrt{2}}[\phi_r(\text{a})\,\phi_s(\text{b}) + \phi_r(\text{b})\,\phi_s(\text{a})] \quad \text{(symmetric)}$$

$$\phi_{\text{nuc}(A)} = \frac{1}{\sqrt{2}}[\phi_r(\text{a})\,\phi_s(\text{b}) - \phi_r(\text{b})\,\phi_s(\text{a})] \quad \text{(antisymmetric)}$$

According to the second expression, the only non-vanishing *antisymmetric* functions are those for which $r \neq s$. There are $(2I + 1)$ ways of choosing ϕ_r and $(2I + 1) - 1 = 2I$ ways of choosing ϕ_s, with $s \neq r$. Hence the number of antisymmetric functions is:

$$\tfrac{1}{2} \times 2I \times (2I + 1) = I(2I + 1).$$

Similarly, $I(2I + 1)$ *symmetric* functions can be formed with $r \neq s$; but $(2I + 1)$ others, with $r = s$, can also be formed. Hence the number of symmetric nuclear spin states is:

$$I(2I + 1) + (2I + 1) = (I + 1)(2I + 1).$$

Suppose that interchanging the nuclei produces the changes

$$\phi_{\text{rot}} \to -\phi_{\text{rot}} \quad \text{and} \quad \phi_{\text{nuc}} \to -\phi_{\text{nuc}}$$

(i.e., both ϕ_{rot} ϕ_{nuc} are antisymmetric in the nuclei). Then, correspondingly, the effect on the product ψ is:

$$\phi_{\text{rot}}\,\phi_{\text{vib}}\,\phi_{\text{el}}\,\phi_{\text{nuc}} \to (-\phi_{\text{rot}})\,\phi_{\text{vib}}\,\phi_{\text{el}}\,(-\phi_{\text{nuc}})$$

if ϕ_{el} and ϕ_{vib} are symmetric. Thus:

$$\psi \to \psi$$

and ψ is *symmetric* in the nuclei.

Now suppose that ϕ_{rot} is antisymmetric, but ϕ_{nuc} is symmetric, i.e., on interchanging the nuclei:

$$\phi_{\text{rot}} \to -\phi_{\text{rot}} \quad \text{and} \quad \phi_{\text{nuc}} \to \phi_{\text{nuc}},$$

$$\phi_{\text{rot}}\,\phi_{\text{vib}}\,\phi_{\text{el}}\,\phi_{\text{nuc}} \to (-\phi_{\text{rot}})\,\phi_{\text{vib}}\,\phi_{\text{el}}\,\phi_{\text{nuc}},$$

$$\psi \to -\psi.$$

Thus, if either one of ϕ_{rot} or ϕ_{nuc} is antisymmetric, then ψ is antisymmetric. Otherwise, ψ is symmetric.

The symmetry of the allowed functions ψ is determined by the general exclusion principle, hence:

if v is odd (ψ is antisymmetric), then *either* ϕ_{nuc} *or* ϕ_{rot}, *but not both* must be antisymmetric:

if v is even (ψ is symmetric), then ϕ_{nuc} *and* ϕ_{rot} must be *both* symmetric or *both* antisymmetric

Case (i): ν *is odd;* e.g., H_2.
Terms in the sum,

$$f_{\text{int}} = \sum_i g_i \, e^{-\epsilon_{\text{int}(i)}/kT}$$

can be divided into two groups. Thus:

$$f_{\text{int}} = \begin{Bmatrix} \text{Terms with} \\ \phi_{\text{nuc}} \text{ antisymmetric} \\ \phi_{\text{rot}} \text{ symmetric} \end{Bmatrix} + \begin{Bmatrix} \text{Terms with} \\ \phi_{\text{nuc}} \text{ symmetric} \\ \phi_{\text{rot}} \text{ antisymmetric} \end{Bmatrix}$$

(para terms) (ortho terms)

$$= f_{\text{el}} f_{\text{vib}} \, I(2I + 1) \sum_{J=0,2,4,\ldots} (2J + 1) \, e^{-J(J+1)\,\Theta_{\text{rot}}/T}$$

$$+ f_{\text{el}} f_{\text{vib}} \, (I + 1)(2I + 1) \sum_{J=1,3,5,\ldots} (2J + 1) \, e^{-J(J+1)\,\Theta_{\text{rot}}/T}$$

Since $I = \frac{1}{2}$ for hydrogen:

$$f_{\text{int}} = f_{\text{el}} f_{\text{vib}} \left[\sum_{J=0,2,4,\ldots} (2J + 1) \, e^{-J(J+1)\,\Theta_{\text{rot}}/T} + 3 \sum_{J=1,3,5,\ldots} (2J + 1) \, e^{-J(J+1)\,\Theta_{\text{rot}}/T} \right] \quad (13.27)$$

Case (ii): ν *is even;* e.g., D_2

$$f_{\text{int}} = \begin{Bmatrix} \text{Terms with} \\ \phi_{\text{nuc}} \text{ antisymmetric} \\ \phi_{\text{rot}} \text{ antisymmetric} \end{Bmatrix} + \begin{Bmatrix} \text{Terms with} \\ \phi_{\text{nuc}} \text{ symmetric} \\ \phi_{\text{rot}} \text{ symmetric} \end{Bmatrix}$$

(para terms) (ortho terms)

$$= f_{\text{el}} f_{\text{vib}} \, I(2I + 1) \sum_{J=1,3,5,\ldots} (2J + 1) \, e^{-J(J+1)\,\Theta_{\text{rot}}/T}$$

$$+ f_{\text{el}} f_{\text{vib}} (I + 1)(2I + 1) \sum_{J=0,2,4,\ldots} (2J + 1) \, e^{-J(J+1)\,\Theta_{\text{rot}}/T}$$

Since $I = 1$ for deuterium:

$$f_{\text{int}} = f_{\text{el}} f_{\text{vib}} \left[3 \sum_{J=1,3,5,\ldots} (2J + 1) \, e^{-J(J+1)\,\Theta_{\text{rot}}/T} + 6 \sum_{J=0,2,4,\ldots} (2J + 1) \, e^{-J(J+1)\,\Theta_{\text{rot}}/T} \right] \quad (13.28)$$

An application of the partition functions 13.27 and 13.28 has been treated in Section 6.2. Expressions of this type also have the following important use in the interpretation of intensities in the spectra of diatomic molecules.

Since transitions of the type ortho↔para are spectroscopically forbidden, the molecular spectrum (electronic–vibration–rotation, or Raman) can be treated as a superposition of two spectra, one arising from ortho→ortho, and the other from para→para transitions.

The intensity distribution can be discussed as in Section 10.1, the intensity of each line being related to the term in the partition function corresponding to the initial state. In the present case, however, the term contains as a factor $I(2I + 1)$ (for para states) or $(I + 1)(2I + 1)$ (for ortho states). Hence the intensities of successive lines are approximately in the ratio:

$$\frac{(I + 1)(2I + 1)}{I(2I + 1)} = \frac{I + 1}{I}$$

the stronger lines being associated with the ortho states. In this way, one can (a) evaluate I, from the measured intensity ratio, and (b) decide upon the parities of the rotational wave functions (i.e., whether J is odd or even) respectively associated with ortho and para states, and hence whether symmetric or antisymmetric molecular wave functions ψ are permitted in a given case; thus (see Table 13.1), experimental confirmation of the general exclusion principle is obtained.

Table 13.1 Nuclear Spins and Allowed Symmetries Determined from Molecular Spectra

Nucleus	H	D	T	^{3}He	^{7}Li	^{12}C	^{14}N	^{16}O	^{35}Cl	^{37}Cl
I	$\frac{1}{2}$	1	$\frac{1}{2}$	$\frac{1}{2}$	$\frac{3}{2}$	0	1	0	$\frac{3}{2}$	$\frac{3}{2}$
ν	1	2	3	3	7	12	14	16	35	37
Symmetry* of ψ	a	s	a	a	a	s	s	s	a	a

*s symmetric, a antisymmetric in the nucleus.

Data from G.H. Herzberg, I, p. 461.

Problems

13.1 Calculate the electronic contribution to the molar entropy or iron at 1000 K. Assume that there are two mobile electrons per atom and that the density of iron is $7{\cdot}5 \times 10^3$ kg m^{-3}.

13.2 From the results of Section 13.4, show that the translational Helmholtz free energy of a Fermi–Dirac gas is $\tilde{A} = \frac{3}{5}\tilde{N}\mu_0$, and, by combining this with Equation 13.9, that $PV = \frac{2}{3}E$.

13.3 By treating liquid ^{3}He as a Fermi–Dirac gas, calculate $\tilde{C}_V$ for liquid ^{3}He at 2.5 K. Compare the result with the corresponding value for ^{4}He shown in Fig. 13.3.

13.4 The vapour pressure of liquid ^{3}He is 0·44 atmosphere at 2.5 K. By means of the Sackur–Tetrode equation (cf. problem 8.5), calculate (a) the enthalpy of evaporation of ^{3}He at 2·5 K, and (b) the normal boiling point of ^{3}He (experimental value = 3·2 K). (Assume Fermi–Dirac entropy for the liquid – no detailed calculation is required – and classical entropy for the vapour, because of the low density of the latter.)

13.5 *Thermionic emission.* A heated metal electrode emits electrons. All the emitted electrons can be drawn to a second electrode, if the latter is at a high enough positive potential relative to the former. The corresponding current divided by the area of the emitting electrode is called the saturation current density $\mathcal{J}$.

Use the transition state theory of Chapter 12 to show that:

$$\mathcal{J} = \frac{4\pi em(kT)^2}{h^3} e^{-\psi/kT}$$

(Richardson's law), where m and e are the electronic mass and charge, and ψ is called the *thermionic work function* for the electrons of the metal, as follows.

The transition state occurs just outside of the metal, where the electronic potential relative to that of the metal is ψ. The coordinate q_* is the direction at right angles to the surface. Determine $\Delta S^{\ddagger}$: S for electrons *inside* the metal is small (page 281); on the other hand, $S^{\ddagger}$ has the classical value (*two*-dimensional translation and $g_{el} = 2$), because the concentration of electrons outside of the metal is small.

13.6 *Thermal conductivity of metals.* According to the kinetic theory of gases (Mayer and Mayer, p.26), the thermal conductivity κ of a gas is:

$$\kappa = \frac{1}{3} l \frac{N}{V} \frac{\tilde{C}_V}{\tilde{N}} \bar{v}$$

where l is the mean free path between collisions and $\bar{v}$ is the average speed of the molecules.

Show that the average speed for electrons in a metal is:

$$\bar{v} = \frac{3}{4} \left(\frac{2\mu_0}{m} \right)^{1/2}$$

and deduce that the thermal conductivity of a metal (due to electrons) is:

$$\kappa = \frac{\pi^2}{4} \frac{N}{V} l k^2 T \left(\frac{1}{2m\mu_0} \right)^{1/2}$$

(a more exact treatment gives $\pi^2/3$ instead of $\pi^2/4$).

CHAPTER FOURTEEN

The Canonical Ensemble

In the previous chapters, discussion has been confined mainly to systems of *independent* particles, i.e., to systems for which the total energy is given as the sum of the energies of the individual particles:

$$E = \sum_i n_i \epsilon_i \qquad (14.1)$$

It will be recalled that this condition led to the introduction of the parameter β (Equations 2.6 and 2.7) in the derivation of the Maxwell–Boltzmann distribution law; again, the condition formed the basis (Equation 2.28 etc.) of our statistical mechanical interpretation of temperature.

For systems in which there are appreciable and varying forces of interaction between the molecules, however, it is impossible to divide up the total energy into the contributions from separate molecules. If the interactions are small, Equation 14.1 can be replaced by:

$$E = U + \sum_i n_i \epsilon_i \qquad (14.2)$$

in which ϵ_i is a possible energy level of an *isolated* molecule and U is the *total energy of interaction* of the assembly of molecules. This interaction energy cannot be treated as a constant parameter; clearly, its value must depend upon the (continually changing) relative positions of all the molecules of the system, and on the occupancy of the molecular quantum states. Again, if the forces of interaction are large (as in a diamond crystal) the term $\Sigma n_i \epsilon_i$ may not have any meaning for the system: only the energy levels of the system *as a whole,* not the ϵ_i, are relevant.

For all such systems, the derivations given in Chapter 2 cannot be used; neither is the statistical interpretation of temperature, given there, any longer appropriate. We therefore discuss, in this chapter, a statistical method, due to J.W. Gibbs, which is much more powerful than that previously described; it is a method which can be used equally well for systems of independent particles and for systems in which intermolecular forces must be taken into account. In Sections 14.4–14.6, three important special applications of this new method will be outlined.

14.1 The canonical ensemble

In deriving the Maxwell–Boltzmann law, we began by considering a system of independent particles having a *fixed volume and fixed total energy.* In practice, however, in so far as experimental measurements are concerned, one is usually interested in the properties of a system *at known temperature* and volume, i.e., a thermostatted system. *One cannot assume, a priori, that systems identical in volume, temperature, and molecular composition necessarily have the same internal energy.* We shall, in fact, show that (to an extremely high degree of probability) identical systems at the same temperature and volume do have essentially the same internal energy; but this is a *deduction* from the statistical treatment, not an initial assumption.

Suppose, then, that is is desired to study the statistical properties of some thermodynamical system A, the temperature, volume, and molecular composition of which are known; for example, the system A might be a vessel containing one mole of hydrogen at 273 K. To do this, we envisage an assembly of *a very large number N* of systems identical, in volume and molecular composition, with A, and in thermal contact with one another (e.g., in a thermostat bath of heat-conducting fluid); and we suppose that the whole assembly is insulated from the outside world. An assembly of this kind is called a *canonical ensemble* of systems of type A.

The existence of thermal contact between the systems implies that the systems can exchange energy with one another. In

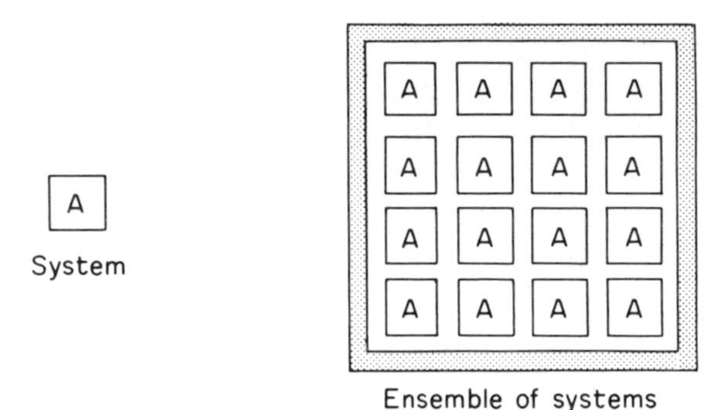

Fig. 14.1 Illustrating the canonical ensemble

consequence, the energies* E_i of the individual systems fluctuate about a mean value, as do the energies ϵ_i of the individual molecules of a gas. We later find that significantly probable fluctuations in the E_i are, in contrast with those of the molecular energies, very small (too small ever to be observed), but this in no way vitiates the mathematical treatment. We therefore assume that, in the ensemble, $\boldsymbol{n}_1$ systems have energy E_1, $\boldsymbol{n}_2$ have energy E_2, $\boldsymbol{n}_i$ have energy E_i, and so on.

The total energy of the ensemble is:

$$\boldsymbol{E} = \sum_{i} \boldsymbol{n}_i E_i$$

and the total number of systems in the ensemble is:

$$\boldsymbol{N} = \sum_{i} \boldsymbol{n}_i$$

The number $\boldsymbol{N}$ can be imagined to be as large as we please. Therefore, for an ensemble in equilibrium, *one can discuss the most probable distribution of energies E_i and quantum states of the systems by a procedure identical with that used in Chapter 2 for the distribution of molecular energies ϵ_i*. Thus:

$$\overline{\boldsymbol{n}_i} = \Omega_i \, e^{-\alpha} \, e^{-\beta E_i} \tag{14.3}$$

where $\overline{\boldsymbol{n}_i}$ is the most probable number of systems in the ensemble with energy E_i. Note that Ω_i (the degeneracy of E_i, or the *number of complexions* associated with E_i) replaces g_i in Equation 2.14. Again, the most probable number $\overline{\boldsymbol{m}_j}$ of systems in the $\boldsymbol{j}$-th

* In this chapter, we use bold face $\boldsymbol{i}$ and $\boldsymbol{j}$ respectively to label an energy level and a quantum state (complexion) of the whole system.

quantum state is:

$$\bar{m}_j = e^{-\alpha} e^{-\beta E_j} \tag{14.4}$$

The arguments used in Chapters 2 and 3 (with E_j replacing ϵ_i) can again be applied to show that β is a *measure of the temperature of the ensemble.* Thus, an interpretation of the temperature of system A is obtained by applying our statistical methods to the canonical ensemble of systems like A. We shall also show how the thermodynamical properties of A can be calculated; for this purpose it is necessary to introduce the concept of the *ensemble average.*

Suppose that the value of some property X of the system is simultaneously determined for each of the N systems in the ensemble. Let the instantaneous values obtained for systems $1, 2, 3, \ldots, N$ be respectively $X_1, X_2, \ldots, X_N$. Then the average value (the ensemble average) of X is:

$$\bar{X} = \frac{X_1 + X_2 + X_3 + \ldots + X_N}{N} = \frac{\sum_{l=1}^{N} X_l}{N}$$

Now, suppose that the value of X for one single system in the ensemble is determined on N different occasions. Since, in general, there will be fluctuations, suppose that the values of X so obtained are $X'_1, X'_2, \ldots, X'_N$. The average of these values is:

$$\bar{X}' = \frac{X'_1 + X'_2 + X'_3 + \ldots + X'_N}{N}$$

Since the systems in the ensemble are identical, it is reasonable to assume* that the two averages $\bar{X}$ and $\bar{X}'$ would be the same, provided that N were large enough. Thus, the value of X obtained by averaging over the systems in the ensemble is just the average value that would be obtained by repeated measurements on one particular system, say A. The measurements on A are measurements made when A is part of the ensemble, i.e., when A gains heat from, and loses heat to, the surrounding systems, whilst remaining at a constant temperature T. In so far as A itself is concerned, however, the surrounding systems could just as well be part of a thermostat bath held at temperature T ("A does not 'know' whence it receives

* If N identical dice are cast together, the average result per throw should be the same as that obtained when one die is thrown N times. But the analogy with the averaging of X is not perfect, because there is no interaction between the N dice. In fact, identification of $\bar{X}$ and $\bar{X}'$ (the ensemble average and the time average) requires a fuller discussion than can be given here.

heat, or to where it emits heat"). Hence, averaging over the canonical ensemble enables one to predict the average of repeated measurements on *one system held at a fixed temperature.* In practice, of course, such repeated measurements always give, within experimental error, the same value; but this presents no problem, because calculations for the ensemble show that the probability of significant deviations from the mean value is extremely small (see Section 14.7).

14.2 Thermodynamic properties

The term $e^{-\alpha}$ in Equation 14.3 can be eliminated by summing over the $\boldsymbol{n}_i$ or $\boldsymbol{m}_j$; as in Section 2.4, we obtain:

$$e^{-\alpha} = \frac{Z}{N} \tag{14.5}$$

where Z is called the *partition function for the system** and is defined as (cf. Equations 2.17 and 2.18:

$$Z = \sum_{\substack{i \\ \text{energy levels} \\ \text{of the system}}} \Omega_i \, e^{-\beta E_i} \tag{14.6}$$

$$= \sum_{\substack{j \\ \text{quantum states} \\ \text{of the system}}} e^{-\beta E_j} \equiv \sum_{\substack{j \\ \text{complexions}}} e^{-\beta E_j} \tag{14.7}$$

The arguments used in Chapters 2 and 3 to relate statistical mechanics to thermodynamics can again be applied, with Z replacing f. Thus:

(i) the total energy of the ensemble is:

$$\boldsymbol{E} = -N\left(\frac{\partial \ln Z}{\partial \beta}\right)_V \tag{14.8}$$

(note again that the volume of the individual systems is assumed to be fixed);

(ii) the entropy of the ensemble is $\boldsymbol{S}$, where:

$$\boldsymbol{S} = k \ln \boldsymbol{\Omega} \tag{14.9}$$

* Mayer and Mayer, and Hill respectively use the symbols $\boldsymbol{Q}$ and Q. Rushbrooke uses P.F.

in which $\mathbf{\Omega}$ is the total number of complexions of the whole ensemble, associated with $\boldsymbol{E}$;

(iii) the temperature of the ensemble is T, where:

$$\beta = \frac{1}{kT} \tag{14.10}$$

(iv)

$$\ln \mathbf{\Omega} = \boldsymbol{N} \ln Z + \beta \boldsymbol{E} \tag{14.11}$$

Substituting Equation 14.10 and 14.11 into 14.8 and 14.9:

$$\boldsymbol{E} = \boldsymbol{N}kT^2 \left(\frac{\partial \ln Z}{\partial T}\right)_V$$

$$\boldsymbol{S} = \boldsymbol{N}k \ln Z + \frac{\boldsymbol{E}}{\mathrm{T}}$$

Therefore, the ensemble averages, which we identify with the experimental (or thermodynamic) values, are:

$$E = \frac{\boldsymbol{E}}{\boldsymbol{N}} = kT^2 \left(\frac{\partial \ln Z}{\partial T}\right)_V \tag{14.12}$$

$$S = k \ln Z + \frac{E}{T} \tag{14.13}$$

The work function is:

$$A = E - TS = -kT \ln Z \tag{14.14}$$

the heat capacity is:

$$C_V = kT^2 \left(\frac{\partial^2 \ln Z}{\partial T^2}\right)_V + 2kT \left(\frac{\partial \ln Z}{\partial T}\right)_V \tag{14.15}$$

and the pressure is:

$$P = kT \left(\frac{\partial \ln Z}{\partial V}\right)_T \tag{14.16}$$

Equations for other thermodynamic properties are correspondingly derived.

If E_j can be expressed as the sum of independent contributions (nuclear, electronic, translational, etc.), then Z can be factorized in just the same way as we discussed in Sections 3.4 and 5.1 for f. Correspondingly, the thermodynamic functions can be expressed as sums of independent terms.

14.3 Systems of independent particles

Equations 14.12 and 14.13, relating energy and entropy (and, through these, all the other thermodynamic properties) to Z, have been derived without any assumptions being made concerning the dependence or independence of the particles: that is, whether or not the interaction energy U in Equation 14.2 is zero. Hence, *these equations will be valid even for those systems in which intermolecular forces are important,* and they are therefore of much wider applicability than the corresponding equations of Chapter 3. However, before discussing more general applications we must show that, in the case of *independent* particles, the present method yields results which are identical with those previously derived on the basis of the assumptions made in Chapter 2.

Consider, firstly, systems of independent *distinguishable* molecules. Comparing the results of the two methods for such systems:

$$E = NkT^2\left(\frac{\partial \ln f}{\partial T}\right)_V \qquad E = kT^2\left(\frac{\partial \ln Z}{\partial T}\right)_V$$

$$S = Nk\ln f + \frac{E}{T} \qquad S = k\ln Z + \frac{E}{T}$$

$$A = -NkT\ln f \qquad A = -kT\ln Z$$

clearly, the two sets of equations agree if:

$$N\ln f = \ln Z \text{ (i.e., if } f^N = Z)$$

To prove that this is indeed the case, we note that:

$$f^N = \left(\sum_{\substack{\text{molecular}\\ \text{quantum states}\\ j}} e^{-\epsilon_j/kT}\right)^N$$

$$= \sum_{j_1, j_2, \ldots, j_N} e^{-(\epsilon_{j_1}+\epsilon_{j_2}+\ldots+\epsilon_{j_N})/kT} \tag{14.17}$$

the summation being over all possible values of the indices $j_1, j_2, \ldots, j_N$. It may clarify the meaning of this identity to consider the following example:

$$\begin{aligned}(e^{x_1} + e^{x_2})^2 &= e^{x_1}e^{x_1} + e^{x_1}e^{x_2} + e^{x_2}e^{x_1} + e^{x_2}e^{x_2}\\ &= e^{x_1+x_1} + e^{x_1+x_2} + e^{x_2+x_1} + e^{x_2+x_2}\\ &= \sum_{j_1, j_2} e^{(x_{j_1}+x_{j_2})}\end{aligned}$$

where j_1 can be 1 or 2 and j_2 can be 1 or 2.

Now:

$$\epsilon_{j_1} + \epsilon_{j_2} + \ldots + \epsilon_{j_N} = E_j$$

the energy of á *particular complexion,* in which molecule 1 has energy ϵ_{j_1}, molecule 2 energy ϵ_{j_2}, and so on. Hence:

$$\sum_{j_1, j_2, \ldots, j_N} e^{-(\epsilon_{j_1} + \epsilon_{j_2} + \ldots + \epsilon_{j_N})/kT} = \sum_{\substack{\text{complexions} \\ j}} e^{-E_j/kT} \tag{14.18}$$

and from Equations 14.17 and 14.18, therefore:

$$(f)^N = Z \tag{14.19}$$

If the molecules are independent *and indistinguishable,* Equation 14.19 must be modified. To illustrate this, consider the term in Equation 14.18 which corresponds to molecule 1 in quantum state 1, molecule 2 in quantum state 2, and so on, so that the first N molecular quantum states are singly occupied. The total energy is then:

$$E_j = \epsilon_1 + \epsilon_2 + \epsilon_3 + \ldots + \epsilon_N$$

There will be other terms in Equation 14.18 with the same total energy:

$$E_{j'} = \epsilon_2 + \epsilon_1 + \epsilon_3 + \ldots + \epsilon_N$$

$$E_{j''} = \epsilon_1 + \epsilon_3 + \epsilon_2 + \ldots + \epsilon_N$$

These correspond to complexions obtained by rearranging the molecules among the N quantum states; clearly, there are $N!$ such terms. If the molecules are distinguishable from one another, such rearrangement produces a new complexion; if they are *indistinguishable,* it does not. Consequently, in Equation 14.18, $N!$ terms appearing on the left-hand side will correspond to only *one* on the right, if the molecules are indistinguishable. It is for this reason that, in the case of indistinguishable, independent particles:

$$Z = \sum_{\substack{\text{complexions} \\ j}} e^{-E_j/kT} = \frac{1}{N!}\left(\sum_j e^{-\epsilon_j/kT}\right)^N = f^N/N!$$

Actually, division by $N!$ in this way slightly *over-compensates* for indistinguishability, because there are also complexions in which *more than one* molecule occupies a given quantum state; this is entirely analogous to the situation discussed on page 52, and, as was there shown, the effect is in most cases completely negligible.

14.4 The Debye theory of crystals

We come now to our first practical application of the ensemble method. In Section 6.4, it was necessary to state Equation 6.17, upon which the Debye calculation of crystal specific heats is based, as an intuitively reasonable result. Now, however, it is possible to derive it rigorously.

According to general theory (cf. Equation 5.15), the vibrational energy of a crystal is:

$$E'_{\text{vib}\,(j)} = \sum_{r=1}^{3N} \epsilon'_r \qquad (14.20)$$

where ϵ'_r is the total energy associated with the r-th vibrational mode of the crystal, i.e.:

$$\epsilon'_r = (v_r + \tfrac{1}{2})\, h\omega_r \qquad (v_r = 0, 1, 2, 3, \ldots)$$

the lowest value of $E'_{\text{vib}\,(j)}$ is obviously the one for which all the v_r are zero. This lowest value is the *zero-point vibrational energy* $E'_{\text{vib}\,(0)}$ of the crystal:

$$E'_{\text{vib}\,(0)} = \sum_{r=1}^{3N} \tfrac{1}{2} h\omega_r$$

As discussed in Section 5.1, we express molecular energies ϵ_i as relative to the lowest molecular energy level; similarly, we adopt the convention that the vibrational energy $\epsilon_{\text{vib}\,(j)}$ of the crystal is the energy expressed relative to the lowest value:

$$E_{\text{vib}\,(j)} = E'_{\text{vib}\,(j)} - E'_{\text{vib}\,(0)} = \sum_{r=1}^{3N} \epsilon_r$$

where $\epsilon_r = v_r h\omega$.

The vibrational partition function for the crystal is then:

$$Z_{\text{vib}} = \sum_j e^{-E_{\text{vib}\,(j)}/kT}$$

Now, $E_{\text{vib}\,(j)}$ is, according to Equation 14.20, the sum of $3N$ independent terms ϵ_r. Therefore, Z_{vib} can be expressed in the usual way, as the product of $3N$ independent factors:

$$Z_{\text{vib}} = f_1 \times f_2 \times f_3 \times \ldots \times f_{3N} \qquad (14.21)$$

where

$$f_r = \sum_{v_r=0}^{\infty} e^{-v_r h\omega/kT} \qquad (14.22)$$

(Alternatively, Equation 14.21 can be derived by reversing the procedure which led to Equations 14.18 and 14.19.)

The sum in Equation 14.22 has the value given in Section 5.4:

$$f_r = \frac{1}{1 - e^{-h\omega_r/kT}}$$

Substitution of Equation 14.21 into 14.15 yields, for the vibrational heat capacity of the crystal:

$$C_{V\,(\text{vib})} = kT^2 \left(\frac{\partial^2 [\ln f_1 f_2 \ldots f_{3N}]}{\partial T^2}\right)_V + 2kT\left(\frac{\partial [\ln f_1 f_2 \ldots f_{3N}]}{\partial T}\right)_V$$

$$= \sum_{r=1}^{3N} C'_{V\,(\text{vib})(r)} \qquad (14.23)$$

where

$$C'_{V\,(\text{vib})(r)} = kT^2 \left(\frac{\partial^2 \ln f_r}{\partial T^2}\right)_V + 2kT\left(\frac{\partial \ln f_r}{\partial T}\right)_V \qquad (14.23a)$$

Summation in Equation 14.23 is over all modes of vibration. Since, in general, g_l modes have the same frequency ω_l, we can sum over the frequencies, to obtain:

$$C_{V\,(\text{vib})} = \sum_{\substack{\text{frequencies} \\ l}} g_l\, C'_{V\,(\text{vib})(i)}$$

Now $C_{V\,(\text{vib})(l)}$ of Equations 6.15 and 6.17 is obviously $3N$ times $C'_{V\,(\text{vib})(r)}$ given by Equation 14.23a, and therefore:

$$C_{V\,(\text{vib})} = \frac{1}{3N} \sum_l g_l C_{V\,(\text{vib})(l)}$$

which is the required Equation 6.17.

14.5 Regular solutions

The system discussed in the preceding section is the first example encountered in this book of a *cooperative assembly*, i.e., an assembly which must be treated as a whole rather than as a collection of independent individuals. There are many such cases, in which intermolecular forces are so important that the behaviour of any one molecule is fundamentally influenced by the quantum states occupied by its neighbours.

A second example, which we shall now briefly consider, is a liquid mixture. In Sections 7.3 and 7.4, we were able to give a simplified account of the Bragg–Williams theory, by considering the constituent molecules as essentially independent particles. Refinement of the theory requires, however, a more powerful method of approach; this was hinted at on pages 135 and 136.

We use the model and notation of Section 7.3. A possible energy of the mixture A + B is (cf. Equation 7.15):

$$E_i = E_{\mathrm{A}(r)} + E_{\mathrm{B}(s)} + N_{\mathrm{AB}}w$$

where $E_{\mathrm{A}(r)}$ and $E_{\mathrm{B}(s)}$ are possible energies of pure A and B, respectively, and N_{AB} is the number of *neighbouring* A–B pairs in the mixture. Hence, the partition function for the system (the mixture) is:

$$Z = \sum_{\substack{\text{complexions} \\ \text{of the mixture}}} \mathrm{e}^{-(E_{\mathrm{A}(r)} + E_{\mathrm{B}(s)} + N_{\mathrm{AB}}w)/kT}$$

As usual, since E_i is the sum of independent terms $E_{\mathrm{A}(r)}$, $E_{\mathrm{B}(s)}$, and $N_{\mathrm{AB}}w$, Z can be factorized:

$$Z = Z_{\mathrm{A}} \times Z_{\mathrm{B}} \times Z_{\mathrm{configuration}}$$

where:

$$Z_{\mathrm{A}} = \sum_r \Omega_r \mathrm{e}^{-E_{\mathrm{A}(r)}/kT}; \quad Z_{\mathrm{B}} = \sum_s \Omega_s \mathrm{e}^{-E_{\mathrm{B}(s)}/kT}$$

and

$$Z_{\mathrm{configuration}} = \sum_{\substack{\text{all spatial} \\ \text{arrangements} \\ \text{of A and B}}} \mathrm{e}^{-N_{\mathrm{AB}}w/kT}$$

The Helmholtz free energy of the mixture is:

$$\begin{aligned} A &= -kT\ln Z \\ &= -kT\ln Z_{\mathrm{A}} - kT\ln Z_{\mathrm{B}} - kT\ln Z_{\mathrm{configuration}} \end{aligned}$$

The first two terms are the free energies of pure A and pure B, and so the free energy of mixing is:

$$\Delta A_{\mathrm{mix}} = -kT\ln Z_{\mathrm{configuration}}.$$

The problem is thus one of evaluating $Z_{\mathrm{configuration}}$.

The Bragg–Williams approximation consists, essentially, of replacing all the N_{AB} values in the exponential terms of $Z_{\text{configuration}}$ by *one* average value, say $\overline{N_{AB}}$. Thus:

$$Z_{\text{configuration}} \approx \sum_{\substack{\text{all spatial}\\ \text{arrangements}\\ \text{of A and B}}} e^{-\overline{N_{AB}}w/kT}$$

$$= \Omega_{\text{config}}\, e^{-\overline{N_{AB}}/kT}$$

where Ω_{config} is just the number of spatial arrangements of A and B in the mixture. This number is (cf. Equation 7.2):

$$\frac{(N_A + N_B)!}{N_A!\,N_B!}$$

and so:

$$Z_{\text{configuration}} = \frac{(N_A + N_B)!}{N_A!\,N_B!}\, e^{-\overline{N}_{AB}w/kT}$$

The Helmholtz free energy of mixing is now:

$$\Delta A_{\text{mix}} = -\,kT \ln \frac{(N_A + N_B)!}{N_A!\,N_B!} + \overline{N}_{AB}w$$

$$= -RT(n_A \ln x_A + n_B \ln x_B) + N_{AB}w$$

When $\overline{N}_{AB}$ is evaluated by the method of Section 7.3, and ΔA_{mix} is identified with ΔG_{mix} (for $V = 0$), we immediately obtain Equation 7.20.

It is not, in principle, necessary to make the above approximation. If terms with the same value of N_{AB} in $Z_{\text{configuration}}$ are collected together, there results:

$$Z_{\text{configuration}} = \sum_{\substack{\text{all possible}\\ \text{values of } N_{AB}}} g_{N_{AB}}\, e^{-N_{AB}w/kT}$$

where $g_{N_{AB}}$ is the number of arrangements with the same value of N_{AB}. Considerable progress has been made in recent years in the involved mathematical problem of calculating $g_{N_{AB}}$ exactly, and evaluating the sum.

14.6 Imperfect gases

The third application of the canonical ensemble method, which we discuss, is in the theory of real (imperfect) gases.

Numerous equations of state have been proposed for real gases. The best-known to physical chemists is the equation of van der Waals:

$$\left(P + \frac{a}{\tilde{V}^2}\right)(\tilde{V} - b) = RT \tag{14.24}$$

where a and b are empirical parameters for the gas. Simple arguments which demonstrate the reasonableness of this equation are given in elementary texts; according to these, b is proportional to the volume of the individual molecules (regarded as hard spheres), and a is a term arising from intermolecular forces.

Statistical mechanical theory does not, however, lead directly to Equation 14.24; instead, the following result is obtained:

$$P\tilde{V} = RT\left(1 + \frac{B}{\tilde{V}} + \frac{C}{\tilde{V}^2} + \frac{D}{\tilde{V}^3} + \ldots\right) \tag{14.25}$$

Equation 14.25 is known as the *virial equation,* and the coefficients B, C, D, etc., as the second, third, fourth, etc. *virial coefficients.*

Actually, within the limits of applicability of the van der Waals equation, Equations 14.24 and 14.25 are equivalent. Thus, Equation 14.24 can be rearranged to give:

$$\begin{aligned} P\tilde{V} &= \frac{RT}{1 - b/\tilde{V}} - \frac{a}{\tilde{V}} \\ &= RT\left(1 + \frac{b}{\tilde{V}} + \frac{b}{\tilde{V}^2} + \ldots\right) - \frac{a}{\tilde{V}} \\ &= RT\left(1 + \left(b - \frac{a}{RT}\right)\frac{1}{\tilde{V}} + \frac{b}{\tilde{V}^2} + \ldots\right) \end{aligned} \tag{14.26}$$

Equations 14.25 and 14.26 are identical up to the term in $1/\tilde{V}$, provided that:

$$B = b - \frac{a}{RT} \tag{14.27}$$

The statistical mechanical derivation of Equation 14.25 proceeds in two stages; firstly, Z is calculated, and secondly, the pressure is derived from Z by means of Equation 14.16.

(i) *The partition function*

As in the case of the molecular partition function f (Section 3.4), Z can be factorized into an internal part Z_{int} and a translational part Z_{trans} where:

$$Z_{\text{trans}} = \sum_{\substack{\text{translational} \\ \text{quantum states } t \\ \text{of the whole system}}} e^{-E_{\text{trans}(t)}/kT}$$

We evaluate Z_{trans} as the classical approximation, by the methods discussed in Chapter 11.

According to Equation 11.14b, the translational kinetic energy of a single molecule can be expressed as the sum of three terms, of the type $p_i^2/2m$. Hence, the total translational kinetic energy of a gas containing N molecules of mass m is:

$$\frac{1}{2m} \sum_{i=1}^{3N} p_i^2$$

the sum being over the momenta conjugate to the $3N$ space-coordinates of the centres of mass of the molecules.

If the gas is imperfect, there is another contribution to the translational energy, namely, the potential energy U arising from the presence of intermolecular forces. If this is regarded as the sum of the potential energies due to the repulsion between *pairs* of molecules, we can write:

$$U = \sum_{\substack{\text{pairs} \\ k,l}} U_{kl} \equiv \sum_{k<l} U_{kl}$$

The symbol Σ with '$k < l$' beneath it means that the corresponding sum is taken over all pairs of indices for which k is less than l. This simply ensures that each pair is counted only once, and that terms such as U_{kk}, with $k = l$, are not present in the sum.

The total translational energy of an imperfect gas is thus:

$$E_{\text{trans}} = \frac{1}{2m} \sum_{s=1}^{3N} p_s^2 + U = \frac{1}{2m} \sum_{s=1}^{3N} p_s^2 + \sum_{k<l} U_{kl}$$

The classical approximation to Z_{trans} is obtained as in Section 11.5; that is:

$$Z_{\text{trans}} = \frac{1}{h^{3N}} \int_{(6N)}^{\dots} \int e^{-E_{\text{trans}}/kT} \, dq_1 \dots dq_{3N} \, dp_1 \dots dp_{3N}$$

$$= \frac{1}{h^{3N}} \int_{(6N)}^{\dots} \int e^{-[(1/2m) \Sigma p_i^2 + U]/kT} \, dq_1 \dots dq_{3N} \, dp_1 \dots dp_{3N}$$

Now, U depends upon the internuclear distances (and therefore upon the q_s, but not upon the momenta p_s. Hence, the right-hand expression of the last equation can be factorized in the usual way.

$$Z_{\text{trans}} = \frac{1}{h^{3N}} \int_{(3N)}^{\dots} \int e^{-(1/2m) \Sigma p_s^2/kT} \, dp_1 \dots dp_{3N}$$

$$\times \int_{(3N)}^{\dots} \int e^{-U/kT} \, dq_1 \dots dq_{3N}$$

Clearly, the first multiple integral can be further factorized into $3N$ integrals of the type:

$$\int e^{-(1/2m) p_s^2/kT} \, dp_s = (2\pi mkT)^{1/2}$$

and therefore:

$$Z_{\text{trans}} = \left(\frac{2\pi mkT}{h^2}\right)^{3N/2} Q_T$$

where

$$Q_T = \int_{3N}^{\dots} \int e^{-U/kT} \, dq_1 \dots dq_{3N} \tag{14.28}$$

is called the *configuration integral.* Q_T can be regarded as the factor in Z_{trans} which corrects for the non-ideality of the gas. To evaluate Q_T, we first introduce a function f_{kl} for each pair of the molecules k,l defined by:

$$f_{kl} = e^{-U_{kl}/kT} - 1 \tag{14.29}$$

Then:

$$e^{-U/kT} = \exp(-\sum_{k<l} U_{kl}/kT) = \prod_{k<l} e^{-U_{kl}/kT}$$

$$= \prod_{k<l} (1 + f_{kl}) \tag{14.30}$$

For example, for an assembly of three molecules, we would have:

$$e^{-U/kT} = (1 + f_{12})(1 + f_{13})(1 + f_{23})$$

There being one bracket for each pair of molecules. Multiplying out the brackets:

$$\begin{aligned} e^{-U/kT} &= 1 + (f_{12} + f_{13} + f_{23}) + (f_{12}f_{13} + f_{12}f_{23} + f_{13}f_{23}) \\ &\quad + f_{12}f_{13}f_{23} \end{aligned}$$

In general, multiplying out the brackets in Equation 14.30 gives:

$$e^{-U/kT} = 1 + \sum_{k<l} f_{kl} + \ldots \tag{14.31}$$

the terms not explicitly written down involving two or more f_{kl} functions.

Q_τ is now evaluated by substituting the expression 14.31 into Equation 14.28. The exact solution of the resulting integral was first achieved in 1927 by H.D. Ursell: a lucid account of this work is given in Chapter 16 of Rushbrooke's book. The derivation outlined below is a grossly simplified treatment, which gives the correct result by the accident of two compensating errors; it is, however, sufficiently similar to the exact theory to form an instructive introduction to the latter.

The simplifying assumption made is that, because the f_{kl} terms are numerically small, products of the type $f_{12}f_{13}, f_{12}f_{13}f_{23}$, etc. can be neglected, and therefore Equation 14.31 can be replaced by:

$$e^{-U/kT} \approx 1 + \sum_{k<l} f_{kl} \,. \tag{14.32}$$

In fact this is not really justifiable, because, although the individual higher products are small, there are many more of them. Nevertheless, we adopt Equation 14.32 in our derivations. Thus:

$$\begin{aligned} Q_\tau &= \int \underset{(3N)}{\cdots} \int 1 \, dq_1 \ldots dq_{3N} + \int \underset{(3N)}{\cdots} \int \sum_{k<l} f_{kl} \; dq_1 \ldots dq_{3N} \\ &= V^N + \int \underset{3N}{\cdots} \int \sum_{k<l} f_{kl} \; dq_1 \ldots dq_{3N} \end{aligned} \tag{14.33}$$

Since $\int (a+b)dx = \int a\,dx + \int b\,dx$, the last multiple integral is obviously the same as the sum:

$$\sum_{k<l} \int \underset{(3N)}{\cdots} \int f_{kl} \, dq_1 \ldots dq_{3N} \equiv \sum_{k<l} \int \underset{(N)}{\cdots} \int f_{kl} \, d\tau_1 \ldots d\tau_N$$

where, in order to simplify the formulae a little, we have used the notation

$$\int\int\int dq_r dq_s dq_t \equiv \int d\tau_k$$

when q_r, q_s, and q_t are the space coordinates of molecule k.

Now, f_{kl} depends only upon the positions of molecules k and l, so the integration over the coordinates of the other molecules can be performed independently, that is:

$$\underset{(N)}{\int \cdots \int} f_{kl}\, d\tau_1 \ldots d\tau_N = \left(\int\int f_{kl}\, d\tau_k\, d\tau_l\right)\left(\underbrace{\int d\tau_1 \int d\tau_2 \ldots}_{(N-2\ \text{integrals})}\right)$$

$$= \left(\int\int f_{kl}\, d\tau_k\, d\tau_l\right)\left(V^{N-2}\right), \qquad (14.34)$$

since $\int d\tau = V$. The double integral can also be simplified. Thus:

$$\int\int f_{kl}\, d\tau_k\, d\tau_l = \int\left(\int f_{kl}\, d\tau_k\right) d\tau_l.$$

The position of k relative to l can be expressed in terms of polar coordinates (r,θ,φ) with l as origin (Fig. 14.2). Integrating over these coordinates:

$$\int f_{kl}\, d\tau_k = \int_0^{2\pi}\int_0^{\pi}\int_0^{\infty} f(r)\, r^2 \sin\theta\; dr\; d\theta\; d\varphi$$

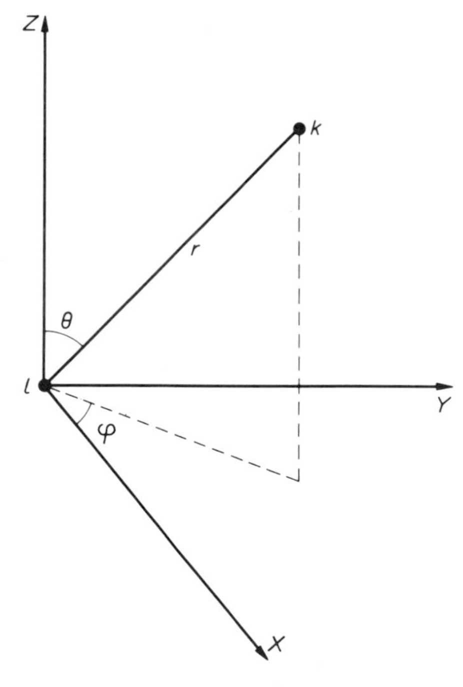

Fig. 14.2

There are two things to note about the second expression.

Firstly, f_{kl} is assumed to depend on the internuclear distance, not on the angles θ and φ (which is strictly correct for monatomic gases and a useful approximation in other cases); hence we now designate it as $f(r)$. Secondly, the limits of integration should be the walls of the vessel containing the gas; since, however, $f(r)$ diminishes rapidly with distance, it makes no difference if the limit ∞ is chosen for r. Integrating over θ and φ:

$$\int f_{kl}\,\mathrm{d}\tau_k = 4\pi \int_0^\infty f(r) r^2\,\mathrm{d}r = A, \text{ say.}$$

Hence:

$$\int \left(\int f_{kl}\,\mathrm{d}\tau_k \right) \mathrm{d}\tau_l = \int A\,\mathrm{d}\tau_l = A \int \mathrm{d}\tau_l = AV,$$

since A is a constant, independent of the position of l. Substituting into Equation 14.34:

$$\int \underset{(N)}{\cdots} \int f_{kl}\,\mathrm{d}\tau_1 \ldots \mathrm{d}\tau_N = A V^{N-1}$$

Since there are $N(N-1)/2 \approx N^2/2$ *pairs* of molecules:

$$\sum_{N<l} \int \underset{(N)}{\cdots} \int f_{kl}\,\mathrm{d}\tau_1 \ldots \mathrm{d}\tau_N = \frac{N^2}{2} A\, V^{N-1},$$

and Equation 14.33 for Q_T becomes:

$$\begin{aligned} Q_T &= V^N + N^2 A V^{N-1}/2 \\ &= V^N (1 + N^2 A/2V). \end{aligned} \qquad (14.35)$$

(ii) *The equation of state*

Since Q_T is the only factor of Z which depends upon volume, Equation 14.16 simplifies to:

$$P = kT \left(\frac{\partial \ln Q_T}{\partial V} \right)_T \qquad (14.36)$$

From Equation 14.35:

$$\ln Q_T = N \ln V + \ln \left(1 + \frac{N^2 A}{2V} \right)$$

In the limiting case of very small deviations from ideality, A is small ($A = 0$ for the perfect gas). If A is sufficiently small that $N^2 A/2V << 1$, the second logarithmic term can be expanded:

$$\ln\left(1+\frac{N^2A}{2V}\right)=\frac{N^2A}{2V}+\ldots,$$

so that:

$$\ln Q_T = N\ln V+\frac{N^2A}{2V}+\ldots. \qquad (14.37)$$

The exact treatment shows that Equation 14.37 is the correct expression for $\ln Q_T$ to within terms of order $1/V$, *whether or not* $N^2A/2V$ *is small*; that is, the error introduced by expanding the logarithm fortuitously cancels with that arising from the neglect of the higher terms in Equation 14.31.

From Equations 14.36 and 14.37 therefore:

$$P = kT\left(\frac{N}{V}-\frac{N^2A}{2V^2}+\ldots\right)$$

i.e.,

$$P\tilde{V} = RT\left(1-\frac{\tilde{N}A}{2\tilde{V}}+\ldots\right)$$

Comparing this with Equation 14.25, we see that the second virial coefficient is:

$$B = -\tilde{N}A/2 \qquad (14.38)$$

Equation 14.38 is of great importance, because it enables one to draw conclusions about the nature of intermolecular forces, from the experimentally determined value of the second virial coefficient. The following general procedure is used.

Suppose that a theoretical treatment leads to an expression $U_{\text{theor}}(r)$ for the potential energy of interaction of two molecules. The corresponding function $f(r)$ is, according to Equation 14.29:

$$e^{-U_{\text{theor}}(r)/kT}-1$$

and the theoretical value of the second virial coefficient:

$$B_{\text{theor}} = -\tilde{N}2\pi\int_0^{\infty} r^2\left(e^{-U_{\text{theor}}(r)/kT}-1\right)dr.$$

Comparison of B_{theor} gives the experimental value B gives an indication of the validity of the expression $U_{\text{theor}}(r)$; even more usefully, if $U_{\text{theor}}(r)$ contains adjustable parameters, the values of these can be chosen so as to give the best agreement between B_{theor} and B.

The simplest example of this approach is obtained by regarding the molecules as hard, impenetrable spheres, of diameter r_0.

Because the spheres are impenetrable, we put:

$$U_{\text{theor}}(r) = \infty, \quad \text{if } r < r_0,$$

so that:

$$e^{-U_{\text{theor}}(r)/kT} = 0, \quad \text{if } r < r_0.$$

If (as is probably the case), for other values of r, $U_{\text{theor}}(r)$ is small compared with kT, the exponential term can be approximated, and:

$$e^{-U_{\text{theor}}(r)/kT} \approx 1 - U_{\text{theor}}(r)/kT \; (r \geqslant r_0)$$

Then

$$\begin{aligned} B_{\text{theor}} &= -\tilde{N} 2\pi \int_0^\infty r^2 (e^{-U_{\text{theor}}(r)/kT} - 1)\, dr \\ &= \tilde{N} 2\pi \int_0^{r_0} r^2\, dr + \tilde{N} 2\pi \int_{r_0}^\infty r^2 \frac{U_{\text{theor}}(r)}{kT}\, dr. \end{aligned}$$

This is the sum of two terms, one of which is constant and the other proportional to $1/T$. According to Equation 14.27, the two terms may be identified with the van der Waals constants b and a, respectively. Thus:

$$b = \tilde{N} 2\pi \int_0^{r_0} r^2\, dr = \tilde{N} \tfrac{2}{3}\pi r_0^3 = 4\tilde{N} v,$$

where v is the volume of one of the hard spheres. Furthermore:

$$a = -\tilde{N}^2 2\pi \int_{r_0}^\infty r^2\, U_{\text{theor}}(r)\, dr$$

i.e., a is dependent upon the potential energy of interaction of the two molecules.

14.7 Further comments on the ensemble average

The ensemble average of any property X of a macroscopic system is calculated in terms of Z in just the same way as the average value of a molecular property p is calculated in terms of f. Thus, in analogy with Equation 2.23b, the ensemble average is

$$\overline{X} = \frac{\sum_j e^{-E_j/kT} X_j}{Z} \qquad (14.39)$$

the sum being over all quantum states (complexions) of the system.

(i) *The Maxwell–Boltzmann Law*

Consider an assembly of independent molecules and suppose that the property X under consideration is, n_i, the distribution number for the i-th molecular level. The ensemble average value of n_i is:

$$\overline{n}_i = \frac{1}{Z} \sum_j n_{ij}\, e^{-E_j/kT} \qquad (14.40)$$

where n_{ij} is the value of n_i for the j-th complexion.

Since the molecules are independent:

$$E_j = n_{1j}\epsilon_1 + n_{2j}\epsilon_2 + \ldots + n_{ij}\epsilon_i + \ldots$$

Hence:

$$\frac{\partial}{\partial \epsilon_i} e^{-E_j/kT} = \frac{1}{kT^2} n_{ij} e^{-E_j/kT}$$

Therefore Equation 14.40 becomes:

$$\begin{aligned}
\overline{n}_i &= \frac{kT^2}{Z}\frac{\partial}{\partial \epsilon_i} \sum_j e^{-E_j/kT} \\
&= \frac{kT^2}{Z}\frac{\partial Z}{\partial \epsilon_i} = kT^2 \frac{\partial \ln Z}{\partial \epsilon_i} \\
&= kT^2 \frac{\partial}{\partial \epsilon_i} \ln (f)^N \qquad \text{(by Equation 14.19)} \\
&= NkT^2 \frac{\partial}{\partial \epsilon_i} \ln f = \frac{NkT^2}{f}\frac{\partial f}{\partial \epsilon_i} \\
&= \frac{N}{f} g_i\, e^{-\epsilon_i/kT}
\end{aligned}$$

Comparing this result with the Maxwell–Boltzmann Equation 2.21, we see that the *ensemble average* of a distribution number is the same as the *most probable value* of that number, derived by the Maxwell–Boltzmann method; this is the result referred to on pages 21 and 59.

(ii) *Fluctuations of Energy in the Canonical Ensemble*

The internal energy of a system in the canonical ensemble is not assumed necessarily to be constant; the thermodynamical internal energy is identified, in Equation 14.12 with *ensemble average* of E. To complete our short account, let us consider the importance of possible departures from the mean value.

We wish to determine how values of E observed for systems of a given kind, at a constant temperature (i.e., members of the ensemble), are likely to differ from the mean energy (now denoted by $\overline{E}$) calculated from Equation 14.39 (or, equivalently, 14.12), i.e., to determine the average of fluctuations of E about the mean.

For this purpose, it is no use evaluating the average value of $(E - \overline{E})$, because, of course, this must be zero. However, a meaningful measure of the departure from $\overline{E}$ will be obtained by calculating $\sigma/\overline{E}$, where σ is the standard deviation. The latter is defined by:

$$\sigma^2 = \text{mean value of } (E - \overline{E})^2$$

Now:

$$(E - \overline{E})^2 = E^2 - 2E\overline{E} + (\overline{E})^2$$

The average of $E\overline{E}$ is simply $\overline{E}$ times the average of E, i.e., it is $(\overline{E})^2$. Therefore, denoting the average value of E^2 by $\overline{E^2}$, we have:

$$\sigma^2 = \overline{(E - \overline{E})^2} = \overline{E^2} - (\overline{E})^2 .$$

The value of $\overline{E^2} - (\overline{E})^2$ is easily derived from known quantities. We observe that:

$$C_V = \left(\frac{\partial E}{\partial T}\right)_V = \frac{\partial}{\partial T}\left(\frac{kT^2}{Z}\frac{\partial Z}{\partial T}\right)$$

$$= \frac{1}{ZkT^2}\sum E_j^2\, e^{-E_j/kT} - \left(\frac{\sum E_j e^{-E_j/kT}}{Z^2}\right)\left(\frac{\partial Z}{\partial T}\right)_V$$

$$= \frac{1}{kT^2}\left[\frac{\sum E_j^2\, e^{-E_j/kT}}{Z} - \left(\frac{\sum E_j\, e^{-E_j/kT}}{Z}\right)^2\right]$$

$$= \frac{1}{kT^2}(\overline{E^2} - (\overline{E})^2),$$

from the averaging formula 14.39. Hence:

$$\frac{\sigma}{\overline{E}} = \frac{[\overline{E^2} - (\overline{E})^2]^{1/2}}{\overline{E}} = \frac{(C_V kT^2)^{1/2}}{\overline{E}} = \frac{C_V^{1/2} R^{1/2} T}{\overline{E}\tilde{N}^{1/2}}$$

Since C_V is of order R, and $\overline{E}$ of order RT:

$$\sigma/\overline{E} \text{ is of order } 1/\tilde{N}^{1/2} \approx 10^{-10}$$

an extremely small number. Thus, for macroscopic systems,

deviations in E from the mean value are physically unimportant. This is the deduction from statistical mechanics which we mentioned in Section 14.1, namely that, to a high degree of probability, systems with the same volume, temperature, and composition have the same internal energy.

Problems

14.1 Derive the Einstein heat capacity equation by applying the canonical ensemble theory to a crystal in which all $3N$ normal modes of vibration have the same frequency ω.

14.2 A general method for evaluating the sum 14.6:

$$Z = \sum \Omega_i \, e^{-E_i/kT}, \qquad \text{(i)}$$

is to use the approximation (cf. page 33) that:

$$\ln Z \approx \ln(\Omega_i \, e^{-E_i/kT})_{\max},$$

where the term in brackets is the largest term in the sum. To illustrate this method, we apply it to the evaluation of the partition function for a system of independent, distinguishable molecules.

(a) Deduce that the same value of Z is obtained if i in equation (i) labels a *distribution*, provided that the sum is taken over *all* distributions irrespective of the value of their energy E_i.

(b) From Equations 1.4 and 4.3,

$$\ln(\Omega_i e^{-E_i/kT}) = \sum n_{ii} \ln n_{ii} - \sum n_{ii} \ln \frac{n_{ii}}{g_{ii}} - \sum n_{ii} \epsilon_i$$

For the largest term:

$$\delta \ln(\Omega_i e^{-E_i/kT}) = 0$$

subject to $\Sigma \delta n_{ii} = 0$ (there is no restriction on the value of $\Sigma n_{ii} \epsilon_i$. Proceeding as in Section 2.2., deduce that:

$$\overline{n}_i = \frac{N}{f} g_i \, e^{-\epsilon_i/kT}$$

and hence (by substitution in the expression for $\ln Z$) that Equation 14.19 is again obtained.

14.3 The following expression for the potential energy of interaction between two molecules was proposed by Sutherland [Hirschfelder, Curtiss, and Bird, *Molecular Theory of Gases and Liquids,* Wiley, New York (1967), p. 158]:

$$U_{\text{theor}}(r) = \infty \quad \text{if} \quad r < r_0$$

$$U_{\text{theor}}(r) = -cr^{-\gamma} \quad \text{if} \quad r > r_0 .$$

(a) Show that in this case the second virial coefficient is

$$B_{\text{theor}} = \tfrac{2}{3}\tilde{N}r_0^3 - 2\pi\tilde{N} \int^{\infty} [\exp(cr^{-\gamma}/kT) - 1]\, r^2\, dr.$$

(b) By series expansion of $\exp(cr^{-\gamma}/kT)$, show that:

$$B_{\text{theor}} = -\frac{2\pi Nr_0^3}{3} \sum_{j=0}^{\infty} \left(\frac{1}{j!}\right)\left(\frac{3}{j\gamma - 3}\right) \left(\frac{c}{r_0^{\gamma}kT}\right)^j .$$

14.4 From Q_T given by Equation 14.37, obtain expressions for E and S for an imperfect gas. Using the result of the preceding example, obtain an expression for the entropy correction referred to in the footnote to Table 8.1.

Appendices

Appendix 1: Approximation of sums by integration

(i) Suppose that $F(x)$ is some function of a variable x. We have, from time to time, encountered sums of the type:

$$\sum_{x=x_0}^{x=x_n} F(x) = F(x_0) + F(x_0+1) + F(x_0+2) + F(x_0+3) + \ldots + F(x_n);$$

i.e., the sum is taken over the F terms calculated for all integer values of x from x_0 to x_n. Examples of such sums are the partition functions for:

rotation ($x \equiv J$) (Section 5.5),

vibration ($x \equiv v$) (Section 5.4),

translation ($x \equiv p$) (Section 4.6).

We wish to discuss the important approximation of replacing such sums by integrals, i.e.:

$$\sum_{x=x_0}^{x=x_n} F(x) \approx \int_{x_0}^{x_n} F(x)\,\mathrm{d}x \tag{A.1}$$

It is necessary to assume that the function $F(x)$ can be defined for non-integral as well as for integral values of x, so that $F(x)$ can be plotted as a smooth curve.

For example, the function

$$F(x) \equiv F(J) = (2J+1)\mathrm{e}^{-J(J+1)\Theta_{\mathrm{rot}}/T} \tag{A.2}$$

has a meaning for non-integral values of J although the only values

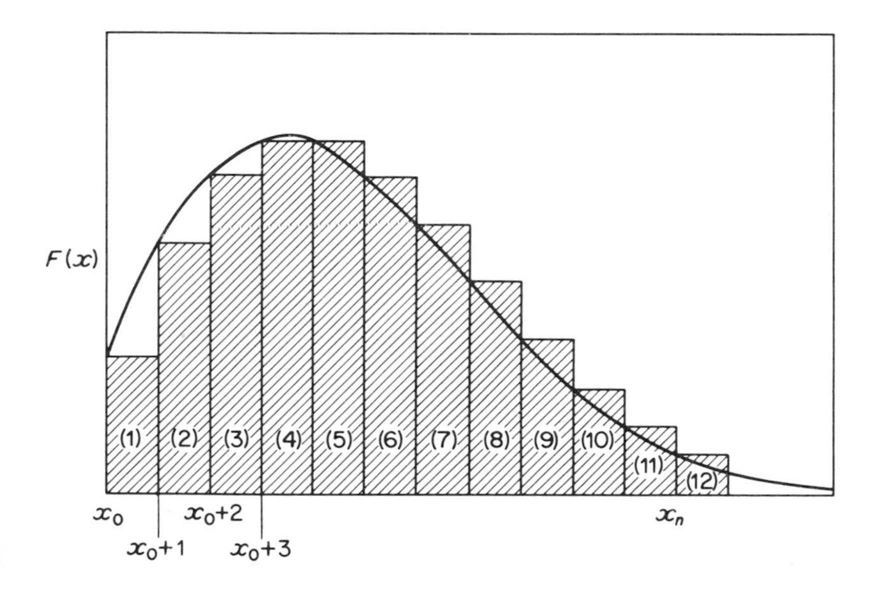

Fig. A.1

of $F(J)$ of physical significance in the rotation of a diatomic molecule are those for which J is zero or an integer. Fig. A.1 shows the curve for $F(J)$.

The shaded area (1) is $F(0) \times 1 = F(0)$, numerically
The shaded area (2) is $F(1) \times (2 - 1) = F(1)$, numerically
The shaded area (3) is $F(2) \times (3 - 2) = F(2)$, numerically
and so on. Hence the total shaded area in Fig. 11.1 is exactly the sum:

$$F(0) + F(1) + F(2) + \ldots \equiv \sum_{x=x_0}^{x_n} F(x)$$

The area under the *curve* between $x = x_0$ and $x = x_n$ is the integral:

$$\int_{x_0}^{x_n} F(x)\,\mathrm{d}x\,.$$

The difference:

$$\sum_{x=x_0}^{x_n - 1} F(x) - \int_0^{x_n} F(x)\,\mathrm{d}x \tag{A.3}$$

is the sum of the areas bounded by the curve and the steps between x_0 and x_n in Fig. A.1. As a first approximation, these areas can be estimated as areas of triangles. The area of each triangle:

$$\begin{aligned} \text{area} &= \tfrac{1}{2}(\text{base} \times \text{height}) \\ &= \tfrac{1}{2}[F(x) - F(x + 1)] \end{aligned}$$

This is positive if the step lies above the curve $[F(x) > F(x+1)]$, and negative if the step lies below the curve.

The sum of the areas of the triangles is:

$$
\begin{aligned}
&\tfrac{1}{2}\sum_{x=x_0}^{x_n-1}[F(x) - F(x+1)] \\
&= \tfrac{1}{2}\left[\sum_{x=x_0}^{x_n-1}F(x) - \sum_{x=x_0}^{x_n-1}F(x+1)\right] \\
&= \tfrac{1}{2}\left[\sum_{x=x_0}^{x_n-1}F(x) - \sum_{x=x_0+1}^{x_n}F(x)\right] \\
&= \tfrac{1}{2}[F(x_0) - F(x_n)]
\end{aligned}
$$

Adding on the final term $F(x_n)$ to the difference A.3, we thus obtain as an estimate of the error of the approximation A.1, $\frac{1}{2}[F(x_0) + F(x_n)]$, and the corresponding *fractional* error is:

$$
\frac{\frac{1}{2}[F(x_0) + F(x_n)]}{\int_{x_0}^{x_n} F(x)\,dx} \qquad \text{(A.4)}
$$

For the terms A.2 of the rotational partition function, substitution of $x_0 = J_0 = 0$ gives $F(0) = e^0 = 1$ and $F(\infty) = 0$. Hence the error is approximately:

$$
\frac{1}{2}\left(\frac{1}{T/\Theta_{rot}}\right) = \frac{\Theta_{rot}}{2T}
$$

In most cases, $\Theta_{rot} < 2$ K, so that the fractional error is less than $1/T\,\text{K} < 0{\cdot}3\%$.

For hydrogen $\frac{1}{2}(\Theta_{rot}/T)$ is $85{\cdot}38/2T$, which is not negligible for temperatures lower than about 300 K; in such cases, approximation A.1 cannot be used.

By applying Equation 5.41 to the case of vibration, the fractional error in A.1 is calculated to be

$$
\frac{\Theta_{vib}}{2T} \approx \frac{10^3}{T\,\text{K}}
$$

according to Table 5.2. This is very large unless T is several thousand degrees, and approximation A.1 is not used.

(ii) More generally, the values of x in the summation may not be integers. If, however, they can be expressed as:

$$
x = r\lambda + x_0 \qquad (r = 0,1,2,3,\ldots)
$$

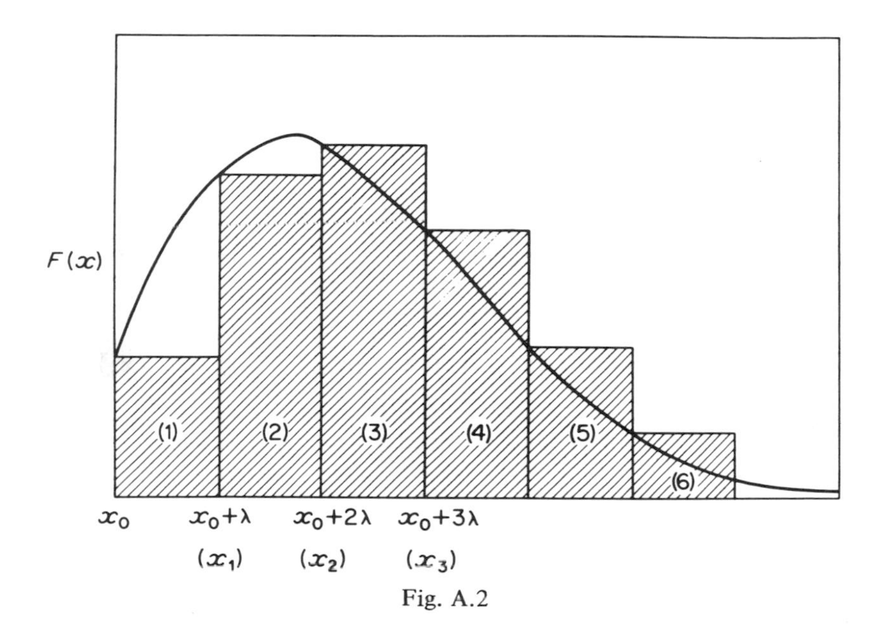

Fig. A.2

where λ is a constant, then:

$$\sum_{\substack{x = x_0 \\ \text{in steps of } \lambda}}^{x_n} F(x) \approx \frac{1}{\lambda} \int_{x_0}^{x_n} F(x)\,dx \tag{A.5}$$

This can be seen from Fig. A.2. The areas of the individual shaded rectangles are now $\lambda F(x)$. Therefore, the total stepped area is:

$$\lambda F(x_0) + \lambda F(x_1) + \lambda F(x_2) \approx \int_{x_0}^{x_1} F(x)\,dx$$

or

$$F(x_0) + F(x_1) + F(x_2) + \ldots \approx \frac{1}{\lambda} \int_{x_0}^{x_1} F(x_1)\,dx$$

which is the approximation A.5.

For example, in evaluating the rotational partition function for parahydrogen when $T >> \Theta_{rot}$, one includes in the sum only terms corresponding to even values of J. Then:

$$J = 2r \quad (r = 0,1,2,3,\ldots),$$

and

$$f_{rot} \approx \frac{1}{2} \int_{0}^{\infty} (2J + 1)e^{-J(J+1)\Theta_{rot}/T}\,dJ = \frac{1}{2}\left(\frac{T}{\Theta_{rot}}\right)$$

the same result is obtained for $J = 2r + 1$, i.e., for orthohydrogen.

Approximation A.5 is also the basis of Equation 11.4. In this case there are two integrations and the steps ($\lambda = \delta p$, $\lambda' = \delta q$) are infinitesimal.

(iii) Approximation A.1 is stated more precisely as the *Euler–Maclaurin summation formula*:

$$\sum_{x=x_0}^{x_n} F(x) = \int_{x_0}^{x_n} F(x)\,\mathrm{d}x + \tfrac{1}{2}[F(x_0) + F(x_n)] + \phi \qquad \text{(A.6)}$$

where:

$$\phi = -\frac{1}{12}[F^{(1)}(x_0) - F^{(1)}(x_n)] + \frac{1}{720}[F^{(2)}(x_0) - F^{(2)}(x_n)] - \frac{1}{30240}[F^{(3)}(x_0) - F^{(3)}(x_n)] + \ldots$$

with:

$$F^{(k)}(x_0) = \left[\frac{\partial^k F(x)}{\partial x^k}\right]_{x=x_0}, \text{ etc.}$$

Applying this to the terms A.2 of the rotational partition function:

$$F^{(1)}(x_0) = 2 - \frac{\Theta_{\mathrm{rot}}}{T}$$

$$F^{(2)}(x_0) = -12\frac{\Theta_{\mathrm{rot}}}{T} + 12\left(\frac{\Theta_{\mathrm{rot}}}{T}\right)^2 - \left(\frac{\Theta_{\mathrm{rot}}}{T}\right)^3$$

Higher derivatives contain only powers of Θ_{rot}/T greater than 1. Therefore, neglecting powers of Θ_{rot}/T higher than the first, and noting that all the derivatives (which contain the factor $\mathrm{e}^{-J(J+1)\Theta_{\mathrm{rot}}/T}$) vanish at $J = \infty$:

$$\phi = -\frac{1}{12}\left(2 - \frac{\Theta_{\mathrm{rot}}}{T}\right) - \frac{12}{720}\frac{\Theta_{\mathrm{rot}}}{T}$$

$$= -\frac{1}{6} + \frac{1}{15}\left(\frac{\Theta_{\mathrm{rot}}}{T}\right)$$

Substituting this value into Equation A.6 together with the results:

$$\int_0^\infty F(x)\,\mathrm{d}x = \frac{T}{\Theta_{\mathrm{rot}}},$$

and

$$F(x_0) = 1,$$

we obtain:

$$f_{\mathrm{rot}} \approx \frac{T}{\Theta_{\mathrm{rot}}} + \frac{1}{3} + \frac{1}{15}\left(\frac{\Theta_{\mathrm{rot}}}{T}\right),$$

the first terms in Mulholland's expansion (page 75).

Appendix 2: Stirling's theorem

Stirling's theorem states that

$$\sqrt{2\pi}\, n^{n+1/2}\, e^{-n} < n! < \sqrt{2\pi}\, n^{n+1/2}\, e^{-n}\left(1+\frac{1}{4n}\right)$$

(see e.g., R. Courant, *Differential and Integral Calculus*, Blackie (1937), *Vol. I* p. 361)

Therefore,

$$\ln n! = (n+\tfrac{1}{2})\ln n - n + C$$

where C is a number, the value of which lies between 0·9189 and $0{\cdot}9189 + \ln\left(1+\frac{1}{4n}\right)$. For large values of n this reduces to the simpler form (1.8),

$$\ln n! \approx n\ln n - n \tag{1.8$'$}$$

the error being approximately $-(1+\frac{1}{2}\ln n)$.

The approximate form (1.8) is easily obtained.

$$\begin{aligned}\ln n! &= \ln[n(n-1)(n-2)\ldots 3.2.1]\\ &= \ln n + \ln(n-1) + \ln(n-2) + \ldots + \ln 3 + \ln 2 + \ln 1\\ &= \sum_{x=1}^{n} \ln x\end{aligned}$$

Applying approximation (A.1),

$$\begin{aligned}\sum_{x=1}^{n} \ln x \approx \int_1^n \ln x\, dx &= [x\ln x - x]_1^n\\ &= n\ln n - n + 1\\ &\approx n\ln n - n,\end{aligned}$$

if n is large: the required result.

Appendix 3: Change of variables in multiple integration

It is necessary to refer elsewhere (e.g., Jeffreys and Jeffreys, p. 184) for an account of this topic. For present purposes, however, only the following result is required.

Suppose that it is required to replace integration over the variables x_1, x_2, x_3 by integration over a_1, a_2, a_3. Then:

$$dx_1\, dx_2\, dx_3 = J\, da_1\, da_2\, da_3$$

where J is called the *Jacobian* of x_1, x_2, x_3 with respect to a_1, a_2, a_3. It is defined as the determinant:

$$J = \begin{vmatrix} \dfrac{\partial x_1}{\partial a_1} & \dfrac{\partial x_2}{\partial a_1} & \dfrac{\partial x_3}{\partial a_1} \\ \dfrac{\partial x_1}{\partial a_2} & \dfrac{\partial x_2}{\partial a_2} & \dfrac{\partial x_3}{\partial a_2} \\ \dfrac{\partial x_1}{\partial a_3} & \dfrac{\partial x_2}{\partial a_3} & \dfrac{\partial x_3}{\partial a_3} \end{vmatrix}$$

J is more usually denoted by:

$$\frac{\partial(x_1, x_2, x_3)}{\partial(a_1, a_2, a_3)} .$$

Examples are:

(i) *Transformation from Cartesian to polar coordinates*

The transformation is

$$\begin{aligned} x &= r \cos\theta \\ y &= r \sin\theta . \end{aligned} \tag{A.7}$$

Then:

$$\frac{\partial x}{\partial r} = \cos\theta , \quad \frac{\partial x}{\partial \theta} = -r\sin\theta , \text{ etc.}$$

and so:

$$\frac{\partial(x, y)}{\partial(r, \theta)} = \begin{vmatrix} \cos\theta & -r\sin\theta \\ \sin\theta & r\cos\theta \end{vmatrix}$$

$$= r\cos^2\theta + r\sin^2\theta = r .$$

Thus:

$$\mathrm{d}x\,\mathrm{d}y = r\,\mathrm{d}r\,\mathrm{d}\theta . \tag{A.8}$$

(ii) *Transformation from Cartesian to spherical polar coordinates*

Proceeding in the same way as above, one obtains:

$$\frac{\partial(x, y, z)}{\partial(r, \theta, \varphi)} = r^2 \sin\theta$$

so that:

$$\mathrm{d}x\,\mathrm{d}y\,\mathrm{d}z = r^2 \sin\theta\,\mathrm{d}\theta\,\mathrm{d}\varphi . \tag{A.9}$$

(iii) *Transformation of momentum variables*

In Section 12.6, we introduced the variables γ and λ defined by:

$$\gamma = \sin\chi\, p_\theta - \frac{\cos\chi}{\sin\theta}(p_\varphi - \cos\theta\, p_\chi)$$

$$\lambda = \cos\chi\, p_\theta + \frac{\sin\chi}{\sin\theta}(p_\varphi - \cos\theta\, p_\chi)\,.$$

Now:

$$\frac{\partial\gamma}{\partial p_\theta} = \sin\chi\,, \quad \frac{\partial\gamma}{\partial p_\varphi} = -\frac{\cos\chi}{\sin\theta}, \text{ etc.}$$

Hence:

$$\frac{\partial(\gamma,\lambda)}{\partial(p_\theta,p_\varphi)} = \begin{vmatrix} \sin\chi & \cos\chi \\ -\dfrac{\cos\chi}{\sin\theta} & \dfrac{\sin\chi}{\sin\theta} \end{vmatrix}$$

$$= \frac{\sin^2\chi}{\sin\theta} + \frac{\cos^2\chi}{\sin\theta} = \frac{1}{\sin\theta}$$

so that:

$$\mathrm{d}p_\theta\,\mathrm{d}p_\varphi = \sin\theta\,\mathrm{d}\gamma\,\mathrm{d}\lambda\,,$$

the result used in obtaining Equation 11.24.

Appendix 4: Some definite integrals

(i) The integral:

$$I = \int_0^\infty \mathrm{e}^{-ax^2} x\,\mathrm{d}x$$

is evaluated by making the substitutions:

$$y = x^2\,, \quad \mathrm{d}y = 2x\,\mathrm{d}x$$

Then:

$$I = \frac{1}{2}\int_0^\infty \mathrm{e}^{-ay}\,\mathrm{d}y = -\frac{1}{2a}[\mathrm{e}^{-ay}]_0^\infty = \frac{1}{2a} \qquad \text{(A.10)}$$

(ii) By differentiation of I with respect to a, other integrals are obtained. Thus:

$$\mathrm{e}^{-ax^2} x^3\,\mathrm{d}x = -\frac{\partial I}{\partial a} = \frac{1}{2a^2}$$

Repetition of this process gives all integrals of the type:

$$\int_0^\infty \mathrm{e}^{-ax^2} x^k\,\mathrm{d}x$$

in which k is an odd integer.

(iii) Consider the integral:

$$I = \int_0^\infty e^{-ax^2}\, dx$$

Then:

$$I^2 = \int_0^\infty e^{-ax^2}\, dx \times \int_0^\infty e^{-ax^2}\, dx$$

$$= \int_0^\infty e^{-ax^2}\, dx \times \int_0^\infty e^{-ay^2}\, dy$$

(changing the name of the variable in the second integral)

$$= \int_0^\infty \int_0^\infty e^{-a(x^2+y^2)}\, dx\, dy\,.$$

Changing to polar coordinates according to Equations A.7 and A.8:

$$x^2 + y^2 = r^2 \quad \text{and} \quad dx\, dy = r\, dr\, d\theta$$

so that:

$$I^2 = \int_0^{\pi/2} \int_0^\infty e^{-ar^2}\, dr\, d\theta = \int_0^{\pi/2} d\theta \int_0^\infty e^{-ar^2}\, r\, dr$$

$$= \frac{\pi}{2}\frac{1}{2a}$$

by Equation A.10. Hence:

$$I = \frac{1}{2}\left(\frac{\pi}{a}\right)^{1/2} \tag{A.11}$$

As in paragraph (ii) above, other integrals can be generated from I by differentiation. For example:

$$\int_0^\infty e^{-ax^2}\, x^2\, dx = -\frac{\partial I}{\partial a} = \frac{1}{4a}\left(\frac{\pi}{a}\right)^{1/2}$$

The results are summarized in Table A.1.

Table A.1

$I = \int_0^\infty x^k e^{-ax^2}\, dx$			
k	I	k	I
0	$\frac{1}{2}\sqrt{\pi/a}$	1	$1/2a$
2	$\frac{1}{4}\sqrt{\pi/a^3}$	3	$1/2a^2$
4	$\frac{3}{8}\sqrt{\pi/a^5}$	5	$1/a^3$

(iv) The integral $I = \int_0^\infty e^{-ax} x^{1/2} \, dx$ can be evaluated from Equation A.11.

Let:

$$y^2 = x \quad , \quad 2y\, dy = dx$$

Then:

$$I = 2 \int_0^\infty e^{ay^2} y^2 \, dy = \frac{1}{2}\left(\frac{\pi}{a^3}\right)^{1/2}$$

a result used in the derivation of Equation 13.17.

Index